AN INTRODUCTION TO X-RAY SPECTROMETRY

TITLES OF RELATED INTEREST

Cathodoluminescence of geological materials
D. J. Marshall

Crystal structures and cation parameters in rock-forming minerals
J. R. Smyth & D. L. Bish

The dark side of the Earth
R. Muir Wood

Environmental magnetism
R. Thompson & F. Oldfield

Geology and mineral resources of West Africa
J. B. Wright *et al.*

Image interpretation in geology
S. A. Drury

The inaccessible Earth
G. C. Brown & A. E. Mussett

The interpretation of igneous rocks
K. G. Cox *et al.*

Komatiites
N. T. Arndt & E. G. Nisbet (eds)

Mathematics in geology
J. Ferguson

Metamorphic geology
C. Gillen

Metamorphic processes
R. H. Vernon

Metamorphism and metamorphic belts
A. Miyashiro

The outcrop quiz
J. B. Wright

Perspectives on a dynamic Earth
T. R. Paton

Petrology of the igneous rocks
F. H. Hatch *et al.*

Petrology of the metamorphic rocks
R. Mason

A practical introduction to optical mineralogy
C. D. Gribble & A. J. Hall

Rheology of the Earth
G. Ranalli

Rutley's elements of mineralogy
H. H. Read

Simulating the Earth
J. Holloway & B. Wood

Volcanic successions
R. A. F. Cas & J. V. Wright

The young earth
E. G. Nisbet

AN INTRODUCTION TO X-RAY SPECTROMETRY

X-ray fluorescence and electron microprobe analysis

K. L. Williams
Department of Geology and Geophysics
The University of Sydney

London
ALLEN & UNWIN
Boston Sydney Wellington

© K. L. Williams, 1987
This book is copyright under the Berne Convention. No reproduction without permission. All rights reserved.

Allen & Unwin, the academic imprint of
Unwin Hyman Ltd
PO Box 18, Park Lane, Hemel Hempstead, Herts HP2 4TE, UK
40 Museum Street, London WC1A 1LU, UK
37/39 Queen Elizabeth Street, London SE1 2QB

Allen & Unwin Inc.,
8 Winchester Place, Winchester, Mass 01890, USA

Allen & Unwin (Australia) Ltd,
8 Napier Street, North Sydney, NSW 2060, Australia

Allen & Unwin (New Zealand) Ltd in association with the Port Nicholson Press Ltd,
60 Cambridge Terrace, Wellington, New Zealand

First published in 1987

British Library Cataloguing in Publication Data
Williams, K. L.
   An introduction to X-ray spectrometry:
X-ray fluorescence and electron microprobe
analysis.
1. X-ray spectroscopy
I. Title
537.5'352   QC482.S6

Library of Congress Cataloging in Publication Data
Williams, K. L.
   An introduction to X-ray spectrometry.
Bibliography: p.
Includes index.
1. Rocks—Analysis. 2. Mineralogy, Determinative.
3. X-ray spectroscopy. 4. Fluorescence spectroscopy.
5. Microprobe analysis.   I. Title.
QE435.W55   1987   552'.06   87–1498
ISBN 0–04–544001–8 (alk. paper)

Typeset in 10 on 12pt Times by Columns, Caversham, Reading
and printed in Great Britain by Biddles Limited, Guildford, Surrey

Preface

Over the past two decades, X-ray spectrometric techniques – particularly X-ray fluorescence and electron microprobe analysis – have emerged from the research laboratory to take their place as versatile and more or less routine methods for chemical analysis of a wide variety of sample materials. The transition was a consequence initially of developments in electronic engineering which greatly enhanced the capabilities and the reliability of the instrumentation, and subsequently of the availability of small but highly efficient computer systems which relieved operators of much of the responsibility for instrument control and data reduction. Few commercial instruments are now sold without some form of computer control system, and it has become standard practice in many laboratories to encourage end users to perform their own analyses with a minimum of training and skilled supervision.

This is a very desirable situation in many ways, but it has one unfortunate corollary. Analysts who do not understand the principles on which a technique is based cannot be expected to maintain properly critical standards of appraisal, particularly when an obliging computer performs concealed 'massaging' of their data. The result is that great faith is sometimes placed in the results of rather poor analyses.

This book has been written primarily for those novice end-user analysts who wish to improve the quality of their work by gaining a better working understanding of the principles and practices of X-ray spectrometric analysis without having to delve too far into the underlying physical theory. It is based on courses I have taught for some years to graduate and undergraduate students who have been regular users of X-ray fluorescence and electron microprobe facilities with which I have been associated. There are some critical differences between the two techniques, but they also have much in common and I have found it convenient to teach them together. The topics covered, and the levels at which they are treated, are those that I have found to be most appropriate to the needs of this group.

I am aware that I will not have satisfied everyone and that I have, for example, treated such subjects as scanning imagery, or related X-ray techniques such as scanning transmission electron microscopy (STEM) and proton-induced X-ray spectrometry (PIXE), very superficially or not at all. However, a more comprehensive review would necessarily have become a massive volume far beyond my original intent, and I have chosen instead to focus on the two X-ray based spectrometric techniques that have already become more or less routine quantitative analytical tools.

PREFACE

There is nothing 'new' in this book – it is no more than an abstraction from the published works of the pathfinders who have developed the science of X-ray spectrometric analysis, and it was written principally as a collation of material that otherwise is scattered in many different sources. The text is not extensively referenced, but a short bibliography is provided, and readers familiar with the writings of Norrish, Norrish and Hutton, Norrish and Chappell, Jenkins, de Vries, Birks, Heinrich, Reed, Goldstein, Springer, Smith, Tertian and Claisse and many others will be well aware of my indebtedness to them as the sources of most of my material. I have attempted throughout to synthesize rather than plagiarize, but it is not always easy to make the distinction, and I apologize if I have accidentally transgressed. Certainly it is my hope that students who begin with this book will be encouraged to read more widely among the original sources – particularly those listed in the bibliography – so that they can resolve the problems that remain unclarified.

Many of the examples used in the text refer to geological applications, which simply reflects my own background and the interests of most of my students. Of course the points that they illustrate are equally applicable to any other appropriate analytical samples.

It is a pleasure to acknowledge the help that many people have given me. I am particularly grateful to the students of Stanford University and the University of Sydney, who have cheerfully tolerated some rather dreary lectures over the years. Charles Taylor showed me how to make an electron microprobe work properly, and also taught me much about how to teach others. My first forays into X-ray fluorescence, in the 1960s, were neither particularly successful nor encouraging; Keith Norrish and Bruce Chappell did much to restore my faith, as they have for many others. Through all the subsequent successes and failures, my family remained tolerant of my eccentricities and frequently unusual working hours, and I am very grateful to them for their support.

The original manuscript for this book could not have been prepared without my venerable Vector Graphic word processor, and the final version, including the text figures, was prepared on an Apple Macintosh. I was fortunate that Michael Hough was able to read much of the manuscript and make many constructive suggestions for improvements. However, neither he nor the authors of my source materials are responsible for my errors and misconceptions.

Roger Jones, the Director of Academic Publishing for Allen & Unwin, has been at all times encouraging and understanding, and it has been a pleasure to work with him.

K.L.W.
Sydney

Acknowledgements

I am grateful to the following individuals and organizations who have given permission for the reproduction of copyright material:
Figures 2.7 & 4.2 reproduced from R. Jenkins, *An introduction to X-ray spectrometry* by permission of John Wiley & Sons Ltd. Copyright 1974; Figures 4.5 & 9.7 reproduced from S. J. B. Reed, *Electron microprobe analysis* by permission of Cambridge University Press; Figures 8.13 & 8.14 reproduced from R. Tertian & F. Classe, *Principles of quantitative X-ray fluorescence analysis* by permission of Cambridge University Press.

Contents

Preface	*page*	vii
Acknowledgements		ix
List of tables		xiii

1	Introduction	1
	1.1 Spectrometric analysis	1
	1.2 The growth of X-ray spectrometry	2
	1.3 The essentials of X-ray spectrometry	8

2	The nature and production of X-ray spectra	10
	2.1 X-rays and X-ray spectra	10
	2.2 Energy transitions of 'inner orbital' electrons	16
	2.3 Energy transitions of 'outer orbital' electrons	27

3	The interaction of X-rays with matter	34
	3.1 X-ray absorption	34
	3.2 Scattering and X-ray diffraction	42

4	X-ray dispersion and detection	47
	4.1 Crystal dispersion	47
	4.2 Detectors and counting equipment	58
	4.3 Gas-filled detectors	60
	4.4 Scintillation detectors	88
	4.5 Solid-state (semiconductor) detectors	91

5	X-ray spectrometers	100
	5.1 Wavelength-dispersive spectrometers	100
	5.2 Energy-dispersive spectrometers	110

6	Summary of instrumentation	121
	6.1 X-ray fluorescence spectrographs	121
	6.2 The electron microprobe	129

CONTENTS

7 Qualitative and quantitative X-ray spectrometric analysis 139

    7.1 Qualitative analysis 139
    7.2 Quantitative analysis 142
    7.3 Sources of error in quantitative analysis 143

8 Composition-dependent errors in X-ray fluorescence analysis 169

    8.1 Interference 169
    8.2 Interelement ('matrix') effects 179
    8.3 Correction procedures 183
    8.4 Examples of matrix correction procedures 209

9 Composition-dependent errors in electron microprobe analysis 222

    9.1 Interference 222
    9.2 Interelement ('matrix') effects 224
    9.3 The 'simplified theory' approach 227
    9.4 The empirical (α-factor) approach 268

Appendix A   Tables of X-ray analysis parameters for XRS data correction 271

Appendix B   Some fundamental statistical concepts 354

Bibliography 364

Index 367

List of tables

2.1	Electronic transition levels	*page* 18
3.1	Absorption filters used in X-ray diffractometry	42
4.1	Characteristics of commonly used dispersion crystals	50
4.2	Inert gas ionization constants	63
4.3	Comparative resolutions of different types of detector	95
6.1	Generalized guide to primary tube selection and operation	125
8.1	Typical spectral interferences in silicate analysis	176
9.1	Selection of parameters for characteristic fluorescence	257
A.1	Polynomial coefficients for calculation of X-ray emission and absorption wavelengths	275
A.2	Polynomial coefficients for calculation of mass absorption coefficients	276
A.3	Wavelengths and energies: K-series	277
A.4	Wavelengths and energies: L-series	278
A.5	Wavelengths and energies: M-series	281
A.6	Mass absorption coefficients for Kα emissions	283
A.7	Mass absorption coefficients for Lα emissions	293
A.8	Mass absorption coefficients for Mα emissions	309
A.9	Mass absorption coefficients for high side of K edge	318
A.10	Mass absorption coefficients for high side of L_I edge	324
A.11	Absorption jump ratios and fluorescence yields for K fluorescence	336
A.12	Absorption jump ratios and fluorescence yields for L fluorescence	337
A.13	Norrish and Hutton coefficients for silicate analysis, normalized to G1:W1	338
A.14	Norrish and Hutton coefficients for silicate analysis, normalized to fusion mix	339
A.15	Sample Pascal program for calculation of mass absorption coefficients	340

1 Introduction

1.1 Spectrometric analysis

The modern analytical chemist has many techniques available for the determination of elemental composition of materials samples. In particular, advances in electronics have led to the development of a wide variety of instrumental methods to supplant many (but not all) of the older classical chemical procedures. Most of the instrumental methods are comparative rather than absolute and the equipment required may be rather expensive, but in general their convenience, sensitivity, accuracy and precision have made them very attractive and they have completely changed the face of both qualitative and quantitative analysis.

Instrumental methods include a variety of *spectroscopic* or *spectrometric* techniques, which depend on the analysis of some portion of the electromagnetic spectrum emitted or absorbed by the sample under the conditions of analysis. The spectra are characteristic of the elements present in the sample (and, in some cases, of the form in which they are present), and analysis is made by comparing results obtained from the analytical sample with those obtained under the same conditions from calibration standards. Measurements are usually made within a discrete range of wavelengths, so that in each case the heart of the instrumentation consists of one or more *spectrometers* – devices to isolate the chosen wavelength or range of wavelengths for measurement.

Wavelengths emitted or absorbed by elements of analytical interest range from the gamma- and X-ray regions (<100 Å, where 1 Å $= 10^{-10}$ m) through the ultraviolet and visible regions to the infrared (*c.* 7000 Å). No single spectroscopic technique can conveniently span this range, which corresponds to a spectral energy range of approximately 1–15 000 electronvolts, and few laboratories can afford the variety of equipment required to encompass all of the various spectroscopic techniques. Most interest therefore attaches to spectrometers which are either relatively inexpensive (flame emission, atomic absorption, colorimeters etc.) but well suited to specific applications, or to instruments which are particularly versatile in terms of good performance over a wide range of applications. X-ray spectrometers, and particularly X-ray fluorescence (XRF) and electron probe analysis (EPA) systems, fall into the second category and are currently in wide use throughout the materials sciences.

X-ray spectra are relatively simple, and therefore interference from one element in the determination of another is not normally a major problem. Reliable XRF and EPA systems are currently available from several commercial manufacturers. They can be used for many different kinds of sample, with analytical detection limits commonly of the order of 1 ppm for XRF. EPA methods are somewhat less sensitive, but have the added advantage of high spatial resolution (i.e. the ability to analyse very small samples, such as dust particles or small inclusions of one phase in another), and they are characterized by high absolute sensitivities – they can detect as little as 10^{-15} g of an element in the small sample volume that they analyse. Both techniques are equally useful in high concentration ranges, which is not always true of other sensitive spectroscopic techniques, and with adequate calibration they meet high standards for accuracy and precision.

Both XRF and EPA are essentially non-destructive, although some form of sample preparation is almost invariably required. They can complete complex, multi-element analyses in a few minutes per sample, so that they are well suited to applications involving large numbers of samples. XRF methods in particular are also widely used for on-line process control – for example in ore-treatment plants.

Their principal disadvantages are the relatively high costs of the necessary instrumentation, and their inefficiency in the light element range. At present EPA systems can be used only for the analysis of elements heavier than boron, and XRF systems are limited to those heavier than nitrogen; in each case performance for the two or three lightest elements in the application range is also somewhat limited.

XRF and EPA differ from each other principally in the method used to produce, or *excite*, X-ray spectra from the samples being analysed. In EPA the excitation energy is obtained by irradiation of the sample with a beam of energetic electrons, whereas in XRF the primary energy source is a flux of X-rays obtained from a conventional sealed X-ray tube. Each method has advantages and disadvantages making it more or less appropriate for various specific applications. Neither can be said to be superior to or more versatile than the other; rather they are complementary techniques which together provide powerful tools for quantitative chemical analysis.

1.2 The growth of X-ray spectrometry

Röntgen announced his discovery of X-rays in 1896, and from his work grew four major fields of research and industrial applications: medical radiography, industrial radiography, X-ray diffraction analysis and X-ray spectrometry. Although the latter is therefore based on principles which have been known and applied in the laboratory for many years, it has only been developed relatively recently into a versatile, more or less routine analytical technique.

By 1911, Barkla had demonstrated that the 'hardness', or energy characteristics, of radiation emitted from an X-ray tube depended on the composition of the tube target. In particular he showed that the emitted spectrum consists of a continuum of X-ray energies on which is superimposed several spectral series of discrete energies characteristic of the target elements (he introduced the nomenclature of K series, L series . . . which is still used to describe the various components of the characteristic spectrum).

By 1913 the Braggs had demonstrated that X-rays could be diffracted by crystalline materials, with the diffraction angles being determined by (and therefore characteristic of) the interplanar spacings of the crystal structures and the wavelengths of the X-rays themselves. This provided a method of using a crystal to disperse X-radiations of different wavelengths, and Moseley was able to construct a crystal-dispersing spectrometer which he used to record the spectra emitted by various elements and to study the relationship between X-ray characteristics and atomic number of the emitting element. He showed that within each of Barkla's spectral series, emission wavelengths (λ) vary systematically with atomic number Z (Fig. 1.1).

The variation can be expressed as:

$$1/\lambda = K(Z - \sigma)^2 \qquad (1.1)$$

where K and σ are constants depending on the spectral series (i.e. the K,

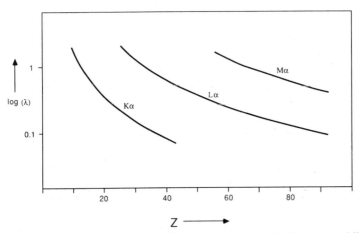

Figure 1.1 Variation of the wavelength (λ) of some characteristic X-ray spectral lines with atomic number (Z) of the emitting atoms, plotted for the strongest lines in each of the K, L and M series.

INTRODUCTION

L and M series of Fig. 1.1). This relationship has become known as 'Moseley's Law'. It implies that determination of wavelengths for the various emissions, or *lines*, in an emitted X-ray spectrum allows identification of the atoms emitting them, and it was soon realized (intuitively at first, by analogy with already developed principles of optical spectra, but subsequently confirmed by experiment) that the relative intensity with which a particular characteristic line is emitted is a function of the concentration of its emitting element in a multi-element target. By 1922 Hadding had developed a method for the analysis of minerals, based on these concepts.

In these early stages, the samples being investigated were fabricated into or were coated on to the surfaces of the targets of so-called 'demountable' X-ray tubes (i.e. tubes constructed to allow the interchange of targets, followed by evacuation and operation under continuous vacuum pumping). The X-ray spectra were then produced by excitation with a beam of accelerated electrons, as in the modern electron microprobe, which is closely analogous in its construction principles to the demountable X-ray tube (Fig. 1.2).

The spectra were dispersed into their component wavelengths by rotatable diffracting ('analysing') crystals. NaCl and $CaCO_3$ were widely used initially, but have now been completely supplanted by other materials that give better results. The spectra were recorded on photographic film, and, from the film record, the wavelength of a spectral line could be calculated from the Bragg diffraction equation:

$$n \lambda = 2d \sin \theta \tag{1.2}$$

where d is the known interplanar distance of the diffracting crystal (discussed in more detail in Ch. 3), θ is the diffraction, or Bragg angle, and n is an integer (1, 2, 3 . . .) expressing the *order* of the diffraction.

Although relatively clumsy by modern standards, this experimental configuration worked well and was responsible for spectacular progress in the development of the periodic classification of the elements. The fundamental significance of the atomic number Z was established, and it was possible to order some elements properly in the Periodic Table that had previously been incorrectly placed on the basis of atomic weight; e.g. Co ($Z = 27$, atomic weight = 58.9) and Ni ($Z = 28$, AW = 58.7). The number of elements from hydrogen to uranium was shown to be 92, which was interesting since six of them had not then been discovered (Tc (43), Pm (61), Hf (72), Re (75), At (85) and Fr (87)). Of these, Hf was first identified in the X-ray spectrum of a sample of zircon, and Re in a columbite X-ray spectrum.

Moseley and others were well aware of the analytical potential of X-ray spectrometry, but were hampered in its routine application by several problems, e.g.:

THE GROWTH OF X-RAY SPECTROMETRY

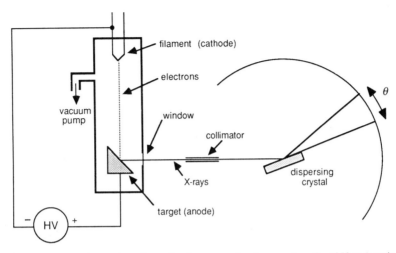

Figure 1.2 An experimental configuration for measuring the wavelengths of X-rays emitted from a selected target material. Electrons from a heated tungsten filament (cathode) are accelerated by a high DC voltage towards the target material, which is coated on the front surface of an anode. With appropriate electron energies, the target material emits its characteristic X-ray spectrum, some of which leaves the tube through a thin window and is dispersed (separated into its constituent wavelengths) by a rotatable crystal (see Ch. 4). The X-ray tube is continuously pumped to low pressures instead of being evacuated and sealed, to allow interchange of target materials. An electron microprobe is, in effect, a large, continuously pumped or *demountable* X-ray tube of this type, with the addition of electromagnetic lenses to focus the electron beam on to a selected small area of the target surface and multiple spectrometers to analyse the emitted spectra (see Ch. 6).

(a) the tube had to be repeatedly evacuated when samples and standards were interchanged, which made it difficult to set-up reproducible analytical conditions;
(b) because of the low pressures required for X-ray generation, volatile samples, including most liquids, could not be analysed;
(c) heat generated by interaction with the electron beam (see later discussion) was likely to damage the sample, despite water cooling of the target or its substrate;
(d) recording and measurement of the films was slow, inconvenient and of limited accuracy (at least in terms of intensity).

Excellent sensitivities were attained under ideal conditions, but in general the method was not well suited to routine analysis. In 1925 von Hevesy suggested that the problems of heat damage and sample volatility (and also of relatively high continuous background, which obscured weak emissions and hence restricted sensitivity) could be overcome by exciting the emitted spectra not with electrons but with a beam of sufficiently energetic X-rays derived from a primary tube which could be sealed rather than made demountable. Use of a primary energetic spectrum to excite a secondary, slightly less energetic spectrum is known as *fluorescence*, and the recording and study of electromagnetic spectra is called *spectrography*. The new technique became known as *X-ray fluorescence* spectrography*.

However, although some progress was made in this direction and the first commercial instruments became available late in the 1930s, difficulties were still encountered in obtaining the stability and sensitivity required to compete with established analytical procedures and successful development had to await the technological advances of the 1940s and later years.

In 1938 Hamos described an X-ray spectrographic system which allowed study of variations in chemical composition over the surface of a solid sample, i.e. a system which provided some degree of the *spatial resolution* which is now regarded as one of the major attributes of the electron microprobe. In the microprobe, resolution of this nature is achieved by focusing the primary electron beam on to a very small portion of the sample; heat problems are minimized by keeping the electron beam current very low – usually of the order of 10–100 nA – but the resulting emitted X-ray intensities are then also very low and could not have been reliably detected or measured by the equipment available in the 1930s.

X-rays cannot be focused in this way, but Hamos managed to achieve a modest degree of spatial resolution by using a special spectrometer to isolate and examine only those X-rays emitted from a small area of the sample surface (of the order of 50 μm diameter, compared with about 1 μm for the electron microprobe); the whole of the sample surface was excited by a beam of primary X-rays, but the spectrometer 'viewed' only the fluorescent X-rays emitted from a small portion of that surface. A picture of the distribution of a particular element over the sample surface could be built up by mechanically translating the sample, on a raster pattern, with respect to the spectrometer and simultaneously recording the variation in intensity of a suitable emitted X-ray line; the same result is now conventionally achieved in electron microprobe analysis by keeping the sample stationary and electronically scanning the electron beam.

An alternative device known as the X-ray macroprobe represented a variant on Hamos's procedure in that the primary X-ray beam was collimated so that it only irradiated a small area on the sample surface. The macroprobe was marketed commercially during the 1960s, but it was

---

* The term 'spectrography' strictly means the recording of spectra; 'spectroscopy' is the generation and observation of spectra, and 'spectrometry' is their measurement. To some extent these three terms are now used more or less interchangeably.

soon supplanted by the markedly superior spatial resolution of the electron microprobe.

Use of a focused beam of electrons to obtain spatial resolution appears to have been first suggested in a 1943 patent application by Hillier, but the concept was not immediately pursued. The very low X-ray intensities were still difficult to measure accurately, even though the original use of film as the recording medium had given way to thin-window Geiger counters and associated electronic counting circuitry. However, progressive improvements continued to be made in detector design and in the development of more efficient focusing spectrometers, and in 1949 Castaing and Guinier described the first workable *'microsonde electronique'*, or electron microprobe. In his doctoral thesis, Castaing (1951) formulated most of the design and operating principles on which modern microprobe analysis is based.

Further improvements have been made in standards of mechanical and electronic performance to bring both XRF and EPA to their present status of routine analytical techniques rather than research innovations. Major emphasis in recent years has been placed on the incorporation of small, dedicated computer systems to control instrumental operating conditions, make the necessary measurements, and process the analytical data on-line to give immediate results. Few commercial instruments of either type are now sold without some degree of mini- or microcomputer automation, and in fact some are so highly controlled that routine operator intervention is virtually limited to maintenance and the loading and unloading of samples (particularly in XRF).

It should be noted, however, that while this situation has obvious advantages, particularly in industrial environments, it has also become regrettably easy to overlook proper standards of analysis appraisal. In particular, the procedures used in data processing may often obscure the fact that the raw analytical data is, for one reason or another, of poor quality (e.g. by automatically totalling all analyses to 100 per cent!). When using highly automated equipment, the analyst must guard constantly against the tendency to relax proper standards of assessment.

There has also been considerable progress in recent years towards the partial replacement of crystal-dispersive spectrometers with so-called *'energy-dispersive'* types which use electronic methods instead of Bragg-type crystal diffraction to isolate the lines of analytical interest from the remainder of the emitted spectrum. They are very effective in many applications, but nevertheless it is unlikely that these 'EDS' systems will completely supplant the older crystal, or *wavelength-dispersive* ('WDS') types. The two systems are again best viewed as complementary, and microprobe laboratories in particular gain great advantage if their instrumentation is equipped with both.

INTRODUCTION

1.3 The essentials of X-ray spectrometry

In its practical analytical applications, such as XRF or EPA, X-ray spectrometry is based on the analysis of the X-ray spectrum generated from a sample under controlled conditions. Elements present in the sample are identified from the wavelengths of characteristic lines present in the emitted spectrum, and their proportions are determined from the relative intensities of a line selected for each element, compared to emission intensities from the same elements in one or more calibration standards of known composition. Before calculation of compositions, measured intensities (and in some cases wavelengths/energies) must be corrected for instrumental errors and for the effects that each element has on its own emission intensities and those of other elements present in the sample.

It is therefore necessary for the X-ray analyst to have at least a basic working knowledge of the physics of X-rays; e.g. their physical nature and their relationships to other regions of the electromagnetic spectrum, their generation and the nature of their interaction with matter, their diffraction by crystalline materials etc. These topics are reviewed in Chapters 2 and 3, at a level which is certainly not comprehensive but should provide an adequate grounding upon which the novice can build as he/she gains practical experience.

Chapters 4 and 5 explore the basis of the devices that are used to isolate X-rays of a chosen wavelength/energy (or, more accurately, of a narrow *range* of wavelength/energy) and to measure the intensity with which they are emitted from a sample under analytical conditions. The dispersion and intensity measurement devices collectively constitute the *spectrometers* which are the heart of the analytical instrumentation. The remaining essential characteristics of the most common XRS instruments (XRF and EPA), including the energy sources used to generate the analytical X-ray spectra, are reviewed in Chapter 6.

Chapter 7 introduces the application of XRS techniques to both qualitative and quantitative chemical analysis, and reviews the principal sources of potential error – some of these are consequences of the quality of construction and maintenance of the laboratory instrumentation, some are functions of operator skill and diligence, and some are inherent in the fundamental physics of the technique. It is important that the analyst learns to distinguish between them and to develop his/her analytical strategies accordingly.

Chapters 8 and 9 treat composition-dependent sources of error in XRF and EPA, respectively, in sufficient detail to provide a basic awareness and understanding that can be used to develop a useful approach to most more or less routine analytical problems.

Illustrative examples are quoted throughout the text, but it is hardly practical in a book of introductory scope to attempt to provide an encyclopaedic summary of solutions to a wide range of specific analytical problems. Rather this book has been envisaged as a reasonably compact introduction to the chemical analytical applications of X-ray spectrometry, to serve as a basis for subsequent wider reading of more comprehensive treatments and particularly of the journal literature in which specific 'case histories' abound. These can be reviewed more critically and more usefully when the beginner has mastered the fundamentals on which Chapters 2–9 have been focused.

2 The nature and production of X-ray spectra

2.1 X-rays and X-ray spectra

X-rays may be defined for our purposes as being that part of the electromagnetic spectrum having wavelengths in the region of 0.1–100 Å (1 Å = 10^{-10} m). X-rays may also be regarded, on a quantum basis, as photons of energy, which is normally expressed in X-ray spectrometry in kilo electron volt (keV) units; energy may also be expressed in joules, with the relationship between the two units being given by

$$1 \text{ keV} = 1.602\ 19 \times 10^{-16} \text{ J (coulomb-volts)} \quad (2.1)$$

Energy and wavelength are related by the fundamental equation

$$E = h\nu = (hc) / \lambda \quad (2.2)$$

where E is the energy, λ the wavelength, ν the frequency, c the velocity of propagation of electromagnetic radiation and h is Planck's constant (6.626×10^{-34}). Thus the energy equivalent of an X-ray photon having a wavelength of 1 Å would be given by

$$\begin{aligned}
E_{(1\ \text{Å})} &= (6.626 \times 10^{-34}) \times (3 \times 10^{8}) / (1 \times 10^{-10}) \\
&= 2 \times 10^{-15} \text{ joules (approx.)} \\
&= (2 \times 10^{-15}) / (1.602 \times 10^{-19}) \text{ eV} \\
&= 12.4 \text{ keV}
\end{aligned}$$

Thus, if E is expressed in keV and λ in Å,

$$E = 12.4 / \lambda \text{ keV} \quad (2.3)$$

If E is expressed in eV and λ in Å,

$$E = 12\ 400 / \lambda \text{ eV} \quad (2.4)$$

These expressions suffice for most practical purposes, but there are actually slight inconsistencies between the energy and wavelength scales due to uncertainty in the accepted values for h and c. Some treatments use the more precise relationship:

$$E = 12.3964 / \lambda \tag{2.5}$$

For many years X-ray wavelengths were expressed in terms of the 'X' unit, which was originally based on the first-order grating constant of calcite and was intended to be numerically equal to 10^{-3} Å. However, calcite was a poor choice of standard, since there are differences from one calcite crystal to another and there were consequently some inconsistencies in early wavelength measurements. When these were resolved the resulting X units were found to differ slightly from their anticipated relationship to Ångström units; in fact

$$1 \text{ Å} = 1.002\,02 \text{ kX}$$

More recently crystallographers and spectroscopists have returned to the Ångström unit as the unit for wavelength measurement; however, some tables still in use in X-ray spectrography laboratories are compiled in X units.

X-rays may be produced in many different ways – for example, the decay of K to Ar involves the capture of an electron by the nucleus and the consequent emission of an X-ray photon. Most commonly, however, they are produced by bombarding matter with appropriately energetic photons, such as accelerated electrons, gamma-rays or other X-rays. In the electron microprobe, the X-rays used for analysis are generated by bombardment of the sample by a beam of accelerated electrons, and in XRF they are produced by a flux of primary X-rays produced by a sealed X-ray tube.

2.1.1 The continuous spectrum

The X-ray spectrum produced by electron bombardment consists of two parts, called the *continuous spectrum* (or *continuum*) and the *characteristic spectrum*. The continuum is always present whenever primary electron radiation has energies within the X-ray band; the characteristic spectrum is only present, however, when the exciting radiation has energies in excess of a critical excitation energy which is uniquely characteristic for each spectral series of each of the elements present in the emitting sample. No continuum is produced under conditions of X-ray fluorescence (except that produced in the primary X-ray tube), and once again the characteristic spectrum is only produced if the primary X-rays include photons with energies higher than the relevant critical excitation energies.

According to modern theory, the continuous spectrum is produced by deceleration of incident electrons through inelastic collisions with atoms of the target material. Energy lost in each such collision (and a single electron may undergo a series of collisions before finally losing all its energy, or 'coming to rest') is emitted as an X-ray photon of corresponding energy. Thus if an electron lost 12.4 keV of energy in a single collision, the result would be the emission of an X-ray photon with an energy of 12.4 keV (or a wavelength of 1 Å).

Clearly the proportion of its total energy that an electron loses in such a collision can range from almost zero in a barely 'glancing' collision up to 100 per cent if the total energy is lost in one collision. The X-ray spectrum produced in this way will thus consist of a continuous range of energies between the limits of zero on one hand and the maximum energy of the electrons on the other. The maximum emitted X-ray energy (corresponding to the minimum wavelength) will be given by

$$E_{max} = 12.4 / \lambda = E_0(kV) \quad \text{(from Eqn 2.4)}$$

where $E_0(kV)$ is the electron accelerating potential in kV.

The distribution of radiation between these limits is given by a formula due to Kramers:

$$I(\lambda) \cdot d\lambda = K\, i\, Z\{(\lambda / \lambda_{min}) - 1\} \{1 / \lambda^2\} \cdot d\lambda \qquad (2.6)$$

where i is the X-ray tube current and Z the mean atomic number of the target; K is a constant. The general form of this distribution is shown in Figure 2.1, which includes two curves – one for the distribution of continuous radiation *generated within* the target, and the other for radiation *emitted from the surface* of the target (after some of the X-rays generated at depth have been absorbed by the target itself).

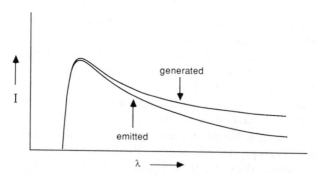

Figure 2.1 The general form of the variation of continuum intensity with wavelength. There is a sharp short-wavelength (high energy) limit which contrasts with a gradual decline to lower intensities towards the longer wavelengths. The upper curve represents the intensities actually generated within the target sample. Some of these, however, are absorbed by the sample itself or by the tube window. The emitted intensities are therefore reduced, or *attenuated*, particularly towards the lower-energy, longer wavelengths.

The wavelength corresponding to the intensity maximum can be obtained by differentiating the Kramers Formula, viz.

$$I(\lambda) = K' \left(\frac{1}{\lambda \cdot \lambda_{min}} - \frac{1}{\lambda^2} \right)$$

and again,

$$\frac{dI}{d\lambda} = -\left(\frac{K'}{\lambda^2 \cdot \lambda_{min}} - \frac{2K'}{\lambda^3} \right)$$

At the point of inflexion, $\lambda = \lambda_{max}$ and $dI/d\lambda = 0$, so that

$$1/\lambda_{min} = 2/\lambda_{max} \qquad (2.7)$$

whence at the intensity maximum $\lambda_{max} = 2\lambda_{min}$. Thus the wavelength of maximum intensity will be approximately twice the minimum wavelength.

This relationship is only approximate because the continuum actually emitted from the target is considerably modified by self-absorption – some of the X-ray photons are generated at depth within the target and may be absorbed in interaction with the target atoms (see later discussion). This is particularly so for the low-energy, long-wavelength, 'soft' X-rays. The nature and extent of self-absorption will depend on the composition of the target and will also be a function of electron energy, since high-energy electrons can generate X-rays at greater depths within the target. The continuum may also be further modified by absorption losses in the window of the X-ray tube, Compton scattering (see Sec. 3.2) etc., but the general form of the curve shown in Figure 2.1 is usually maintained.

Increasing the accelerating potential V on the electrons will have two principal effects:

(a) the short-wavelength limit of the continuum will shift to shorter wavelengths, in accordance with the relationship

$$\lambda_{min} = 12.4/V \qquad \text{(Eqn 2.3)}$$

(b) the wavelength of maximum intensity will also shift towards lower wavelengths.

Since each electron also has more energy to lose in collisions, the intensity of any particular wavelength will also increase as V is increased. These relationships are shown in Figure 2.2.

The Kramers formula also indicates that the continuum intensity for any particular wavelength increases linearly as a function of Z. The

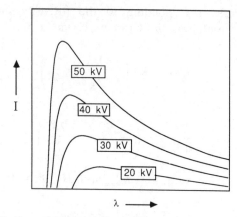

Figure 2.2 The distributions of continuum intensities produced in an X-ray tube operated at different electron accelerating potentials. At higher potentials the short wavelength limit moves towards lower wavelengths (higher energies), and the intensities increase at all wavelengths.

integrated continuum intensity is thus greater for heavy target atoms than for light. Since the continuum from the primary tube is often used as the excitation source in X-ray fluorescence, general-purpose XRF tubes are often fitted with heavy-metal anodes (e.g. W, Au).

The total energy of the continuous spectrum, expressed as a proportion P of the total energy of the incident electrons, is given by the empirical relationship

$$P = 1.1 \times 10^{-9} \times Z \times V$$

in which Z is the mean atomic number of the target and V is the accelerating potential (in volts). This confirms that the integrated continuum intensity increases linearly with increasing Z and/or V.

It is also interesting to note that in a Cr-target X-ray tube operated at 50 kV only 0.13 per cent of the incident electron energy is converted to continuous radiation; even in an Au-target tube at 100 kV this figure rises only to 0.87 per cent. A very small amount of additional energy may be converted to characteristic radiation (see below), and most of the remainder is dissipated as heat. This explains why the early investigators encountered difficulties due to heat damage of samples mounted on demountable tube targets. Modern X-ray tubes operated at ratings of up to 4 kW must provide for very efficient cooling of the targets, usually by water flow, and the targets should ideally be made of high-melting materials of good thermal conductivity.

2.1.2 Characteristic spectra

As noted earlier, the continuous spectrum is produced whenever energetic electrons interact with matter, so long as the electrons have energies in the X-ray range. In addition, characteristic, or *line*, spectra are produced whenever the electrons have energies in excess of a critical excitation energy which varies from element to element in the target. Thus in Figure 2.3, which shows the spectra produced from a molybdenum target tube at a range of electron accelerating voltages from 5 to 25 kV, it can be seen that only the continuous spectrum is excited below 20 kV. At 25 kV two sharp peaks appear, superimposed on the continuum at wavelengths below 1 Å. These are called the Kα and Kβ *lines* of molybdenum; they always appear at identical wavelengths in the spectra from any Mo-bearing samples, providing the accelerating potential exceeds 20.0 kV, which is called the *critical excitation potential* for the Mo K spectrum. The critical excitation potential for the longer-wavelength, lower-energy Mo L spectrum is 2.9 keV, and the Mo L lines appear in the longer-wavelength region, beyond 3 Å, if the tube

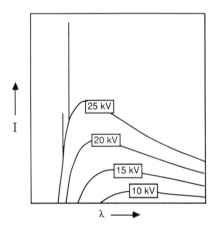

Figure 2.3 The X-ray spectra produced from a molybdenum target at electron accelerating potentials of 10–25 kV. Below 20 kV, only the continuum is in the wavelength range of this figure, but at all higher potentials sharp, relatively intense characteristic emissions appear at wavelengths of 0.632 Å (19.6 keV) and 0.709 Å (17.5 keV). These are called the Mo Kβ and Mo Kα lines respectively. All elements other than H and He have similar critical excitation potentials at which groups of characteristic line spectra appear, and all except the lightest elements have several groups in their line spectra which appear at different potentials and are called the K, L, M. . . spectra. The actual critical excitation potentials and the wavelengths of the characteristic lines are unique to each element, so that an element can be identified from the 'fingerprint' of its characteristic spectrum, and its concentration in the target can be estimated from the relative emission intensities of one or more of the characteristic lines. This is the basis of X-ray spectrometry.

potential exceeds 2.9 kV. No other element emits identical lines at identical wavelengths.

The characteristic spectrum is produced whenever the incident electrons have sufficient energy to excite electrons from the 'inner' sub-shells of the target atoms to higher-energy orbital levels; such excitation can take place only to a vacant energy level, which for a ground state atom means beyond the Fermi level (which means, effectively, removal of the electron to beyond the valence band – see Section 2.3).

The excited atom is unstable, and regains its initial ground state by transfer of high-energy, outer orbital electrons to the temporarily unfilled inner level. Each such transition releases energy, which is then emitted as an X-ray photon whose energy in turn is equal to the difference between the initial and final energy states of the transferred electron:

$$E_{photon} = E_{initial(electron)} - E_{final(electron)} \qquad (2.8)$$

Thus the wavelength of the emitted photon depends on the energy levels of the electron orbitals of the excited atom, which are quantized and characteristic of each atomic species. Since the transitions primarily involve inner orbital levels rather than valency or conduction band electrons, in general the energy or wavelength of the emitted photon is not dependent on the valency or bonding state of the atom (in fact under certain circumstances small variations in the emitted energies can occur, and they can actually be used to derive valuable information about valencies and bonding states – see Section 2.2.2).

2.2 Energy transitions of 'inner orbital' electrons

2.2.1 Quantum theory of electron transitions

In simple terms, atoms consist of nuclei surrounded by electrons. Each electron may be considered as existing in a particular energy state which can be described by four parameters – the so-called 'quantum numbers' of the electron. These are as follows:

(a) n, the *principal quantum number*, which can have only positive integer values 1, 2, 3. . . . Electrons for which $n = 1$ are said to belong to the K shell, for $n = 2$ the L shell, for $n = 3$ the M shell and so on.

(b) l, the *angular quantum number*, which can have values ranging from 0 to $(n - 1)$. Thus for K shell electrons, the only permissible value of l is zero. For L shell electrons, l can be 0 or 1, for M shell 0, 1 or 2 and so on. Spectrographers, following the early conventions of optical spectrography, refer to electrons with $l = 0$ as 's' ('sharp') orbitals, with $l = 1$ as 'p' ('principal') orbitals, with $l = 2$ as 'd' ('diffuse') orbitals and with $l = 3$ as 'f' ('fundamental') orbitals.

(c) m, the *magnetic quantum number*, which can take values from $-l$ through 0 to $+l$. Thus for K shell electrons, m is restricted to 0; for L shell electrons, in which $l = 0$ or 1, $m = 0$, $+1$ or -1, and so on.

(d) s, the *spin quantum number*, which only has values of $\pm 1/2$.

The Pauli exclusion principle states that no two electrons in the same atom can have the same four quantum numbers. Hence there can be no more than 2 electrons in the K shell, 8 electrons in the L shell, 18 in the M shell or 32 in the N shell (i.e. the maximum number of electrons in the nth shell is equal to $2n^2$). The Periodic Table of the elements can be neatly explained on the basis of progressive filling of energy levels, from the K shell (lowest energy) through L, M, N . . . shells, and numerous texts give details of the electronic structures of the various elements.

The possibility of interaction between the electrons must also be considered, particularly when electronic transitions are being reviewed. It is important to consider the coupling of the l and s quantum numbers, in terms of j, the vector sum of the l and s moments:

$$j = l + s$$

where j is the total moment which, in conjunction with l, determines the so-called *transition levels* between which electrons may be transferred. Where $l = 0$, the orbit is spherically symmetrical and there is only one possible value of j. For each value of l greater than or equal to 1 there are two possible values of j. The resulting transition levels are listed in Table 2.1, from which it is evident that there is only one transition level associated with the K shell, three with the L shell, five with the M shell and so on. These are conventionally designated the K, L_I, L_{II}, L_{III}, M_I, M_{II}, M_{III}, M_{IV}, M_V, N_I . . . levels.

Transitions between these various levels are permissible, according to quantum theory, as long as certain selection rules are satisfied, namely that

(a) $|\Delta n| \geq 1$;

(b) $\Delta l = +1$; and

(c) $\Delta j = +1$ or 0.

Thus $L_{II} \to K$ or $L_{III} \to K$ transitions are permitted, but $L_I \to K$, for which $\Delta l = 0$, is not (Table 2.1).

This very brief summary of quantum theory is sufficient to provide an explanation of the relationships observed in the characteristic spectra

Table 2.1 Electronic transition levels.

Transition level	l	j
K	0	1/2
L_I	0	1/2
L_{II}	1	1/2
L_{III}	1	3/2
M_I	0	1/2
M_{II}	1	1/2
M_{III}	1	3/2
M_{IV}	2	3/2
M_V	2	5/2
N_I	0	1/2
N_{II}	1	1/2
N_{III}	1	3/2
N_{IV}	2	3/2
N_V	2	5/2
N_{VI}	3	5/2
N_{VII}	3	7/2
.	.	.
.	.	.

produced by either electron or X-ray excitation, and to provide a basis for the conventional nomenclature applied to X-ray spectral lines. It is important to note that the nomenclature was developed from the work of Barkla and others, long before the quantum theory of the atom had been derived, and while the observed relationships are explained adequately by quantum theory, the older nomenclature does not always appear now to be entirely logical.

When an inner orbital electron (e.g. a K shell electron) is displaced from the target atom, the latter can return to its stable ground state by any of a number of processes, among which two predominate.

First, an electron from an 'upper' or 'outer' level (i.e. a transition level for which n is greater than that of the ionized level) may transfer to the excited level. In so doing it will lose energy, which will be emitted as a photon or succession of photons whose characteristic energies will be determined by the quantized energy levels concerned. Return to the ground state may be by a single electron transition from the conduction or valency band, but will much more probably be by a series of transitions, each involving transference between adjacent or nearly adjacent transition levels, and each resulting in the emission of a characteristic photon.

Thus if the initial excitation involved removal of a K shell electron, the atom would be excited to the K^+ state. Transference of an electron from, say, the L_{III} level to the K level would change the state of the atom from K^+ to L^+, thereby reducing its energy and resulting in the emission of a spectral photon identified as a $K\alpha_1$ photon. The 'K' indicates that it was produced by a transference to the K level, the 'α' that it came from the L level, and the subscript '1' that it came from L_{III} rather than L_{II} (note that the $L_I \rightarrow K$ transition is forbidden).

If the transition were from M_{III} to K the resulting emission would be identified as $K\beta_1$, which is quite logical; unfortunately for most of the other possible transitions the logic of conventional identification is more difficult to perceive, although this comment applies only to the *nomenclature* and not to the transition principles, which remain the same. Figure 2.4 is an electron energy level diagram for a heavy atom (i.e. one with a large number of possible transitions), in which the nomenclature conventionally adopted for the more important transitions is shown. The L^+ state is still an excited state, and a further electron transfer might then take place from, say, the M_V level, changing the atom from the L^+ to the M^+ state and resulting in the emission of an $L\alpha_1$ photon. Alternatively, a change from L^+ to N^+ would produce an $L\beta_2$ photon.

Such sequences will continue until the atom regains its stable ground state; obviously the last electron transferred will come from the energy 'extremities' of the atom where electron energy levels are almost continuous (e.g. the conduction or valence bands). Although it would be possible to de-excite by a single transition of a conduction band electron direct to the K level, this would be statistically improbable in all but the lightest atoms and the resulting spectral emission would be very weak. The Al $K\beta_{1,3}$ emission is produced by a 3p $\rightarrow$ K transition involving valence electrons in the unfilled 3p band; although very weak it is usually visible in the Al K spectrum as a broad line whose shape is strongly dependent on the chemical bonding of the aluminium atoms. The $K\beta_{1,3}$ emission from heavier atoms (e.g. Sn) is much stronger relative to the $K\alpha_{1,2}$, but in such cases it does not represent a transition from the valence band, and it is much less sensitive to bonding or chemical state.

The relative probabilities of the various de-excitation routes can be predicted; they are constant for any one element, for which a reproducible relative intensity scale can therefore be established, but they vary somewhat from one element to another. There are some regularities that can be simply explained on at least a qualitative basis. Thus L $\rightarrow$ K transitions ($\Delta n = 1$) are more probable than M $\rightarrow$ K ($\Delta n = 2$) or N $\rightarrow$ K ($\Delta n = 3$), and $K\alpha$ emissions are therefore always more intense but less energetic than $K\beta$ emissions from the same element. The $K\alpha : K\beta$ intensity ratio is larger for light elements (about 25 : 1 for Al compared with 3 : 1 for Sn), but the $K\alpha_1 : K\alpha_2$ intensity ratio is always close to 2 : 1

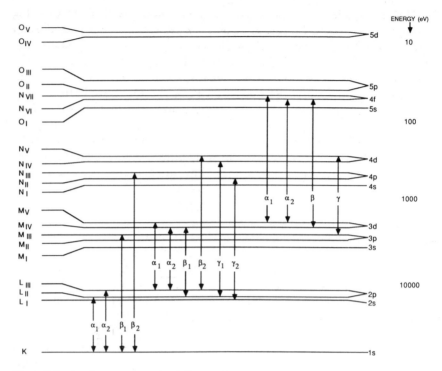

Figure 2.4 An electron energy level diagram for a hypothetical heavy atom, i.e. one with a relatively large number of electrons and hence of electron energy levels. Each 'level' represents the potential energy of the atom with one (and only one) of its electrons removed; if two or more electrons are removed the energy levels are perturbed to different values. The arrowed Greek letters show the nomenclature conventionally applied to the most probable electron transitions to the K, L and M shells, and also to the resulting characteristic X-ray emissions. The nomenclature was applied to the X-ray emissions before the concept of energy level transitions was developed, so that it now appears to have little logical basis.

(these lines are always very close to each other in energy and are usually unresolved or only partly resolved by spectrometers built for routine analytical applications).

Secondly, the radiation produced following the transfer of an upper level electron to an unfilled state is itself energetic and potentially capable of producing a further ionization, so that it may not in fact be emitted from the atom as an X-ray photon. Instead its energy might be consumed in further adjustments to the electron distribution, leading ultimately to the ejection of an electron called a *photoelectron*. For example, initial excitation may produce a K level vacancy which is filled by transfer of an L electron. Instead of emission of a K photon with energy equal to $(E_K - E_L)$, the resulting energy might be used to expel a photoelectron from the M level; the photoelectron will be expelled with an energy of $\{(E_K - E_L) - \Theta_M\}$, where Θ_M is the binding energy of the M level, or the energy required to expel an M electron beyond the Fermi level. This is known as the *Auger process*, and the resulting photoelectron is known as an *Auger electron*.

The Auger Process can cause the production of two or more simultaneous vacancies in the upper levels, e.g. L^+ and M^+ in the example cited, and can thus contribute to the formation of *satellite lines* (see below).

When these two processes are taken into account, it is evident that the intensity of emitted X-radiation depends on the relative efficiencies of the photon emission and Auger emission processes. The ratio of X-ray photon emission to photoelectron emission is called the *fluorescent yield* (ω), and is in practice defined as the ratio of the number of emitted X-ray photons in a given spectral series to the total number of vacancies formed in the same transition level in the same time increment; i.e.

$$\omega_K = (\Sigma(n)_K) / N_K$$
$$= (nK\alpha_1 + nK\alpha_2 + nK\beta_1 + nK\beta_2 + \ldots) / N_K \qquad (2.9)$$

Obviously, the fraction of vacancies filled by the Auger process will be equal to $(1 - \omega)$.

For a given spectral series, ω varies systematically with Z (Fig. 2.5); it decreases markedly with decreasing Z in both K and L spectral series,

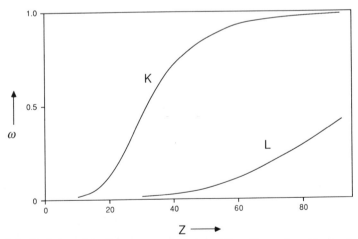

Figure 2.5 Variation of the fluorescence yields (ω) of K and L spectra with atomic number (Z) of the emitting element. The low yields in the low-Z range for each spectral series constitute one of the major factors that limit the sensitivity of X-ray spectrometry in the analysis of light elements.

and the low fluorescent yields for values of $Z < 10$ place fundamental constraints on the use of X-ray spectrographic techniques for light-element analysis, even if various instrumental problems in this region could be overcome. On the other hand, the light elements are obviously ideally suited to Auger electron spectroscopy.

Observed X-ray spectral emissions belong to one of three major groups:

(a) Normal transitions, which can be defined by the simple selection rules given earlier.
(b) A few 'forbidden' transitions, which are observed only as weak emissions, at wavelengths which do not accord with the selection rules.
(c) Lines which arise from simultaneous double ionizations, e.g. K^+L^+. The second vacancy perturbs the single-vacancy energy levels of the atom, resulting in transitions between energy levels that are not those characteristic of the singly ionized state. It has already been noted that the Auger effect provides a partial explanation of these *satellite*, or *non-diagram*, lines. However, satellite lines are most significant in the K spectra of the light elements – the $K\alpha_3,\alpha_4$ satellite doublet of Al, for example, commonly has an intensity of as much as 10 per cent of the $K\alpha_1,\alpha_2$ doublet. This implies dual ionizations by simultaneous double electron impacts.

Although satellite lines are always weak because of the low statistical probabilities inherent in their generation, they may be of analytical consequence if they are emitted by a major element at wavelengths close to the analytical lines being used for trace elements. A classic example is encountered in the determination of trace levels of cobalt in a high-iron matrix, where the satellite Fe $K\beta_0$ line interferes with the Co $K\alpha$ doublet at 1.79 Å (6.924 keV): the most sensitive cobalt line and hence the one that would normally be preferred for trace analysis.

Tables of wavelengths and relative intensities of X-ray emission lines, including both diagram and non-diagram lines, are available and must be consulted whenever analyses are being made in low-concentration ranges. The tables of White and Johnson (1972) are particularly comprehensive and convenient to use.

Features of the characteristic spectrum may be illustrated by reference to the example of barium ($Z = 56$), a relatively heavy atom with filled K, L and M shells and partially filled N and O shells. The transition levels and their associated binding energies are shown in Figure 2.6. In the barium atom, binding energies for the K and L transition levels are as follows:

K: 37.44 keV

ENERGY TRANSITIONS OF 'INNER ORBITAL' ELECTRONS

L_I: 5.99 keV

L_{II}: 5.63 keV

L_{III}: 5.25 keV

M shell binding energies are of the order of 1 keV, and N shell of the order of 0.1 keV.

The energy levels shown in diagrams such as Figure 2.6 are equivalent to the energies of the atom when one electron at a time is removed from each transition level – *and when only one such electron is removed*. As noted above, the single-ionization energy levels are significantly perturbed by double or multiple ionizations.

If electrons with energies of, say, 3 keV strike a barium atom, only the M and N spectra will be excited (and the latter, with wavelengths in excess of 100 Å, is beyond the practical range of X-ray spectrometry). If the incident photon energy rises to 5.25 keV, the L_{III} spectrum will be excited, and the Ll, $L\alpha_2$, $L\alpha_1$ and $L\beta_2$ lines will appear in the characteristic spectrum, in addition to the M spectrum. However, L_{II} and L_I spectra will not be excited because the incident photons have insufficient energy to remove electrons from these transition levels.

The highest energy in the Ba L_{III} characteristic spectrum is that of Ba $L\beta_2$, which has an energy of 5.156 keV (equivalent to a wavelength of 2.404 Å). This emphasizes the point that the critical excitation energy of

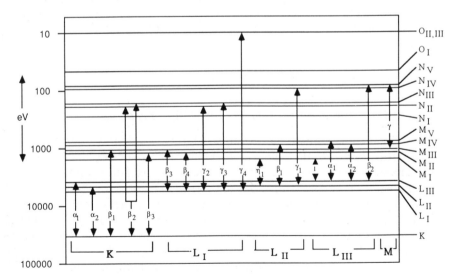

Figure 2.6 Electron energy level diagram for barium ($Z = 56$), showing details of the observed L spectrum (cf. the more generalized diagram of Fig. 2.4).

a given spectral series is always higher than that of the highest energy line in the series.

At 5.63 keV the L_{II} spectrum is excited, and at 5.93 keV the L_I spectrum appears. Considerably more energy is required to displace a K shell electron, so that the critical excitation potential for the Ba K spectrum is 37.44 keV, equivalent to a wavelength of 0.332 Å. The shortest wavelength emission in the Ba K spectrum is Ba $K\beta_2$, produced by the $N_{III} \rightarrow K$ transition, with a wavelength of 0.333 Å.

If the excitation is achieved with primary X-rays derived from a Cr-anode tube, then the incident radiation will consist of the tube continuum, the weak Cr L spectrum (at wavelengths just above 20 Å), and the Cr $K\alpha$ and Cr $K\beta$ lines at 2.29 Å (5.41 keV) and 2.08 Å (5.96 keV) respectively – providing the tube is operated, as it normally would be, above 5.99 keV, the critical excitation potential for the Cr K spectrum.

The Cr $K\alpha$ line thus has sufficient energy (5.41 keV) to excite the Ba L_{III} spectrum (5.25 keV), but not the Ba L_{II} or Ba L_I. The Cr $K\beta$ line (5.96 keV) is of lower intensity, but has sufficient energy to excite Ba L_{II}, but not Ba L_I (5.99 keV). Hence if a Cr-anode tube is to be used for the analysis of Ba, and if it is desired to use the Ba L_I spectrum, then the tube operating conditions must be adjusted to ensure adequate continuum intensity in the vicinity of the Ba L_I critical excitation energy of 5.99 keV.

The energy scale of Figure 2.6 is logarithmic, which may tend to obscure the fact that the transition energies for any element are greatest for transitions to the K shell. Thus the K spectrum always has the highest energies, or shortest wavelengths, of any of the spectral series of a particular element. For example, Ba $L\gamma_4$ ($O_{III} \rightarrow L_I$) has an energy of only 5.98 keV (2.075 Å), well short of the least energetic line in the Ba K spectrum (Ba $K\alpha_2$ at 31.8 keV, or 0.39 Å).

2.2.2 Wavelength shifts in the characteristic spectrum

It has already been noted that the energy levels of a doubly ionized atom differ from those of the same atom in the singly ionized state. Since elements in compounds have different electronic structures from those in the free state, particularly in so far as valence electrons are concerned, it is possible that the consequent shifts in energy levels might result in shifts in energy, or wavelength, of characteristic spectral emissions. This would appear to be particularly likely for line emissions produced by transitions involving electrons from the valence or near-valence bands.

The lines commonly employed in X-ray spectrometry lie mostly between 1 and 10 Å, and most of them are produced by transitions from L or M levels to K or L levels (i.e. most analytical lines are selected from

ENERGY TRANSITIONS OF 'INNER ORBITAL' ELECTRONS

the K or L spectra of the various elements). For the heavier elements, valence electrons are not involved in these transitions, and the inner transition levels are 'shielded' from the effects of small energy shifts in the valence band. Wavelength shifts in analytical lines of the K and L spectra are thus usually too small to be of practical consequence.

For light elements, however, the situation may be significantly different. For example, the elements between $Z = 13$ and $Z = 17$ have unfilled 3p orbitals, which are therefore involved in bonding with other atoms. Since the $K\beta$ line is produced by a transition from the 3p orbital, it is to be expected that $K\beta$ emission wavelengths from these elements (Al, Si, P, S, Cl) should show some dependence on chemical state. The same effect would be expected in the $K\alpha$ emissions of elements with unfilled 2p orbitals (B, C, N, O, F).

Ground state aluminium atoms, for example, have filled K, L and 3s levels, together with a single 3p electron. The $K\beta_{1,3}$ emission is thus weak and broad, and the profile of the peak depends very much on the nature of the chemical bonding of the Al atoms (Fig. 2.7). In Al_2O_3, for example, the Al $K\beta$ line has a band-like structure since it arises not from the discrete energy level characteristic of an aluminium atomic orbital but instead from the energy *band* of the molecular orbital of aluminium and oxygen.

Sulphur occurs naturally in a wide range of oxidation states, ranging from -2 in the sulphide ion to $+6$ in the sulphate ion. Both of these extremes are found in the thiosulphate ion $(S_2O_3)^{-2}$, and if the S $K\beta$ line emitted by a sample of sodium thiosulphate is scanned with an X-ray spectrometer of sufficiently good resolution it is in fact found to consist of two peaks approximately 0.0015 Å (2 eV) apart (more sensitive electron spectroscopic techniques have indicated as much as 6.5 eV difference between the peaks). Chappell and White (1968) took advantage of this phenomenon to establish that sulphur occurs in the mineral scapolite in the form of $(SO_3)^{-2}$ rather than as S^{-2}, which had previously been postulated.

These relationships (including corresponding shifts in the L spectra of some of the heavier elements) can therefore be used to derive similar information concerning chemical bonding, although for this purpose X-ray methods have been largely supplanted by more sensitive and more versatile electron spectroscopic techniques. Of more direct significance to X-ray spectrometry are the errors that can arise in XRF and EPA analysis if wavelength shifts are not recognized and properly compensated.

It will be recalled that XRF and EPA are both *relative* analytical methods, i.e. they depend on comparison of the intensities of identical lines, emitted under identical conditions, from analysis samples on one hand and calibration standards on the other.

It is often tacitly assumed that an X-ray spectrometer tuned, or

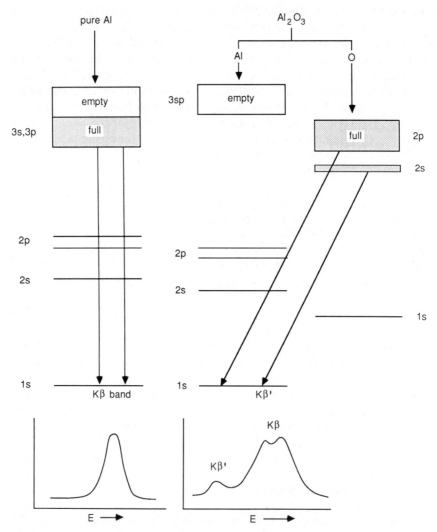

Figure 2.7 Dependence of the profile of the Al Kβ peak on the involvement of the Al valence electrons in chemical bonding. The Kβ emission, or *peak*, is produced by transitions from the 3s and 3p levels to the 1s level. In Al metal, the 3s and 3p electrons are not involved in chemical bonding and the transitions produce a relatively narrow band of energies and hence a sharp, well defined peak. In Al_2O_3, the valence electrons combine with the oxygen 2s and 2p electrons to fill molecular orbitals which span a wider energy band. The Kβ emission becomes broader, asymmetric, and resolved into separate Kβ and satellite Kβ' peaks (after Jenkins 1974).

'peaked', to a particular emission from a calibration standard will be equally well peaked for the analysis samples (in particular, this assumption is often made when computer-controlled systems are used). However, if there is a significant wavelength shift between the two, this will not be the case. To obtain accurate analysis, it will then be necessary:

(a) to repeak the spectrometer between sample and standard, or vice versa, or
(b) to deliberately sacrifice some spectrometer resolution (e.g. by using a coarse collimator, see below), to ensure that the maximum necessary *range* of wavelengths is measured, or
(c) to apply a mathematical correction to measured intensities, based on calibration of the peak shapes.

Of these, (a) is possibly the most widely used, since it can be easily adapted to computer-controlled systems (which can rapidly scan the analytical peak on each sample or standard and ensure that the intensity measurement is made on the true peak centroid).

Whenever there is doubt concerning the possible magnitude of wavelength shifts, it is essential to determine accurate peak profiles on both samples and standards and derive an appropriate strategy before making the analysis.

For electron excitation, either in a primary tube or in the electron microprobe, the intensity of any characteristic line (expressed as the rate of production of photons per unit time) is given by the empirically derived expression

$$I = C i (E_0 - E_c)^p \qquad (2.10)$$

where C is a constant, i is the electron current, E_0 is the accelerating potential, E_c is the critical excitation potential for the appropriate spectral series and the exponent p has a value of about 1.7 when E_0 is less than about $1.7E_c$ and tends to smaller values at higher accelerating potentials. In other words, the efficiency of the characteristic spectrum excitation process depends on the extent to which the critical excitation potential is exceeded (but not indefinitely); in microprobe analysis this is called the *overvoltage effect*, and it partially explains why the microprobe is typically operated at voltages equal to or greater than $2E_c$ (although other factors, such as background intensity or the magnitude of composition-dependent errors, may also have to be considered; see Chs. 8 & 9).

2.3 Energy transitions of 'outer orbital' electrons

The logarithmic scale of Figure 2.6 emphasizes another point – that the 'gap', or energy difference, between successive transition levels is very much smaller for the outer orbitals (e.g. those that include the valence electrons) than it is for the inner. Hence photons emitted as a result of transitions between outer orbital energy levels have much less energy – typically of the order of 1–100 eV, or approximating to the thermal, visible and ultraviolet regions of the electromagnetic spectrum. Similarly, the amount of energy required to eject an outer orbital electron is in the same low-energy range, and an X-ray photon may have sufficient energy to ionize tens to hundreds of atoms if its energy is consumed in the ejection of outer orbital electrons. Ionization of this kind forms the physical basis for various devices used to detect X-rays (e.g. gas-filled and scintillation detectors, which are described in more detail in Ch. 4).

The electron energy relationships so far described mostly refer only to isolated atoms, as, for example, in a monatomic gas. In solid compounds, on the other hand, adjacent atoms influence each other. The energy levels of their outer, or valence, bands are perturbed because of the Pauli exclusion principle. Since no two electrons in a common population can occupy precisely the same quantum state, the discrete energy levels of the isolated atom may be modified to almost continuous bands of closely spaced energies. Electrons in the 'inner' orbitals are not significantly affected by the interaction because they are shielded by the 'outer' orbitals and particularly by the outermost valence electron orbitals. However, the effects on the latter are more pronounced and account for marked variations in some key physical properties between various elements and compounds. In particular, they may have considerable effects on the electrical properties (e.g. the electrical conductivity) of solid materials. Conductivity varies by as much as 23 orders of magnitude between, say, pure copper on one hand and polystyrene on the other.

These effects constitute the physical basis for so-called *solid-state* or *semiconductor* X-ray detectors, which are further described in Chapter 4. The following brief review is intended only to provide a working basis for understanding the use of such detectors in analytical X-ray spectroscopy; more comprehensive reviews are provided in specialized texts on electronics and on the physical properties of solid materials.

2.3.1 Electrical conductivity

Electrical conduction, or flow of electrons in a material under the influence of an applied field, requires physical movement of *charge carriers*, such as electrons or ions. In a solid material, in which ions are relatively immobile, this implies that electrons must be excited into orbitals that are relatively loosely bound to their nuclei, thus allowing them to move more or less freely through the solid material.

In some atoms, the valence band is only partly filled with electrons; that is, unfilled energy states are available within the valence band and

ENERGY TRANSITIONS OF 'OUTER ORBITAL' ELECTRONS

relatively little additional energy is required to 'promote', or raise a valence electron to a higher energy level and free it to act as a charge carrier (Fig. 2.8). The highest ground state (i.e. non-excited) energy level occupied by the electrons is called the *Fermi level*, which is defined as the energy level at which the probability of finding a conduction electron is exactly one half. At absolute zero, all available energy states below the Fermi level are filled and all states above it are empty. Materials in which the Fermi level lies within the partially filled valence band are electrical *conductors*, because at any temperature above absolute zero there is enough thermal energy to ensure the presence of some charge carriers by the promotion of some of the valence electrons to energy states above the Fermi level (i.e. the probability of finding one or more conduction electrons is greater than one half).

In *semiconductors* and *insulators*, the valence band is filled and there is a 'forbidden' *band gap* between the valence band and the first permissible energy levels in the next, or conduction, energy band. This band gap is equivalent to the forbidden energy gap between successive discrete energy levels in the inner orbitals. The Fermi level lies within the band gap for these materials, and their conductivity then depends on the availability of sufficient energy to promote valence electrons across the band gap to the conduction band. Insulators are those materials for which the band gap is large, so that sufficient energy cannot easily be obtained. Semiconductors form the interesting group of materials that have finite but relatively small band gaps (Fig. 2.8).

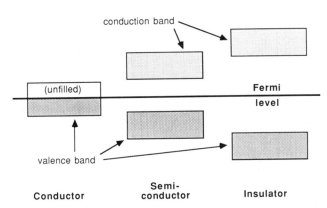

Figure 2.8 Band gaps between the valence and conduction bands in conductors, semiconductors and insulators. In conductors the band gap is negligibly small, and electrons are easily promoted to the conduction band, in which they are free to act as charge carriers. As the band gap increases, promotion of a valence electron becomes progressively more difficult, and electrical conductivity is reduced. Semiconductors form the transitional group in which low conductivity can be enhanced by providing sufficient external energy (thermal, X-ray, photon etc.) to promote electrons across a moderate band gap.

2.3.2 Semiconductors

When a valence electron is promoted to the conduction band, it leaves behind an unfilled level, or *hole*, in the lower energy orbitals of the valence band. Holes can also carry electrical current (since they are charge carriers, with charges opposite to those of the removed electrons), but they 'move' with different speeds or mobilities, depending on the actual material. In fact, holes do not 'move' at all; they have the effect of moving as electrons promoted out of one orbital fall back into another, just as the marquee lights outside a theatre appear to move across or around a sign (Fig. 2.9). If electrons are promoted in this fashion directly from the valence band to the conduction band (e.g. by absorption of thermal energy or energy from incident electromagnetic radiation), there will be an equal number of conduction electrons and holes and the material is said to be an *intrinsic semiconductor*.

Since thermal vibration of atoms can provide sufficient energy to promote electrons across at least small band gaps, the electrical conductivity of intrinsic semiconductors will be temperature-dependent. This property is utilized in devices called thermal resistors, or *thermistors*, which use variations in thermal conductivity of intrinsic semiconductors to measure or control temperature.

The conductivity of a semiconductor also increases when it is exposed to electromagnetic radiation which has sufficient energy to promote valence electrons across the band gap. Band gap widths are typically of the order of a few electronvolts, so that light in the visible and near-visible regions of the spectrum has sufficient energy to promote electrons in many semiconductors. The latter can therefore also function as photodetectors; for example, the band gap for pure silicon is 1.1 eV,

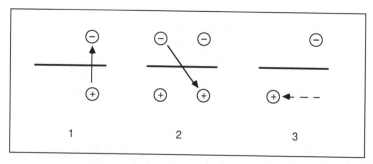

Figure 2.9 The 'marquee' effect, simulating apparent mobility of an electron vacancy, or *hole*. A positively charged hole is first created by promotion of an electron to the conduction band (1). An electron promoted from a second hole (2) may 'fall' into the hole produced in (1) and neutralize it (3). The result thus simulates physical movement of the hole, in the same way that flashing lights appear to move around a theatre marquee.

corresponding to the infra-red energy range, and that of CdS is 2.4 eV, in the visible range. CdS in particular has been widely used as the light-sensing agent in photographic exposure meters.

Promotion of electrons across the band gap can be facilitated by adding small amounts of an impurity whose valence or conductivity bands lie at least partly within the band gap of the host and thus effectively reduce the magnitude of the minimum promotion energy. Modification of the host in this way produces a *doped* or *extrinsic semiconductor*.

The impurity element may affect the electrical properties of the host by either

(a) providing an unfilled electron energy band which lies between the host valence band and its Fermi level, or
(b) providing a partly filled electron energy band which lies above the host Fermi level but below its empty conduction band.

In the first case (Fig. 2.10), the impurity band provides a lower energy 'sink' into which host valence electrons can be readily promoted (i.e. more readily than into the higher energy host conduction band). Electrons promoted in this fashion do not become charge carriers, but each leaves behind an electron hole in the valence band. The electron holes are free to act as charge carriers and, since they are positively charged, this type of extrinsic semiconductor is described as *p-type*.

In the second case, the impurity band provides a 'source' of electrons which can be relatively readily promoted into the host conduction band. The electrons themselves are now free to act as charge carriers; since they are negatively charged, this type of semiconductor is described as *n-type*.

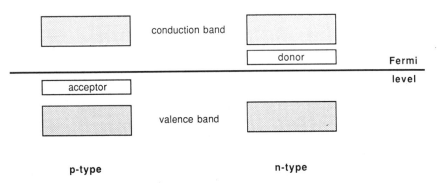

Figure 2.10 Depending on its energy relationship to the band gap and Fermi level of a host element, an impurity band can function either as an electron acceptor or 'sink' to produce a p-type extrinsic semiconductor, or as an electron donor or 'source' of easily promoted electrons to produce an n-type semiconductor. In either case, the effect of the impurity band is to reduce the band gap across which an electron must be promoted to produce either an electron hole (p-type) or an electron (n-type) charge carrier.

Note that in this case no (or at least relatively few) electron holes are generated in the host valence band, since this is not the source of (most of) the promoted electrons.

For example, silicon belongs to Group IV of the Periodic Table, and has four valence electrons available for bonding. In a single crystal of pure silicon, each atom is surrounded by and bonded to four nearest neighbour atoms arranged in a tetrahedral configuration. With the 'sharing' of valence electrons, the valence band of each atom is filled, and conductivity of the crystal depends on promotion of electrons across the 1.1 eV band gap to the conduction band. Pure crystalline silicon is therefore an intrinsic semiconductor.

If a silicon crystal is doped with small amounts of a Group V element, such as arsenic, then the arsenic atoms occupy the places of some of the silicon atoms in the crystal structure. Arsenic, however, has five valence electrons, of which only four are required to satisfy the bonding requirements and fill the valence bands of the four adjacent silicon atoms. The fifth electron will be relatively loosely held and can readily become a mobile charge carrier. The energy required to promote this electron to the conduction band is only a few per cent of the normal silicon band gap energy. When such an electron is promoted, the valence band remains filled because of the 'sharing' of valence electrons with adjacent atoms. In this case, therefore, no corresponding electron hole is produced in the valence band. Silicon doped with arsenic thus has an excess of negative charge carriers, and it is therefore an n-type semiconductor. Because there is an imbalance in the numbers of positive and negative charge carriers it is an extrinsic semiconductor. This does not mean that it has an overall negative charge, in the sense of having more electrons than protons, but merely that it has more negative than positive charge carriers.

Alternatively, silicon may be doped with a small proportion of a Group III element, such as gallium, with only three valence electrons. Again the gallium atoms occupy silicon sites in the crystal structure, but they cannot provide enough electrons to satisfy bonding requirements by filling the valence bands of four adjacent atoms. Instead extra electrons have to be 'taken from the surroundings', and particularly from electrons that have been promoted (e.g. by thermal activation) from the host valence band – possibly via the conduction band.

When such electrons are promoted, electron holes are created in the valence band. Hence absorption of the conduction electrons into the gallium valence band leaves a residual electron hole, and the process leads to an overall surplus of positive charge carriers. Gallium-doped silicon is thus a p-type extrinsic semiconductor. Again, the energy required to generate charge carriers in extrinsic semiconductors may be obtained from incident electromagnetic radiation, such as X-rays. Such materials can therefore be used in solid-state devices designed to detect X-rays by monitoring changes in their electrical properties.

Other similar materials include intrinsic or extrinsic compound semiconductors consisting of two or more elements combined in more or less equal proportions (e.g. GaAs or GaAsP), with or without doping with small amounts of other elements. Single crystals can be doped with different elements in different regions so that they may be, for example, p-type at one end of the crystal and n-type at the other. Such materials have interesting electrical properties that are utilized in a variety of devices (rectifiers, voltage regulators etc.) among the electronic components of X-ray spectrometers, but their characteristics are beyond the scope of this discussion.

3 The interaction of X-rays with matter

3.1 X-ray absorption

When an X-ray beam passes through any matter it is progressively *attenuated*, or reduced in intensity, as a consequence of a complex series of interactions between the X-ray photons and the atoms of the attenuating medium. Such *absorption* effects are of major significance in both XRF and EPA analysis, since in both cases a portion of the analytical spectrum is generated beneath the sample surface and is therefore partly absorbed by the sample itself. The extent of the absorption is a function of sample composition, and will generally differ from samples to standards (except in the rare cases where analytical samples and calibration standards are of identical composition). It is therefore necessary to understand the phenomenon and to apply corrections for its effects.

The primary exciting radiation, whether X-rays or an electron beam, is similarly absorbed as it penetrates the sample, and again the absorption effects may differ appreciably from samples to standards and hence will require appropriate correction procedures.

The physical basis of X-ray absorption is reviewed in this chapter, and the methods used to correct for its effects are discussed in Chapters 8 and 9. The absorption of an electron beam is also discussed in Chapter 9. Electron absorption phenomena are somewhat different from those that affect X-rays, which suffer intensity losses as a combined result of several different processes, as follows (Fig. 3.1):

(a) They may be *photoelectrically* absorbed by interaction with absorbing atoms, leading to either:

   (i) the emission of *photoelectrons* (e.g. Auger electrons), or
   (ii) the emission of *X-rays* characteristic of the absorbing atoms, i.e. secondary fluorescence. The wavelengths of the secondary X-rays depend on the composition of the absorbing medium, and are not directly related to those of the absorbed radiation.

(b) They may be *scattered*, either:

   (i) *coherently*, with no loss in energy, or
   (ii) *incoherently*, with a small energy loss and hence a slight increase in wavelength.

X-RAY ABSORPTION

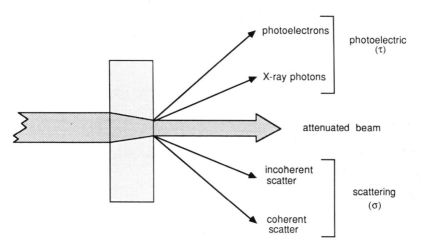

Figure 3.1 Schematic representation of the major processes by which X-rays interact with the atoms of an absorbing medium.

3.1.1 Linear absorption coefficients

Consider an infinitely thin layer, of thickness dl, of the absorbing material, and suppose that the intensity I_0 of a monochromatic incident beam of X-rays is reduced by dI on passing through dl (Fig. 3.2). The incremental intensity loss dI will be proportional to the absorbing thickness dl, i.e.

$$dI \propto -I \cdot dl$$

whence

$$dI = -\mu I \cdot dl$$

where μ is a proportionality constant called the *linear absorption coefficient*. It incorporates the combined effects of all photoelectric and scattering processes, i.e.

$$\mu = \tau + \sigma$$

Hence

$$dI / I = -\mu \cdot dl$$

On integrating over a finite thickness l (Fig. 3.3),

$$\ln I \Big|_{I_0}^{I_l} = -\mu l \Big|_0^l$$

whence

$$\ln I_l - \ln I_0 = -\mu l$$

or, in more familiar form,

$$I_l = I_0 \, e^{-\mu l} \qquad (3.1)$$

3.1.2 Mass absorption coefficients

As well as path length in a given system, the extent of absorption will also depend on the *density* of the absorbing medium; for example, for a particular X-ray wavelength, absorption along a path of 1 mm of liquid mercury will obviously be much greater than along the same path in mercury vapour. It is therefore more convenient to express the thickness

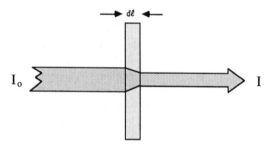

Figure 3.2 Incremental loss of intensity of an X-ray beam transmitted through a thin layer dl of an absorbing medium.

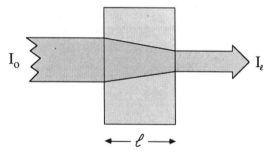

Figure 3.3 Integrated intensity loss of an X-ray beam transmitted through a finite thickness l of an absorbing medium.

of a finitely thick absorber as the product of its linear thickness and its density, i.e. as ϱl, which has the dimensions of mass/unit area. The absorption equation can then be written

$$I \, (= I_t) = I_0 \, e^{-(\mu/\varrho)\varrho l} \tag{3.2}$$

in which the constant (μ/ϱ) – constant for a given wavelength and a given absorbing medium – is known as the *mass absorption coefficient*.

Strictly speaking, the full symbol '(μ/ϱ)' should always be used to distinguish a mass absorption coefficient from the corresponding linear absorption coefficient 'μ'. However, linear absorption coefficients are not often used in XRS calculations, and it has now become common practice to use the shorter symbol 'μ' to denote a *mass* absorption coefficient. *This practice will be followed in the remainder of this book*; wherever the same symbol is used to refer to a linear absorption coefficient, this will be clearly indicated in the text.

Mass absorption coefficients are convenient to use for two major reasons:

(a) They depend only on the atoms present in the absorbing medium and the wavelength of the X-rays being absorbed. Because of the incorporation of the density component they are independent of the physical state of aggregation of the absorber.
(b) It is simple to calculate the mass absorption coefficient, for a particular wavelength, of a mixture of elements if the mixture composition and the individual element coefficients are known. If the mixture consists of n elements with weight fractions $c_1, c_2 \ldots c_n$, and with respective mass absorption coefficients $\mu_1, \mu_2 \ldots \mu_n$ then

$$\mu_{\text{mixture}} = \Sigma \, \{c_n \, \mu_n\} \tag{3.3}$$

For example, the mass absorption coefficient of the mineral chalcopyrite ($CuFeS_2$) for Ni Kα radiation may readily be calculated as follows, utilizing data from wavelength and mass absorption coefficient tables such as those of White and Johnson (1972) and Heinrich (1966) or the summary tables in the Appendix:

$$\text{Ni K}\alpha = 1.658 \, \text{Å}$$

Mass absorption coefficients for Ni Kα:

$$Cu = 64 \qquad Fe = 382 \qquad S = 108$$

Atomic weights:

THE INTERACTION OF X-RAYS WITH MATTER

$$Cu = 63.54 \qquad Fe = 55.85 \qquad S = 32.06$$

$$\Sigma \text{ (formula weights) for } CuFeS_2 = 63.54 + 55.85 + (2 \times 32.06)$$
$$= 183.51$$

Composition of chalcopyrite:

$$Cu = 63.54 / 183.51 = 34.62 \text{ (wt) \%}$$
$$Fe = 55.85 / 183.51 = 30.43 \text{ (wt) \%}$$
$$S = (2 \times 32.06) / 183.51 = 34.94 \text{ (wt) \%}$$

$$\mu_{Ni\ K\alpha}(CuFeS_2) = (0.3462 \times 64) + (0.3043 \times 382) + (0.3494 \times 108)$$
$$= 176.135$$
$$= 176$$

(since mass absorption coefficients are mostly not known with sufficient precision to warrant the decimal places).

3.1.3 Absorption edges

The mass absorption coefficient of any element or compound also varies with the wavelength or energy of the absorbed X-rays. If μ is plotted against wavelength for any chosen absorber it will show a general increase towards the longer, or 'softer' wavelengths, as might be expected. However, the variation is not continuous – instead it is marked by a series of abrupt discontinuities called *absorption edges* (Fig. 3.4). Between the edges, the mass absorption coefficients vary with wavelength approximately according to a relationship of the form of

$$\mu = K \lambda^u Z^v \qquad (3.4a)$$

where K, u and v are constants. The value of K changes at each absorption edge; u and v are to some extent functions of λ and Z, but approximate to 3 and 4 respectively. Several empirically derived modifications to this general expression have been proposed; a convenient form due to Dr Keith Norrish is

$$\mu = (A + BZ + CZ^2)^3 \lambda^n \qquad (3.4b)$$

in which the values of A, B, C and n change at each absorption edge and n is also a slowly varying function of Z (see Table A.2 in Appendix A).

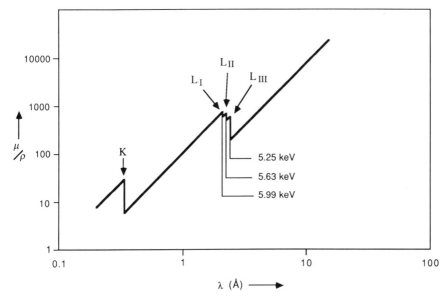

Figure 3.4 Plot of the mass absorption coefficients of barium for wavelengths between 0.2 Å and 11.0 Å. The marked discontinuities are the *absorption edges*; each corresponds to a critical excitation potential in the K, L_I, L_{II}, ... spectral series.

It is difficult or impossible to measure some mass absorption coefficients directly, particularly for long-wavelength X-radiation or for very heavy absorbers in which incident radiation is completely attenuated in very short path lengths. It is then necessary to calculate the absorption coefficients by extrapolation from regions where experimental measurement is more feasible. It should not be forgotten, however, that the expressions used for extrapolation, such as the one quoted above, are themselves empirically derived by extrapolation from a limited number of experimental observations and hence represent a potential source of error (particularly when they are hidden away in a general-purpose computer-based correction procedure).

There is an obvious relationship between the absorption edges of each element and the characteristic spectrum of that element. For each of the K, L, M ... spectral series and sub-series there is an absorption edge at a wavelength just short of the shortest wavelength present in that series, and the correlation allows the edge to be labelled K, L_I, L_{II}, L_{III}, M_I ... etc. Thus the K absorption edge of an element is always found at a wavelength just short of (or an energy just higher than) that of the $K\beta_2$ emission of that element – the highest energy line normally present in the K spectrum ($K\beta_4$, produced by the N_{IV} or 4d to K transition, is a 'forbidden' line which is always very weak in intensity and is very close in energy to, and hence not normally resolved from, $K\beta_2$).

Likewise the L_I absorption edge is just short of the $L\gamma_4$ emission ($O_{III} \rightarrow L_I$), the L_{II} edge is just short of the $L\gamma_6(O_{IV} \rightarrow L_{II})$ line, and so on.

Furthermore, the energy corresponding to the absorption edge is found to be exactly equivalent to the critical excitation potential for that particular spectral series (cf. Figs 2.6 & 3.4). Clearly the energies of the absorption edges are the equivalents of the electron binding energies for each of the transition levels.

Photons with energies just higher than that of an absorption edge will be heavily absorbed by the photoelectric absorption process, whereas those with energies just lower than the edge do not have sufficient energy to effect an ionization at that transition level and hence will not be as heavily absorbed. They will still be subject to scattering losses, however, and they may undergo photoelectric absorption at edges of lower energy, so that absorption will not decline to zero on the long-wavelength, low-energy side of an absorption edge.

It should be emphasized that to excite a characteristic emission line a photon must have an energy *higher* than that of the appropriate absorption edge, which in turn is always higher than that of the emission line itself. Thus Co Kα, with a wavelength of 1.79 Å, cannot excite Fe Kα, even though the latter has a longer wavelength of 1.94 Å. The Fe K absorption edge is at 1.74 Å and therefore has a higher energy than Co Kα.

It has been noted that absorption at any particular wavelength is the sum of scattering (σ) and photoelectric (τ) components. Each varies with wavelength, though the scattering component is relatively constant and usually relatively minor (except at short wavelengths, below the range of normal analytical interest).

If the scattering component is thus neglected, the absorption curve may be seen to be made up of the additive effects of photoelectric absorption due to each of the absorption edges (Fig. 3.5). At wavelengths shorter than the K edge,

$$\tau_{total} = \tau_K + (\tau_{L_I} + \tau_{L_{II}} + \tau_{L_{III}}) + (\tau_{M_I} + \tau_{M_{II}} \ldots) + \ldots$$

As the K edge is passed towards lower energies or longer wavelengths, the τ_K component disappears. The ratio r_K of absorptions immediately on either side of the edge, viz.

$$r_K = \frac{\tau_K + \tau_L + \tau_M + \ldots}{\tau_L + \tau_M + \ldots} \tag{3.5}$$

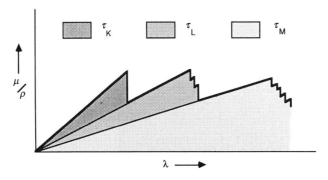

Figure 3.5 The mass absorption coefficient for any wavelength is the sum of components due to photoelectric processes at all absorption edges of longer wavelengths. At wavelengths shorter than the K edge of the absorber, the total absorption is thus the sum of all K, L, M... absorptions. For a longer wavelength, with insufficient energy to eject K-shell electrons, the K absorption component disappears but is replaced by more severe L, M... absorption as the X-ray wavelengths approach closer to the L, M... edges.

is called the *absorption jump* or *absorption jump ratio*, in this case for the K absorption edge. It varies from about 2 for heavy elements with abundant L, M... electrons (e.g. W) to about 20 for light elements. For any pure element, the absorption jump ratio corresponding to a particular shell (K, L, M...) determines the proportion of total primary ionizations that are ionizations of that shell. For example, the absorption jump ratio r_K for silicon is approximately 13. The fraction of total Si ionizations that are K shell ionizations is given by the ratio $(r_K - 1) : r_K$, which for silicon is $(13 - 1) / 13$, or 0.92. This means that approximately 92 per cent of all Si photoelectric ionizations are K shell ionizations.

In a multi-element sample, the height of the absorption jump characteristic of a particular element is proportional to the concentration of that element in the absorber, and this relationship can be used analytically in a technique known as *X-ray absorptiometry*. For example, the U L_I edge is at 0.722 Å; for a uranium-bearing solution the jump ratio can be established by measuring the attenuation of Mo Kα (0.707 Å) and of Nb Kα (0.748 Å), and the uranium concentration can be established with reference to one or more calibration standards. This technique, however, is also of limited value in the soft X-ray region.

In X-ray diffractometry, advantage is often taken of absorption edges to obtain sources of nearly monochromatic X-rays. Thus the K spectrum from a copper-anode tube, for example, contains both Kα and Kβ emissions at 1.54 and 1.39 Å respectively. The Ni K absorption edge lies at 1.49 Å, so that a thin sheet of nickel foil strongly absorbs and effectively filters out most of the Cu Kβ radiation passing through it. The attenuation of Cu Kα is much less severe, since it lies on the long-

Table 3.1 Absorption filters used in X-ray diffractometry.

Tube anode	Kα(λ)	Kβ(λ)	Filter	Absorption edge of filter (Å)
Cr	2.29	2.08	V	2.27
Mn	2.10	1.91	Cr	2.07
Fe	1.94	1.76	Mn	1.90
Co	1.79	1.62	Fe	1.74
Cu	1.54	1.39	Ni	1.49
Mo	0.71	0.63	Zr	0.69

wavelength side of the absorption edge and does not have sufficient energy to eject Ni K electrons; the result is the production of a nearly (effectively) monochromatic beam of Cu Kα X-rays (superimposed on unfiltered continuum). Table 3.1 lists several other commonly used anode/filter combinations.

3.2 Scattering and X-ray diffraction

As noted earlier, the scattering coefficient includes two terms, corresponding to coherent (Rayleigh) and incoherent (Compton) scattering respectively. In fact,

$$\sigma = Zf^2 + (1 - f^2) \tag{3.6}$$

in which the first term describes the extent of coherent scatter and the second the incoherent scatter. f is the atomic scattering factor, or the ratio of the amplitude of the wave scattered by an atom to that scattered by a free electron. Scattering by an aggregate of atoms, as in a crystal structure, is expressed by the structure factor F, which is the integral of the atomic scattering factors (taking phase differences into account). These aspects are reviewed comprehensively in discussions of X-ray crystallography; for the present it is sufficient to note that scattering efficiency depends on Z and on the additional factors that determine f (e.g. θ, λ) and F (crystal structure). Although scattering generally constitutes only a minor component of the absorption mechanism, there are circumstances under which it becomes extremely important, for it is the basis of the X-ray diffraction phenomenon on which wavelength-dispersive X-ray spectrometers are based.

3.2.1 Coherent and incoherent scattering

Coherent scattering arises when an X-ray photon 'collides' with an atom and is deviated without loss in energy. An electron in an alternating electromagnetic field, such as is generated by an X-ray photon, will oscillate at the same frequency as the field. The oscillating electron then re-emits electromagnetic radiation with the frequency of its oscillation, i.e. an X-ray photon with the same energy as the incident photon, and for scattering by a single atom there are no constraints on the direction of re-emission (Fig. 3.6).

Alternatively, the photon may lose part of its energy to the scattering electrons, particularly if some of the latter are loosely bound. Since total momentum must be maintained, the wavelength of the scattered photon must change, increasing in accordance with the expression

$$\Delta\lambda = 0.0243 \, (1 - \cos \phi) \quad (\text{Å}) \quad (3.7)$$

where ϕ is the angle of scatter (Fig. 3.7). This is the physical basis of *incoherent scattering*.

Incoherent scattering is often observed in XRF analysis because the spectrometer 'views' the sample at an angle of approximately 90° to the primary X-ray beam direction. Both characteristic and continuum radiation from the primary tube can therefore be scattered by the sample into the spectrometer. Much of it will be coherently scattered without change in wavelength, but some will also be incoherently scattered, with increases in wavelength of about 0.024 Å, since the scattering angle ϕ is close to 90° (there is always some finite beam divergence about the mean scattering angle). As a result, peaks in the spectrum due to scattered

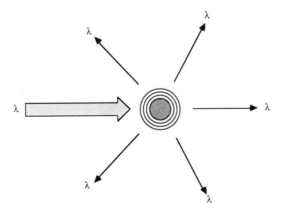

Figure 3.6 Coherent scattering: X-rays may be scattered by an atom in any direction, with no loss in energy and hence no change in wavelength.

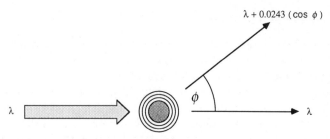

Figure 3.7 Incoherent scattering: a scattered X-ray photon may lose part of its energy to the scattering electrons, so that its wavelength increases by an amount which depends on the scattering angle φ.

anode characteristic radiation will be asymmetric, with subsidiary Compton-scattered peaks centred around wavelengths 0.024 Å higher than the coherently scattered peaks. The effect is generally observed as an asymmetry or a splitting of the coherently scattered and Compton-scattered peaks as they are recorded on a spectrum profile (e.g. the profile of a strip-chart recorder operated in synchronization with a spectrometer scanning through a range of Bragg angles). By contrast, coherently scattered peaks and those due to lines emitted from the sample typically show only the very slight asymmetry that arises from incomplete resolution (separation) of closely spaced lines such as the $K\alpha_1, \alpha_2$ doublet.

3.2.2 Diffraction

X-rays are a form of electromagnetic radiation, and their behaviour can be described effectively in terms of wave phenomena. In particular they exhibit *interference* effects: that is, a coherent group of X-rays travelling along a common path are reinforced in amplitude (intensity) if they are in phase with each other, or the overall intensity is reduced by interference if they are not. However, the directions in which coherent scatter can take place become severely limited if scattering of a beam of X-rays by a *number* of atoms is considered.

In the simplest instance, suppose a beam of X-rays interacts with two atoms, A and B, with θ being the angle between the incident beam direction and the line AB. Consider scattering at an angle φ to AB (Fig. 3.8). AY is the wave front prior to scattering; for the scattered waves to be in phase, and hence not reduced in intensity, ab must be the wave front after scattering. This can only be the case if the path lengths Aa and YBb are equal or if the difference between them is an integral multiple of the X-ray wavelength ($n\lambda$).

It is evident that the path lengths will only be equal if the scattered rays

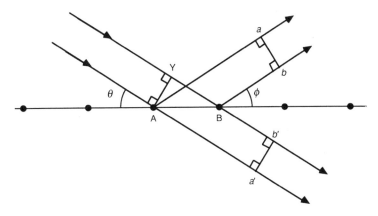

Figure 3.8 Scattering and diffraction of a coherent beam of X-rays from a row of equally spaced atoms. The condition for diffraction is that of coherence of the diffracted beam; that is, the rays scattered from adjacent atoms must remain in phase with each other. This requires either that they have identical path distances (transmission or reflection, as in this figure) or that any path difference be a whole (integral) multiple of the X-ray wavelength.

are undeviated, or if $\theta = \phi$. The number of possible situations where the path difference ($YBb - Aa$) is equal to $n\lambda$ (n must be an integer if there is to be no phase difference) is limited by the relative dimensions of λ and the distance AB. It is therefore obvious that there are limitations on the number of directions in which X-radiation can be coherently scattered from a row (or, by simple extension of this argument, from a plane) of atoms.

When a three-dimensional atomic structure is considered, the limitations become even more severe. For example, coherent scattering from a row of atoms can take place whenever $\theta = \phi$ (i.e. conditions of reflection). However, in a three-dimensional structure, similar scattering can also take place from the next atomic plane beneath the surface, and a beam can be propagated without intensity loss in the direction of reflection only if the two waves (that scattered from the surface layer and that scattered from the next layer beneath the surface) are in phase (Fig. 3.9).

In this case it is evident that the path difference CBD can only be zero if $\theta = 0$ (or, unrealistically, if $d = 0$). Otherwise, for the two scattered waves to be in phase, CBD must be equal to an integral number of wavelengths, or $n\lambda$. If d is the distance between the planes,

$$CBD = 2CB = 2d \sin \theta \; (= n\lambda) \tag{3.8}$$

which is the familiar statement of Bragg's Law. This is a relatively simple

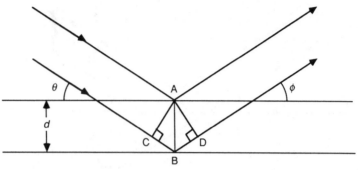

Figure 3.9 Scattering and diffraction of a coherent beam of X-rays from parallel planes of atoms, spaced a distance d apart. The diffraction condition is again that of coherence of the diffracted beam (cf. Fig. 3.8). The path lengths of the two rays shown differ by the distance CBD, which must therefore be equal to a whole (integral) number of wavelengths ($n\lambda$) for the rays scattered at the angle ϕ to be in phase with each other.

but adequate explanation which can easily be extended to the more general case which does not assume $\theta = \phi$. For given values of d and λ there may (depending on their relative magnitudes) be more than one possible solution in θ for the Bragg Equation, according to whether $n = 1, 2, 3. \ldots$ Such successive directions of reinforced coherent scattering are known as *first-, second-, third- . . . order reflections*, and they are encountered frequently in XRF and EPA analysis where crystals of high d (relative to λ) are often used.

The reflecting power of the crystal usually diminishes with successively higher orders of reflection, at least in a general way. In some cases the differences from one order to the next are not extreme (e.g. with an LiF analysing crystal, second-order reflections typically have intensities as high as 10 per cent of those of the corresponding first-order reflections). In other cases (e.g. topaz) there may be marked changes; the systematic nature of reflections and extinctions is in fact used by crystallographers to derive fundamental information about crystal structures.

In X-ray spectrometric analysis, the diffraction phenomenon has a very practical application in that crystals of known interplanar spacing (d) may be used to *disperse* the X-radiation emitted from an excited sample; that is, to isolate chosen wavelengths of analytical interest from the rest of the emitted spectrum. The dispersive function is performed within a *spectrometer*, a common type of which is based on Moseley's original design in which a crystal can be rotated about an axis such that the X-radiation falls on the crystal surface at an angle θ, which in turn can be selected to provide a solution of Equation 3.8 for any chosen wavelength within the range permitted by the crystal interplanar spacing d. The principles of crystal-dispersive spectrometer design and operation are reviewed in more detail in Chapters 4 and 5.

4 X-ray dispersion and detection

4.1 Crystal dispersion

4.1.1 'Wavelength-dispersive' spectrometers

The spectrometers used in XRF and EPA analysis have two major objectives:

(a) they must *disperse* the X-ray spectrum emitted by the sample under analysis conditions, permitting the isolation of individual wavelengths or at least narrow ranges of wavelength for measurement, and
(b) they must make provision for measurement of the *intensity of emission* of the selected analytical wavelength.

In so-called 'wavelength-dispersive' spectrometers the first task is accomplished by using a *diffracting crystal* to disperse different wavelengths to different Bragg angles, depending on the appropriate solutions to the Bragg Equation, and the relevant intensities are measured with an electronic *X-ray detector* equipped with timing and counting circuitry. Such spectrometers are also often called *Bragg spectrometers* because of their dependence on the Bragg relationship.

Although details of the construction and geometry of Bragg spectrometers vary considerably, they all include one or more crystals carefully cut or cleaved parallel to a system of diffracting planes of known interplanar (d) spacing. The spectrometer components are arranged so that radiation emitted from the sample falls on the crystal surface at known (and variable) θ angles, or *Bragg angles*. Since d is known and θ can be set to any desired value between 0° and 90° (at least in theory if not in practice – see below), the spectrometer can therefore be adjusted so that the Bragg condition is met for any desired wavelength within the range of the crystal (Fig. 4.1).

Under the appropriate conditions, X-rays of the desired wavelength will be diffracted at the required Bragg angle θ towards the detector used to measure their intensity. X-rays of other wavelengths will be absorbed by the crystal or will interfere destructively with each other when they are scattered at the set diffraction angle.

Qualitative analysis can be performed by *scanning* such a spectrometer through the complete range of Bragg angles, noting those at which the detector records above-background intensities indicating that a wavelength in the emitted spectrum is satisfying the Bragg condition. These wavelengths can then be calculated from the observed θ angles and the known d-spacing, and the emitting elements can be identified from tables of characteristic spectral emissions. Alternatively, computer-controlled spectrometers can be slewed rapidly between a series of predetermined Bragg angles, one or more for each element sought in the qualitative analysis, pausing at each to check the X-ray intensity level and to write an identification message if emissions from the appropriate element are detected. Automated multi-spectrometer microprobes driven in this way can check a sample for each of the elements within the probe range (normally $Z = 8$ or higher) in only a few minutes.

In recent years, the performance of 'energy-dispersive' X-ray spectrometers (see Sections 4.5 & 5.2), which use electronic techniques to isolate X-rays of the desired energy and hence have no need of dispersive

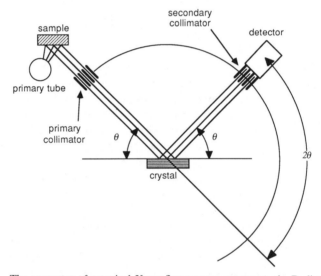

Figure 4.1 The geometry of a typical X-ray fluorescence spectrograph. Radiation from a primary X-ray tube produces fluorescent X-ray emissions from atoms in the sample. The fluorescent X-rays, and some scattered tube X-rays, are passed through a primary collimator to reduce their divergence so that they fall at an angle θ on a dispersing crystal of known interplanar spacing d. Rotation of the crystal allows the selection of any angle θ between 0° and 90° (in theory; in practice the working range is about 10°–70°), to provide an appropriate solution of the Bragg Equation for a chosen emission wavelength λ. X-rays diffracted at this angle pass through a secondary collimator to a detector. The crystal, secondary collimator and detector are mounted on the arms of a *goniometer*, which is a mechanical device that maintains Bragg geometry by rotating the collimator–detector assembly at twice the angular rotation rate of the crystal. For qualitative analysis, the goniometer can be scanned continuously through all or part of its angular range while the detector output is monitored (e.g. with a chart recorder) to establish the Bragg angles at which the diffraction of above-background intensities indicate the emission of a characteristic wavelength from the sample.

crystals or mechanical scanning devices, has improved considerably, to the point where these spectrometers effectively complement, and for some purposes have largely replaced, the mechanically more complex Bragg spectrometers. However, the latter have superior *resolution* in the energy range of analytical interest (i.e. they have superior ability to separate and isolate closely spaced energies or wavelengths, particularly where the emitted intensities are low), and they are still considered essential for many applications.

4.1.2 Dispersing crystals

The first crystals to be used in X-ray spectrometry were those of natural materials, such as rock salt, mica, gypsum, calcite and quartz, whose structures had been among the first to be determined by diffraction methods. However, these have since been largely replaced by a group of synthetic crystals which give improved performance over a wider range of operating conditions.

The number of high-order reflections that can be diffracted by a particular crystal is limited by the relationship between d and λ. $n\lambda$ cannot exceed $2d$, since this would require a value of $\sin \theta$ greater than unity. Since we are concerned in X-ray spectrometry with analytical wavelengths between 1 and 100 Å, it would appear that the prime requirement is for a crystal of large d-spacing, necessary to diffract the long wavelengths. Such a crystal (e.g. lead stearate, d = approximately 50 Å), however, can diffract the *short* wavelengths only in the very inefficient high orders (i.e. for relatively high values of n), and most of the elements emit their strongest characteristic radiations in the shorter wavelength portion of the X-ray band. Hence it is apparent that a range of crystals is required for efficient cover of the complete energy range of interest, and most spectrometers are designed so that up to six or more dispersing crystals may be interchanged; most modern designs provide for interchange during the course of an analysis as successive elements are determined, albeit with varying degrees of convenience. Microprobes in particular, but also some XRF spectrographs, are also commonly equipped with three or more spectrometers so that several different wavelengths can be measured simultaneously. Complete wavelength coverage, and hence maximum analytical flexibility, is obtained by equipping each of these spectrometers with at least two or three interchangeable crystals.

The crystals most commonly used in Bragg spectrometers, together with some of their more important characteristics, are listed in Table 4.1. Some of these can be cleaved relatively easily parallel to the required reflecting planes (e.g. LiF, mica); others have to be carefully cut (e.g. quartz). For focusing spectrometers (Sec. 5.1.2), the dispersing crystals

Table 4.1 Characteristics of commonly used dispersion crystals.

Crystal	Abbrev.	(hkl)	2d (Å)	Lowest Z detectable (max $\Theta = 70°$)		Efficiency
				K	L	
topaz	—	(303)	2.712	23(V)	59(Pr)	average
lithium fluoride	LiF	(220)	2.848	23(V)	57(La)	high
lithium fluoride	LiF	(200)	4.028	19(K)	49(In)	intense
germanium	Ge	(111)	6.532	16(S)	40(Zr)	average
quartz	—	(10$\bar{1}$1)	6.686	15(P)	40(Zr)	high
quartz	—	(10$\bar{1}$0)	8.50	14(Si)	37(Rb)	average
pentaerythritol	PET	(002)	8.742	14(Si)	37(Rb)	high
ethylene diamine d-tartrate	EDDT	(020)	8.804	14(Si)	37(Rb)	average
ammonium hydrogen phosphate	ADP	(101)	10.648	12(Mg)	33(As)	low
mica (muscovite)	—	(002)	19.8	9(F)	26(Fe)	low
potassium acid phthalate	KAP	(10$\bar{1}$0)	26.6	8(O)	23(V)	average
rubidium acid phthalate	RAP, RbAP	(10$\bar{1}$0)	26.1	8(O)	23(V)	good
thallium acid phthalate	TAP, TlAP	(10$\bar{1}$0)	25.75	8(O)	23(V)	high
lead stearate decanoate	PbSD	—	100	5(B)	20(Ca)	average
lead melissate	—	—	160	4(Be)	20(Ca)	average

have to be bent and ground to appropriate radii of curvature, for which some crystals are better suited than others. Mica, for example, can be easily bent and is often used in flat-crystal or semi-focusing spectrometers; however, because of its perfect cleavage it cannot be as easily ground and it is not normally used in fully focusing spectrometers.

4.1.3 Selection of dispersing crystals

Selection of a crystal for a particular application must be based on several considerations (often conflicting and often requiring some measure of compromise), of which the following are the most important.

d-SPACING

It is nearly always most desirable, other things being equal, to use first-order reflections of relatively strong spectral lines, since these yield the highest intensities and hence the best sensitivity and statistical precision. This requirement alone demands a range of analytical crystals whose first-order reflections cover the range of analytical interest.

For general purpose analysis, the LiF_{200} crystal is probably the most widely used. With a $2d$-spacing of 4.028 Å, however, its use is obviously restricted to wavelengths of less than 4 Å. Even this limit is mechanically impractical, since for a first-order emission of 4 Å

$$\sin \theta = n\lambda / 2d$$

$$= 4 / 4.028$$

whence $\sin \theta \approx 1$ and $2\theta \approx 180°$. This implies linear coincidence of sample and detector, which is mechanically impossible, so that the maximum wavelength that can be usefully dispersed by any crystal must be significantly less than $2d$. The useful low wavelength limit of a crystal also depends partly on geometrical features of the spectrometer design (including such factors as the sample area from which radiation is excited and the size and shape of the crystal itself). Thus the useful θ range of Bragg spectrometers is seldom more extensive than 10°–70°, and may be appreciably less, particularly at the low-angle end. Performance in this respect varies appreciably from one design to another; small differences may be quite significant for certain applications and should be carefully examined when evaluating instrument specifications.

For LiF_{200}, the Bragg angle range of 10°–70° corresponds to a wavelength range of 0.7–3.8 Å in the first order, which encompasses the $K\alpha$ emissions of elements between potassium ($Z = 19$) and molybdenum ($Z = 42$) and the $L\alpha$ emissions of elements between indium ($Z = 49$) and uranium ($Z = 92$); however, spectrometer performance in the low-wavelength, or heavy-element, portions of these ranges is usually relatively poor and LiF crystals would not often be used for the determination of Mo or U, for example.

The Mα lines of all natural elements are of wavelengths greater than 3.8 Å (U Mα = 3.91 Å) and are therefore beyond the operating range of LiF_{200} crystals.

For analyses based on the K spectra of elements lighter than potassium but heavier than sodium (i.e. the Z range from 12 to 18, which is of considerable geological interest), the spectroscopist can choose between ADP ($2d$ = 10.64 Å) and PET ($2d$ = 8.742 Å) dispersing crystals. The 10°–70° θ wavelength ranges of these crystals are 1.5–8.2 Å for PET and 1.8–10.0 Å for ADP. There is little practical difference at the lower ends of the ranges, which both overlap the useful range of LiF_{200} or LiF_{220} crystals. However, the ADP crystal has a significantly greater long-wavelength range which encompasses Mg Kα (9.89 Å) and Al Kα (8.34 Å), neither of which can be diffracted by a PET crystal in spectrometers of typical θ ranges.

It might seem, therefore, that of these two crystals ADP would be the better all-purpose choice, particularly since PET crystals are soft and easily damaged, tend to deteriorate with time and exposure to X-radiation, and have undesirably large thermal expansion coefficients which make them much more sensitive than most crystals to temperature fluctuations within the spectrometer (equivalent to 30″ drift in θ per °C at θ = 45°, which is quite severe). Despite these limitations, however, PET is usually favoured in this medium-wavelength range, which illustrates the importance of some of the other factors involved in crystal selection. Specifically, PET has a much higher reflection efficiency leading to improved analytical sensitivity, it has slightly better dispersion characteristics (see below), and it is free of disturbing enhanced-background effects arising from X-ray fluorescence of the crystal itself (see below).

For even lighter elements, with K spectra in the long-wavelength range, KAP and its analogues RbAP and TlAP extend the useful working range down to oxygen (Z = 8; O Kα = 23.6 Å), although the lower end of the range is not usually practicable in XRF systems which lack the high vacuum necessary to minimize absorption of the soft X-rays by air molecules (XRF spectrographs usually incorporate relatively simple vacuum systems designed to maintain pressures of the order of 10^{-1} torr, whereas microprobe operating pressures are of the order of 10^{-5} torr).

In the relatively low pressures of the microprobe it is possible, under ideal circumstances, to work with the K spectra of elements as light as beryllium (Z = 4; Be Kα = 114.0 Å) or boron (Z = 5; B Kα = 67.6 Å). For wavelengths of this magnitude no suitable natural dispersing crystals are available; however, it is possible to prepare 'pseudocrystals' consisting of layered soap films which have effective $2d$-spacings of the order of

100 Å or more. The soap films consist of metal salts of straight-chained fatty acids, such as

(a) stearic acid, $C_{18}H_{36}O_2$, $2d = 100$ Å;
(b) lignoceric acid, $C_{24}H_{48}O_2$, $2d = 132$ Å;
(c) melissic acid, $C_{30}H_{60}O_2$, $2d = 160$ Å etc.

The diffracting ability of such 'crystals' depends on scattering by layers of metal atoms which are separated by the long fatty acid chains. It is obviously desirable to use a metal of high scattering power, and lead ($Z = 82$) is the most widely used.

With recent technical developments it is now possible to manufacture synthetic dispersing 'crystals' of almost any desired interplanar spacing and with high reflection efficiencies. These seem likely to find many practical applications.

For the present, the full analytical range of wavelengths can be usefully encompassed by a minimum of four crystals: LiF_{200}, PET (or ADP), KAP (or RbAP or TlAP) and one of the pseudocrystals. For some specific applications better results may be obtained from one of the other crystals listed in Table 4.1 (or even from one of many others not included in this list; for example, clinochlore, $2d = 28.39$ Å, which is particularly well suited to the dispersion of O Kα or V Lα, but is not much used for other applications). In such cases the crystal d-spacing, though always important, may be deemed less significant than some other parameter, such as dispersive or resolving power.

DISPERSION

Differentiation of the Bragg Equation yields

$$\frac{d\theta}{d\lambda} = \frac{n}{2d} \cos \theta \qquad (4.1)$$

Thus at a given Bragg angle the angular separation of adjacent wavelengths is greatest for crystals with the smallest d-spacings. Such crystals are said to have better *dispersion* or *resolving power*. In cases where spectral interference from an overlapping line (i.e. one with almost but not quite the same wavelength) is a problem it may therefore be advantageous to choose a dispersing crystal with the lowest possible d-spacing in order to minimize the effects of the interference.

In the determination of vanadium in titanium-bearing samples, difficulty is often encountered because of interference with the most sensitive V line (V Kα, $\lambda = 2.505$ Å) by Ti Kα ($\lambda = 2.514$ Å); these lines are only partly resolved by the LiF_{200} crystal normally used in this wavelength region. If the vanadium concentration is relatively low, it is not feasible to use another (weaker) vanadium line such as V Kβ or

V Lα. However, the magnitude of the interference can be reduced – though not entirely eliminated – by using a LiF$_{220}$ crystal, which has approximately 40 per cent better angular dispersion because of its lower d-spacing (Table 4.1).

Note that dispersing power also increases with θ and with n. Interference problems thus tend to be most acute at low Bragg angles, which is a major reason for relatively poor spectrometer performance in the low-angle region. Furthermore, if concentrations and hence intensities permit, it may be advantageous to work with a higher-order reflection than might otherwise have been chosen, because of the improved resolution. This may then create some problems with automated XRF systems which use a device called a *sin-θ potentiometer* in their pulse height discrimination circuitry (see Sec. 4.3). However, such systems usually make provision for automatic adaptation, if specified, to at least second-order reflection conditions.

REFLECTION EFFICIENCY

The reflecting power of a crystal depends primarily on the atoms of which it is composed and the ways in which they are arranged in its crystal structure. Heavy atoms scatter X-rays more efficiently and hence generally produce higher intensities in the diffracted beams. Thus RbAP crystals give small but significant increases in intensity compared to KAP crystals as a consequence of the difference in scattering ability between K atoms ($Z = 19$) and Rb atoms ($Z = 37$); the crystal structures and d-spacings are otherwise almost identical. In turn the reflecting power is even further enhanced in TlAP, in which the K or Rb atoms are replaced by thallium atoms ($Z = 81$) – again with only minor effects on the crystal structure and d-spacings. The much heavier Tl atoms lead to a considerable improvement in reflection efficiency – by a factor of almost 100 per cent compared to KAP.

However, crystals containing heavy atoms also produce higher backgrounds due to radiation scattered from the crystal surface rather than diffracted, and the improvement in sensitivity, as measured by peak-to-background ratios rather than absolute peak intensities, is not always as pronounced as the difference in absolute peak intensities might suggest. The heavy atoms may also emit fluorescent radiation, some of which will reach the detector regardless of the Bragg angle setting of the spectrometer and will therefore contribute further to elevated background intensities (see below).

The reflecting power of a crystal varies to some extent with the wavelength of the diffracted radiation, because the atomic scattering factor f (Eqn 3.6) is wavelength-dependent. The wavelength dependence of reflectivity varies from one crystal to another; thus the reflectivity of topaz$_{303}$ is approximately 20 per cent of that of LiF$_{200}$ at 1.54 Å, 27 per cent at 0.71 Å and 39 per cent at 0.39 Å. Hence the superior dispersion of a topaz crystal is outweighed at wavelengths greater than about 0.7 Å by its relatively poor reflection efficiency, and LiF_{200} (or, better, LiF_{220}, which has a dispersion comparable with that of topaz) will usually be preferred in the longer wavelength range.

A perfect crystal – even if it could be prepared – would actually be a poor reflector of diffracted rays, because of the extinction effects illustrated in Figure 4.2. A perfect crystal may be regarded as being made up of brick-like blocks, perfectly oriented with respect to each other. Incident radiation falling on the surface layer of blocks at an appropriate Bragg angle θ will be diffracted in the normal way. However, radiation penetrating the crystal and diffracted from deeper layers of blocks will, after diffraction, strike the undersurface of the overlying block layers at the same Bragg angle θ and some of it will be diffracted downwards again; there will be a consequent loss of total diffracted intensity in the direction of the detector.

By contrast, an 'ideally imperfect' crystal is made up of a mosaic of small blocks, or block-like domains, of the order of 10^{-5} cm in size and misoriented by a few minutes of arc with respect to each other. In such crystals (Fig. 4.2) diffraction loss is considerably reduced because the Bragg condition will be met much less frequently by rays incident on the upper surfaces of the lower blocks. This misalignment cannot be too substantial, or the Bragg condition will be met in some of the blocks over too wide a range of crystal angles and the resulting peak will be broad

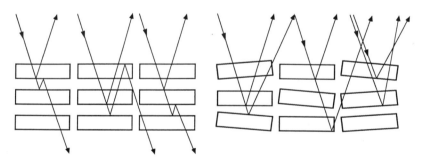

Figure 4.2 Enhanced reflection efficiency from a mosaic of small, slightly misaligned blocks ('domains') in a dispersing crystal. If the blocks are perfectly aligned, some rays scattered through the first layers and reflected back from others beneath will be lost by further reflection at the lower surfaces of the upper layers. This effect is minimized by slight misalignment of the blocks in successive layers, at the cost of some divergence in the diffracted beam, leading to 'peak broadening' and some loss in the ability of the crystal to separate closely spaced wavelengths. The divergence can be countered to some extent by using a secondary collimator (Fig. 4.1), but this itself reduces the measured peak intensity. Optimum compromise is achieved with a crystal that is slightly but not excessively imperfect (after Jenkins 1974).

and ill-defined, and hence of limited analytical use. Adjacent peaks will be poorly resolved from each other, or perhaps not resolved at all. Even 'ideal' misalignment – the minimum necessary to reduce the diffraction loss problem substantially – must necessarily detract from the resolution of the crystal by broadening the diffraction peaks. The extent of broadening (which also depends on such factors as the size of the sample emission surface, the effective surface area of the crystal and the degree of collimation) is expressed as a factor known as the *rocking curve* of the crystal, which can be thought of in terms of the width : height ratio, or 'sharpness', of the resulting peak. A good crystal should yield a peak width, at half maximum intensity, of the order of 0.3° 2θ or better (while still reflecting at adequate intensity levels).

Mosaic imperfections of this type are often deliberately introduced into the crystal during manufacture by abrasion or etching of the crystal surface. Alkali halide crystals, such as LiF or NaCl, can be prepared in this condition relatively easily. Quartz crystals, on the other hand, tend to be more perfect, but their efficiency can be improved considerably by abrasion or by other forms of elastic or plastic strain; elastic straining of quartz crystals has been shown to produce up to 20-fold increases in reflection efficiency. Controlled surface abrasion of LiF crystals can improve reflecting power by factors of up to 10.

Crystals used for longer wavelengths should not be abraded, as they nearly always give better results with polished surfaces. This is because rough-surfaced crystals tend to absorb long-wavelength radiation, particularly at low Bragg angles.

Considerations such as these explain why some crystals perform better than others of the same type; for example, ADP crystals obtained from the same manufacturer, and presumably produced by the same procedures, can vary in their reflectivities by factors of as much as three or four. It is not uncommon for crystal manufacturers to select unusually good crystals from their production runs and sell them at a premium price; the relatively small additional investment on the purchaser's part is usually amply justified by the enhanced performance. Crystal performance is a factor that should be carefully considered when preparing specifications for XRS instrumentation.

CRYSTAL FLUORESCENCE EFFECTS

Since the analysing crystal necessarily lies in the path of radiation emitted from the sample, it is constantly exposed to all the wavelengths in the sample spectrum and the atoms of the dispersing crystal are therefore potential sources of secondary fluorescent radiation. Some of this secondary radiation will reach the detector, regardless of wavelength or the spectrometer Bragg angle, because there are no further dispersing crystals or other devices to 'filter' it (the same considerations also apply to radiation which is scattered from the surface of the crystal rather than diffracted by it). The result will be an increased background intensity over the whole angular range of the spectrometer.

This effect is seldom serious when radiation of less than about 3.5 Å is being measured, since most of the crystals used in this range are composed of light elements which can emit only soft, long-wavelength radiation which will be absorbed either along the path from the crystal to the detector (by residual air molecules) or in the window of the detector itself; in EPA analysis, for example, the sealed gas detectors used for short-wavelength radiation have relatively thick windows, and in XRF analysis the vacuum conditions employed are less stringent and air absorption of soft radiation is correspondingly more pronounced.

The problem can become significant, however, in the longer-wavelength ranges in which crystals such as ADP, KAP, RbAP, TlAP and mica are normally used. Each of these crystals contains one or more elements (P, K, Rb, Tl, Si etc.) which can emit fluorescent radiation at wavelengths within the general analytical range of the crystal itself. Although some of the fluorescent background can be removed by pulse height discrimination (Section 4.3), it is usually difficult to eliminate the problem altogether, and some measure of sensitivity is nearly always lost under these conditions. For major-element analysis the problem is usually not serious, but it may become significant in the analysis of very low concentrations, when any enhancement of the background constitutes a serious difficulty.

An example of this effect is afforded by the interference due to fluorescent Tl Mα radiation from TlAP crystals used for the determination of Mg. Tl Mα has a wavelength of 5.46 Å, and can be only partially resolved from Mg Kα (9.89 Å) by pulse height discrimination; both have relatively broad pulse amplitude distributions (see below). More obviously, fluorescent P emissions from an ADP crystal will present problems if an ADP crystal is to be used for the measurement of low concentrations of phosphorus.

TEMPERATURE EFFECTS

Mention has already been made of the difficulties that may arise from the high thermal expansion coefficients of some crystals, particularly PET but also, to a lesser extent, LiF_{200} and LiF_{220}. Angular shifts due to temperature fluctuations are most pronounced at high Bragg angles, although in this region they are partially compensated by a broadening of the diffraction peaks that is normally observed at high goniometer angles as a result of increased dispersion (the goniometer is the mechanical device which simultaneously rotates both crystal and detector in such a fashion that the Bragg Equation is preserved at all wavelengths).

Despite the temperature problem, the reflection efficiencies of PET and the LiF crystals are so high that X-ray spectroscopists are reluctant to discard them. Instead, some provision is normally made for maintaining uniform spectrometer operating temperatures. This can be more difficult than it sounds, particularly in the more massive XRF systems, because of the large thermal inertia of the spectrometer which makes it difficult to maintain constant temperatures with short-period thermostatic devices. Air conditioning of the laboratory helps to some extent, but is not usually sufficient. Spectrometers may be water cooled, using closed circuit, constant temperature coolant systems, or in some instances good results have been obtained by equipping them with thermostatic devices that maintain their temperatures at set levels 5–10 °C above ambient laboratory temperatures. In either case, temperature stabilization after start-up from cold may take 24 hours or even more.

Even with efficient temperature stabilization, it is good practice to recalibrate crystal angles frequently to guard against possible error (particularly in PET-based Si and Al determinations, which are made at relatively high Bragg angles).

ABNORMAL REFLECTIONS

Reflections which are not apparently attributable to the sample may sometimes cause difficulty in the interpretation of X-ray spectra, particularly in trace element analysis. Such anomalous reflections may include satellite lines (which are not always included in wavelength tables), primary tube radiation scattered by the sample, Laue-type peaks due to diffraction by large crystals in the sample, or abnormal reflections produced by the dispersing crystal.

The last are not often encountered, and even when present are usually (but not necessarily) of relatively low intensity; they are typically broad and irregular in profile. If necessary they can usually be eliminated by improving the degree of secondary collimation (which may be accomplished on many systems, for example, by using the scintillation detector instead of the flow detector, since the scintillation detector usually has an auxiliary collimator). They are most likely to be observed when using dispersing crystals of relatively low symmetry (e.g. topaz, mica).

4.2 Detectors and counting equipment

The basic function of an X-ray *detector* is to convert the energy of the X-ray photons into forms which can be recognized, visually or electronically, and counted over a finite period of time to provide an estimate of the intensity of the detected radiation. A secondary, but extremely useful, characteristic of the detector is that its response should be in some way proportional to the energies of the detected X-ray photons; it then becomes possible to discriminate electronically against some unwanted interfering responses (e.g. crystal fluorescence or scattered continuum) to ensure that only X-rays in a selected energy/wavelength band are measured.

There are many different types of X-ray detectors, all of them depending basically on the ability of X-rays to ionize atoms by photoelectric processes. For convenience, four major groups of detector can be distinguished, depending on the way in which the electrons and ions produced by photoelectric interaction are used to produce a detectable signal:

(a) *Photographic film* types, in which the incident X-rays reduce silver halides to metallic silver by ionization processes. The integrated intensity of the X-rays detected over a finite time interval can be estimated from the degree of blackening of the film. This is the oldest and simplest technique of X-ray detection, and it is still widely used in X-ray crystallographic applications (e.g. Laue, Weissenberg, powder diffraction methods). However, it is inconvenient, lacks precision for intensity measurements (even if optical densitometers are used to measure the degree of film blackening) and does not have good proportional response (i.e. it is not easy to relate the strength of the detector 'signal' – the degree of blackening – to the energy of the X-ray photon that produced it). Hence film measurement is no longer used in quantitative X-ray spectrometry and will not be further considered in this discussion.

(b) *Gas-filled detectors*, in which inert gases such as xenon or argon are ionized by the X-rays in a strong magnetic field produced by a high DC voltage applied across a cathode and anode. Electrons produced by the ionization process migrate to the anode, where they produce voltage pulses which can be amplified and fed to electronic counting circuits.

(c) *Scintillation detectors*, in which the X-ray photons interact with a phosphor material in such a way as to promote valence-band electrons to higher energy levels. When these electrons revert to their stable valence-band state, energy is emitted in or close to the visible light range and is converted into voltage pulses by a photomultiplier. These pulses may be amplified and passed to counting or integration systems.

(d) *Semiconductor ('solid-state') detectors*, in which electrons produced by X-ray ionization of solid-state semiconductor atoms are promoted into conduction bands, thus causing the semiconductor (e.g. silicon) to become more conductive; a 'hole', equivalent to a free positive charge, is left in the valence band. Under the influence of an applied bias voltage, the charge carriers (electrons and holes) migrate to the bias electrodes where they produce small voltage pulses for subsequent amplification and counting. The proportional response of these detectors is better than that of either gas-filled or scintillation detectors, and in fact they have sufficiently good inherent resolution to allow the electronic separation of output pulses produced by X-rays of different energy, with a resolution adequate for many analytical applications. Hence detectors of this type form the basis of energy-dispersive spectrometers which need no dispersing crystals (see later discussion). On the other hand, the energy resolution of a gas-filled or scintillation detector *used in conjunction with a dispersing crystal* is better than that of the solid-state detector alone, and hence is preferred in many quantitative analytical situations, particularly where X-ray intensities are low.

In most XRF and EPA systems, gas-filled or scintillation detectors predominate; XRF systems commonly use both, but the relative bulk of the photomultiplier system usually precludes the use of scintillation detectors in the more compact spectrometers used on microprobes. It is becoming increasingly common to equip microprobes with auxiliary solid-state detector systems, and some commercial XRF systems are also designed around energy-dispersive spectrometers.

4.3 Gas-filled detectors

4.3.1 Principles of operation

A gas-filled detector consists essentially of a hollow metal cylinder fitted with a thin wire along its radial axis (Fig. 4.3). This wire forms the anode of the detector, and is held at a high positive potential, usually in the range of 1.3–2.0 kV, with respect to the grounded metal case, which therefore functions as a cathode.

The anode wire is tensioned to prevent sagging, and the detector is filled with a suitable gas or gas mixture, at pressures normally between 0.5 and 1 atm. In theory any gas can be used, but some give a much better performance than others. Electronegative gases such as oxygen are avoided because the operation of the detector depends on the formation and migration of electrons, and electronegative gas molecules tend to form attachments to the electrons, inhibiting their migration and resulting in signal losses. Inert gases are preferred because they provide adequate signals at lower potentials than are required for most other gases, and their lack of chemical reactivity minimizes problems arising from corrosion of the anode wire. Argon and xenon are the most widely used filling gases.

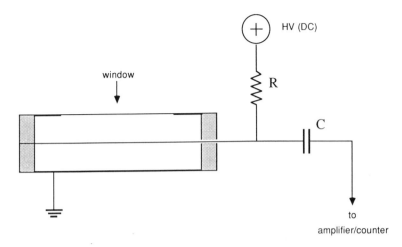

Figure 4.3 A gas-filled X-ray detector. An insulated thin wire is stretched axially in a hollow, grounded, gas-filled, metal cylinder and held at a high positive DC potential (c. 2 kV) which can be controlled with a variable resistor R. An X-ray photon entering the detector through a thin window on the side of the metal casing produces a train of ionized gas atoms; the electrons produced by the ionizations are accelerated towards the wire anode, producing additional electron-ion pairs by collision with other gas atoms along the way. Ultimately, all the electrons are collected at the anode, producing a short-term voltage drop at the capacitor C. This voltage pulse is amplified and passed to counting circuitry which records the detection of the X-ray photon.

X-rays enter the detector through a thin, low-absorbance window in the wall of the cathode casing. The window must be thin enough to transmit a sufficient proportion of the incident X-rays, but it must also be thick enough to prevent excessive leakage of the detector gas into the spectrometer vacuum chamber. This is another area in which some compromise is obviously essential, and several different types of detector are used for different applications. A sealed detector is permanently sealed after filling, and the window in this type of detector is usually made of beryllium, approximately 100 μm thick. Beryllium ($Z = 3$) is a low absorber of X-rays, and this thickness of window gives adequate transmission of X-ray wavelengths shorter than about 4 Å (transmission is approximately 75 per cent at 4 Å and better at shorter wavelengths).

For longer wavelengths, thinner windows must be used, but these are usually significantly permeable to the counter gas, which therefore gradually diffuses into the spectrometer vacuum chamber. Thin beryllium could be used, but it is fragile and withstands poorly the pressure differential between the inside of the detector (0.5 to about 1 atm.) and the spectrometer vacuum (10^{-5}–10^{-1} torr). Better results are obtained from windows made of Mylar (polyethylene terephthalate). 6 μm Mylar is very strong and has adequate transmission for X-ray wavelengths up to about 7 Å. For longer wavelengths 1 μm Mylar has better transmission characteristics, but it is comparatively weak and the so-called 'thin windows' need careful treatment and maintenance, and relatively frequent replacement. In some designs, thin Mylar windows are supported with a strengthening grid (e.g. nylon mesh); the grid fibres absorb some of the X-rays, but a loosely woven mesh has sufficient interstitial space to ensure only small losses in transmitted intensities while improving the mechanical stability of the thin window.

Other window materials (e.g. polypropylene, which contains no oxygen and therefore has no absorption edge at 23.3 Å, or Teflon, which has good transmission for the F Kα line) may be used for special purposes. It is interesting to note that microprobe spectrometers usually have poor performance for N K radiation (N Kα = 31.6 Å), which lies in the wavelength range just short of the C K absorption edge at 43.8 Å. Organic window materials are thus poor transmitters of nitrogen X-rays; they are, however, relatively transparent to carbon K radiation and hence give better performance for the 'softer' carbon X-rays than for those of nitrogen.

The problem of gas leakage in thin window detectors is overcome by continuously supplying the detector with a slow stream of gas to compensate for the diffusion loss. Detectors of this type are known as *gas flow detectors*. They give good results with low-energy radiations, but they require special precautions to maintain constant filling gas density, and it is practically impossible to prevent the introduction of small amounts of contaminants (oxygen, water vapour, dust particles etc.) which produce a gradual deterioration in detector performance. It is therefore necessary to monitor the detector performance characteristics closely, and to remove the detector periodically from the spectrometer for maintenance (cleaning or replacement of the anode wire, replacement of the window etc.).

Apart from beryllium, most window materials are electrical insulators, and their presence disturbs the symmetry of the electric field within the detector. This effect seriously detracts from the proportional characteristics of the detector (Section 4.2), but it may be overcome by coating the inside surface of the window with a thin layer of conducting material (e.g. vacuum evaporated, or 'flashed', aluminium). The conductive layer must of course be thin enough not to cause serious absorption of X-rays. In microprobe applications the outside surface of the window should also be flashed with a conducting layer to prevent charging effects due to absorption of electrons scattered from the sample surface.

The basic principle of the gas-filled detector is that an X-ray photon entering the chamber will produce ionizations of the filling gas; in the case of an argon-filled detector,

GAS-FILLED DETECTORS

$$Ar + h\nu \rightarrow Ar^+ + e^-$$

The positive ion and negative electron so produced are called an *ion pair*. The average amount of energy required to remove an outer electron from an inert gas atom is relatively small – of the order of 30 eV (Table 4.2).

Note that the *average* energy required to produce an ion pair is significantly higher than the first ionization potential, because of the loss of a significant proportion of the X-ray photon energy (almost half in argon detectors) in non-ionizing collisions. Collision with gas molecules is a statistical process, which explains why some photons produce more ion pairs than others of the same energy, and hence why the energy required to produce an ion pair is expressed as an average.

If the average energy required to produce an ion pair is e eV, an X-ray photon will produce an average of N ion pairs, such that

$$N = E / e \qquad (4.2)$$

where E is the energy of the photon. Thus in an argon-filled detector, an Fe Kα photon ($E = 6.398$ keV) will produce an average of 6398/26.4, or 242.3, ion pairs (the decimal value being justified as a statistical average; it is of course not possible to produce 0.3 of an ion pair). Some Fe Kα photons will produce more ion pairs than this, if they undergo less non-ionizing collisions than average; others with more than average non-ionizing collisions will produce less than average ion pairs. The distribution of observed ion pair numbers around the mean, or average, value is quasi-Gaussian (see Appendix B).

Without the applied electric field, the ion pairs would be rapidly destroyed by simple recombination processes; the end effect would simply be the scattering and absorption of the incident X-ray photons (including, under appropriate circumstances, some generation of X-radiation characteristic of the gas atoms). The detector would not produce any signal as the result of absorption of an X-ray photon.

Application of a low potential (less than about 500 V) does not greatly alter this situation. There will be some tendency for electrons to drift

Table 4.2 Inert gas ionization constants.

Element	Z	First ionization potential (eV)	Average energy to produce an ion pair (eV)
Ar	18	15.8	26.4
Kr	36	14.0	22.8
Xe	54	12.1	20.8

towards the anode and for the positive ions to migrate to the cathode; the drifts, however, will be slow and the probabilities of recombination high. Only a small proportion of the charges (positive or negative) generated by the X-ray photon will actually be collected at the cathode or anode, the voltage change ('pulse') produced at the anode will be very small, and the quantum efficiency of the detector will thus be low.

However, as the applied potential is increased it will at some stage reach a level where the rate of drift of electrons towards the anode or ions towards the cathode becomes sufficiently high that *unit efficiency* is achieved; the number of charges collected will be equal to the number produced by the photon-induced ionizations. Recombination is not prevented, but at unit efficiency it is exactly counterbalanced by additional ionizations produced by electrons which are accelerated towards the anode and attain sufficient kinetic energy to generate further ion pairs when they collide with other gas atoms. The total effect at unit efficiency is thus the collection of no more and no less charges than were produced by the original X-ray photon, and the effective amplification of the detector, usually called the *gas amplification*, becomes unity. Detectors operating at or close to unit gas amplification are said to be operating in the *ionization chamber region*.

At progressively higher applied voltages, the energies of electrons accelerated towards the anode are further increased, allowing each electron to produce progressively more ionizations; the new electrons thus produced can themselves induce further ionizations in a *cascade*, or *avalanche*, process. The result is that the charges finally collected at the anode and cathode become greater than those originally produced by the X-ray photon, and the detector is thus functioning as an *amplifier*. Gas amplification increases rapidly as the applied voltage is increased, and gas amplification gains of 10^2 to 10^7 are commonly observed under these conditions.

The magnitude of the charge collected then becomes a function of

(a) the energy of the X-ray photons, which determines the average number of ion pairs produced before gas amplification, and
(b) the degree of gas amplification, which is a function of the applied voltage, the gas density, the electric field characteristics etc.

If the factors controlling the gas amplification are held constant, the magnitude of the charge collected will thus be *proportional to the energy of the X-ray photon*, and the detector is said to be operating in the *proportional region*. Detectors designed to operate under these conditions are called *gas* (or *gas flow*) *proportional detectors*. (They are in fact often called *proportional counters*, although they do not themselves do any counting.) The actual voltage pulse produced at the anode by collection of the generated electrons is typically of the order of a few millivolts and requires additional electronic amplification to voltage levels sufficient to trigger counting circuits (approximately 1 V).

At still higher applied potentials (of the order of 1800 V or more, depending on detector design, gas density etc.), gas amplifications of the order of 10^9 or more can occur. Under these conditions, however, the cascades, or avalanches, of successive ionizations produced by the accelerated electrons become limited by the low gas density and by the more or less permanent fields that build up around the anode wire. Proportionality of response is thus lost, and at such high gains the magnitude of the charges collected becomes virtually independent of the energy of the detected X-ray photons. This is known as the *Geiger region*, since it is the region in which Geiger counters were originally designed to operate.

Geiger detectors have some practical advantages, and they are still widely used for X-ray detection applications where proportional response is not required. Because they do not have proportional response they do not require expensive, highly stabilized detector high-voltage supplies necessary to ensure constant gas gain. Furthermore, their output signals (which have been gas amplified by several orders of magnitude more than the signals from a proportional detector) are of sufficient amplitude to trigger electronic counting equipment, without need of expensive amplifiers and preamplifiers. Since field symmetry is of no great consequence the detector window can be placed in the end of the cathode tube rather than in its side, increasing the photon path length in the detector gas and hence improving the quantum counting efficiency. Geiger counters can be made physically rugged, and hence are ideally suited to certain field applications. They are no longer used, however, for quantitative X-ray spectrometry.

Finally, at applied potentials somewhere in the 2–3 kV region or higher, *corona* and ultimately *glow discharges* are induced. In these regions the detector becomes useless for X-ray measurement since the generation of electron avalanches is sustained indefinitely.

The behaviour of the positive ions accelerated towards the cathode is also significant, since they play important rôles in determining the characteristics of the detector. Most of the ion pairs are actually produced in the vicinity of the anode (within a distance equivalent to only a few diameters of the anode wire, which is only a small fraction of the detector radius). The electric field strength is highest in this region of the detector, so that this is where most of the avalanching takes place. Drift of the positive ions towards the cathode is thus a much slower process than the collection of the electrons at the anode, because they have a longer distance to travel and because the field strength diminishes rapidly away from the anode. As long as the positive ions remain in the vicinity of the anode the field strength is lowered; initiation of further avalanches is then inhibited and the detector does not respond to the absorption of further X-ray photons. This gives rise to the so-called *dead time* of the detector (Section 4.3.7).

When the positive ions reach the cathode, they can be neutralized by transfer of electrons from the cathode surface. However, this process will itself produce a new photon whose energy will depend on the difference in energy between the positive ion and the work function of the cathode surface. The new photon can eject further electrons from the cathode and initiate new avalanches; under some circumstances the initial avalanche could be sustained indefinitely and the detector would become useless.

It is therefore customary to add small proportions of a second gas to the detector to *quench* the cathode neutralization process. The quench gas is usually a diatomic molecular gas or a hydrocarbon which easily produces free radicals. The latter freely donate electrons to the inert gas ions, producing inert gas atoms plus another quench gas species which can react with electrons at the cathode surface without producing photons capable of initiating new avalanches. Commonly used quench gases include methane, alkyl halides, halogens and carbon dioxide. 'P10' gas, widely used in flow detectors, consists of 90 per cent argon and 10 per cent methane; higher methane contents may be used to advantage in soft X-ray detection, although the reduced argon density then necessitates higher applied potentials to maintain a given level of gas amplification.

4.3.2 *Formation of a detector pulse*

Reaction of an X-ray photon with the counter gas initiates an avalanche sequence, with the elapsed time between photon entry and avalanche production being of the order of 10^{-7} to 10^{-6} seconds (the period AB in Fig. 4.4). The electrons produced in the avalanche move towards the positively charged anode, where their collection produces a relatively sharp voltage drop (BC in Fig. 4.4). The positive ions move towards the cathode casing, first rapidly but then more slowly because of the progressive weakening of the field intensity outwards from the anode wire (CD in Fig. 4.4).

The total integrated charge collected by the anode in the time interval BC is thus relatively small, and the output pulse from the detector is actually produced largely by the outward migration and collection of positive ions. As noted earlier, during migration of the positive ions the detector is effectively 'dead', or insensitive to further X-ray photons. However, the contribution of the last stages of positive ion migration to the total charge collected is relatively small; the dead time of the detector can therefore be reduced by using an electronic device called an 'RC clipping circuit' in the amplification system, with a time constant of about 1 µs. Without clipping of the pulse 'tail', the effective dead time of the

GAS-FILLED DETECTORS

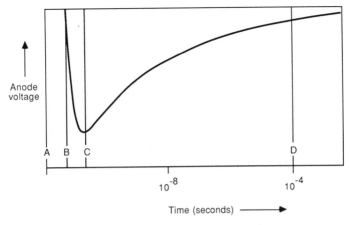

Figure 4.4 Photon-induced voltage pulse produced at the output capacitor C (Fig. 4.3) of a gas-filled X-ray detector. The voltage drop produced by electron collection is relatively rapid (time interval B–C) because most of the electrons are produced close to the anode wire where field strengths are highest. Restoration of the baseline voltage requires slower collection of the positive ions at the detector casing (C–D). To reduce the total collection time, during which the detector is insensitive to the arrival of further X-rays, the baseline voltage is 'clipped', or electronically restored to its original value, when the pulse decay is almost complete.

detector system would be about 10^{-4} s (100 μs), which is unacceptably high because of the restrictions it would impose on the maximum count rates at which the detector could function.

The integrated amplitude of the pulse produced at the output capacitor (Fig. 4.3) by the charge collection process is equal to Q/C volts, where Q is the total charge collected and C is the capacitance of the detector system. Thus a Cu Kα photon (1.54 Å, 8.04 keV) entering an argon-filled detector produces about 8040 / 26.4 = 300 ion pairs, and with a gas gain (A_g) of 10^4 a total of 3×10^6 effective ion pairs will be produced. The total charge produced by collection of each ion pair is equal to the charge on a single electron (1.6×10^{-19} coulombs), so that the total charge produced per photon event is 4.8×10^{-13} C. For a detector capacitance of 10^{-10} F (= 100 picofarads) the voltage produced will thus be

$$(4.8 \times 10^{-13}) / (10^{-10}) = 4.8 \times 10^{-3} \text{ volts}$$

$$= 4.8 \text{ mV}$$

Pulses of this amplitude are too small to trigger conventional electronic counting equipment and therefore require additional amplification, which must preserve the proportional relationship between voltage response and X-ray photon energy. Hence high-quality linear amplifying systems are required, with overall electronic gains (A_e) of the order of 1000×, to produce counting pulses with amplitudes in the range of 1–10 V.

The amplifier is usually mounted in a rack physically remote from the detector and connected to it with a length of shielded (coaxial) cable. Such cable, however, has a capacitance of about 1 pF cm^{-1}, so that a 2 m length of cable would add 200 pF to the capacitance of the detector system and reduce the amplitude of the input pulse to the amplifier to 1.6 mV in the above example. It is not mechanically feasible to mount the main amplifier directly on the detector or even close to it, so it is customary to use a small, low-noise preamplifier on the detector to produce output pulse amplitudes that are sufficiently large for cable transmission to the main amplifier. In earlier, vacuum tube designs the preamplifiers were relatively large and in some cases contributed substantially to mechanical limitations on the Bragg angle range of the spectrometer. However, modern solid-state systems are much smaller and cause much less difficulty in this respect. Even so, the preamplifiers used in some microprobe spectrometers are still connected to the detector by lengths of coaxial cable which are long enough to add significantly to the capacitance of the detector system.

The main amplifier uses a differentiating circuit to convert the counter output into a short, clipped pulse (Fig. 4.5).

4.3.3 *Pulse production statistics*

The output pulse of the gas-filled detector is proportional to the charge collected, provided that the detector is operated in its proportional region. The charge collected is in turn proportional to the number of ion pairs produced by the X-ray photon, and this is proportional to the energy of the photon. The final voltage produced at the amplifier output is thus proportional to E, A_g and A_e; for a given detector operating at fixed applied voltage and constant gas density, A_g and A_e will be constant, so that the pulse voltage, or *pulse amplitude*, will be proportional to E. The reasons for meticulous stabilization of detector voltage and gas flow rate in flow detectors are thus evident.

N, the average number of ion pairs produced by an X-ray photon of energy E, is given by

$$N = E / e$$

It has been noted that e, the *average ionization energy*, exceeds the theoretical first ionization potential of the detector gas because some of the energy of the X-ray photon is dissipated in non-ionizing collisions. If this were not the case, there would be no variation in the number of ion pairs produced by photons of a given energy, and all such photons would

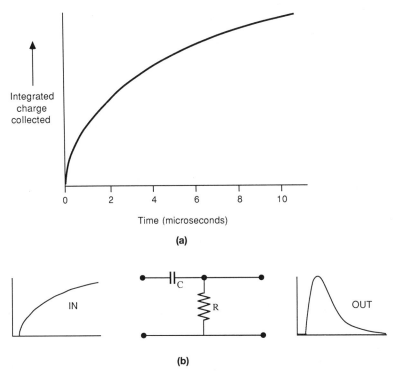

Figure 4.5 (a) Integrated charge collected on the detector anode, as a function of time. The rate of charge collection is initially high, but declines rapidly (Fig. 4.4), so that a differentiating RC circuit (b) can be used to convert the detector output to a pulse whose shape and amplitude are determined by the form of the charge-collection curve and by the values of R and C. If R is a variable resistor, the RC circuit can be 'tuned' to optimize the pulse shape for subsequent amplification and counting, and to adjust the dead time of the detector system (after Reed 1975).

result in output pulses of exactly the same amplitude. Electronic discrimination between pulses produced by photons of different energies – variously called *pulse height analysis* (PHA) or *pulse height discrimination* (PHD) – would then be relatively easy, and there would in fact be little need for spectrometers equipped with dispersing crystals.

On the other hand, if ionizing collisions were only a small fraction of all collisions in the detector, they would occur randomly and the resulting pulse amplitudes would show a *normal*, or *Gaussian*, distribution around a mean amplitude, such that the standard deviation (σ) about the mean N would be approximately equal to $\sqrt{N}$ (Fig. 4.6; see Appendix B for a review of elementary statistics).

In the case of argon-filled detectors, roughly half of the collisions are non-ionizing – that is, the actual situation is intermediate between the two

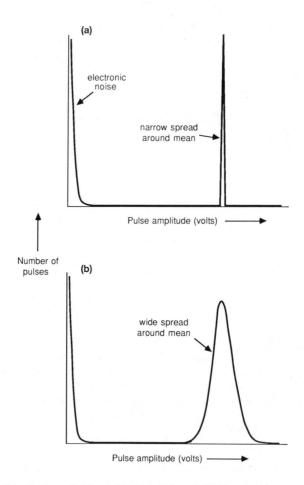

Figure 4.6 Distribution of the amplitudes of the amplified pulses entering the counting circuit. Low-voltage electronic 'noise', consisting of random pulses with amplitudes mostly less than about 1 V, is always present. If it is not filtered, it constitutes a background which may partly or wholly obscure the desired higher-amplitude detector pulses. Because of statistical variations (see text), the latter are distributed in quasi-Gaussian fashion around a mean amplitude; the relative width of the distribution is a function of the energy of the detected X-ray photon and of the detector/amplifier characteristics. Broad distributions imply poor *resolution*, i.e. poor ability of the system to separate the pulse distributions produced by X-ray photons of slightly different energies.

statistical extremes of Figure 4.6, and $\sigma < \sqrt{N}$. The distribution of the pulse amplitudes is then better described by the relationship

$$\sigma = \sqrt{(FN)}$$

in which *F* is known as the *Fano factor*. The *relative standard deviation* (or *coefficient of variation*) of the distribution is given by (σ / N), which will be equal to $\sqrt{(Fe / E)}$ or $\sqrt{(FN)} / N$, usually expressed as a percentage.

The production of secondary ion pairs in the avalanches is also a collision process subject to additional statistical fluctuations which can be included in the distribution expression by taking a larger *effective Fano factor*, which for argon in a gas-filled detector is usually about 0.8.

The result of these statistical considerations is that the *distribution* of pulse amplitudes produced by X-ray photons of a given energy will have the quasi-Gaussian appearance shown in Figure 4.7. Apart from low-voltage electronic noise, the pulse amplitudes are symmetrically distributed around a mean voltage, which will obviously be determined by the combined effects of E, e, A_g and A_e.

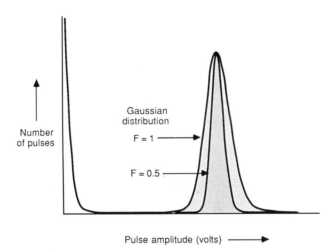

Figure 4.7 If ionizing collisions in the detector were only a very small fraction of all collisions they would occur rarely and randomly, and the resulting pulse amplitude distribution would be a true Poisson distribution (Appendix B), with its standard deviation equal to the square root of its mean. On the other hand, if *all* collisions were ionizing, all pulses would have exactly the same amplitude and there would be no spread about the mean. In practice the true situation invariably lies somewhere between these hypothetical extremes, as indicated by the *Fano factor (F)* which indicates the extent to which the standard deviation departs from the square root of the mean.

4.3.4 Pulse amplitude discrimination

Consider the output pulses from a gas-filled proportional detector which is receiving radiation of two different wavelengths (energies), e.g. Si Kα (7.125 Å, 1.739 keV) and Fe Kβ (1.757 Å, 7.057 keV). At first sight it might seem that these wavelengths could be easily and efficiently dispersed by a suitable analysing crystal, but closer examination shows that some of the Fe Kβ emissions will be diffracted in the fourth order at an apparent wavelength of $4 \times 1.757 = 7.028$ Å. Though its intensity would be relatively low in the higher-order reflections, nevertheless the detection of small concentrations of Si in the presence of large amounts of Fe would become difficult because the most sensitive Si line, Si Kα, will suffer some interference from the adjacent, incompletely resolved Fe Kβ(4).

However, suppose we examine the characteristics of the counting pulses produced by each of these radiations under the same detector/amplifier conditions. In each case the pulse amplitude V, which is proportional to E, e, A_g and A_e, will in fact be proportional to E, since all the other variables are held constant. Hence

$$V_{Fe\ K\beta} : V_{Si\ K\alpha} = E_{Fe\ K\beta} : E_{Si\ K\alpha}$$
$$= 7.057 : 1.739$$
$$\approx 4$$

In other words, the mean amplitude of the Fe Kβ pulses would be about four times that of the Si Kα pulses. If the gas and electronic gains are adjusted to amplify the Si pulses to a mean amplitude of 2 volts, then the mean amplitude of the Fe pulses will be about 8 volts (Fig. 4.8).

It now becomes feasible to apply electronic *discrimination* by setting up an electronic '*window*' (in no way related to the physical window of the detector) around the Si Kα pulses, and passing to the counter only those pulses having amplitudes which fall within this window, e.g. 2 V ± 1 V (Fig. 4.8). The range of the window is conventionally defined in terms of two parameters, viz. the *threshold voltage* (sometimes called the base, baseline, lower level, or simply just E), and a *window width* (or ΔE). In the example of Figure 4.8, the threshold voltage is 1 V and the window width is 2 V. Such a window will filter out any pulses having amplitudes less than 1 V (e.g. electronic noise, which typically consists of random pulses with low amplitudes – mostly less than 0.5 V under normal operating conditions) or more than 3 V. Only pulses with amplitudes between 1 and 3 V will be passed to the counting circuitry.

In the Si Kα – Fe Kβ example, if the amplifier gain is adjusted so that the mean amplitude of the pulses produced by Si Kα photons is about 2 V, these pulses will be passed to the counter. The mean amplitude of

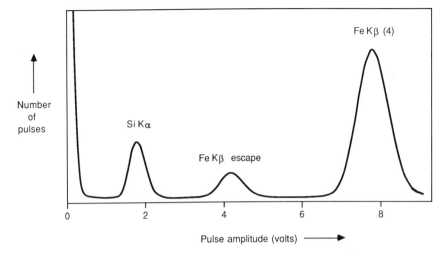

Figure 4.8 Portion of the distribution of amplitudes of the pulses from a detector/amplifier receiving Si Kα and Fe Kβ X-rays. The Fe Kβ photons have approximately four times the energy of the Si Kα photons, so that they produce about four times as many ion pairs and their amplified output pulses are centred about a mean amplitude approximately four times that of the Si Kα distribution. The additional distribution identified as 'Fe Kβ escape' is discussed in Section 4.3.6.

the Fe Kβ pulses, however, will be about 8 V; these pulses will be filtered because they lie outside the window, and the interference of Fe Kβ with the measured Si Kα intensities will be eliminated. This is the principle of pulse amplitude (or 'height') discrimination.

In setting up a pulse discrimination system, limits are obviously imposed on the window width and height settings by the need on one hand to include all the desired pulses within the window, regardless of their spread around their mean, and on the other to reject all the unwanted pulses, although they too will be spread around their means and it is possible that two or more distributions may overlap or come very close to overlapping. Clearly, the minimum window width required is determined by the actual extent of *spread of amplitudes around the mean* or means. Good *resolution* of the discriminant system, i.e. good ability to separate the unwanted amplitudes, implies relatively narrow spreads around each of the amplitude distribution means. Specifically, the resolution of the system R (usually referred to as the *detector resolution*, and expressed as a percentage) may be conveniently defined as

$$R = W / V \qquad (4.3)$$

where W is the width of the amplitude distribution at half its maximum and V is the mean amplitude (Fig. 4.9). Note that 'resolution' as defined in this way is quite different from the resolution of the spectrometer, which refers to the ability of the spectrometer to disperse closely spaced wavelengths.

For a Gaussian distribution, which is closely approached by the pulse amplitude distribution, $W = 2.36\sigma$ (see Appendix B). A pulse distribution of $\pm 0.5W$ ($= \pm 1.18\sigma$) about the mean will include 76.2 per cent of all the pulses. The resolution R then becomes

$$R = W / V$$
$$= 2.36\, \sigma / V$$
$$= 236\, \sigma / V\%$$

But (σ / V) is the *relative standard deviation*, or coefficient of variation, which has already been defined in terms of F, E and e; hence

$$R = 236\, \sqrt{(Fe / E)} \qquad (4.4)$$

For argon, $e = 26.4$ volts and F is approximately 0.8, so

$$R = 236 \times \sqrt{(0.8 \times 0.0264)} / \sqrt{e}\, \% \quad (E \text{ in keV})$$
$$= 34.3 / \sqrt{E}\, \%$$

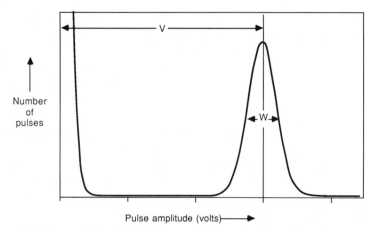

Figure 4.9 The resolution R of a detector and its associated amplifier and pulse-processing circuitry, for a particular X-ray photon energy, is expressed as W / V, where W is the width of the pulse amplitude distribution at half its maximum and V is the distribution mean. W and V are usually measured in volts, and R is conventionally expressed as a percentage.

Hence for Fe Kα ($E = 6.40$ keV), $R = 13.6$ per cent, and for Si Kα ($E = 1.74$ keV), $R = 26.0$ per cent. Note that resolution is poorer (amplitude distributions are more widely spread) for lower energy (longer wavelength) X-rays.

Alternatively,

$$R = 34.3 / \sqrt{E}$$
$$= 34.3 / \sqrt{(12.4 / \lambda)} \quad (E \text{ in keV})$$
$$= 9.74 \times \sqrt{\lambda}$$

These expressions have been calculated for argon-filled detectors with $F = 0.8$. For $F = 1.0$, for example, $R = 10.87 \sqrt{\lambda}$.

Resolutions calculated in this way are theoretical values (note that optimum resolution is indicated by a low value of R), but they can be approached closely in practice with good equipment and skilled operation. Experimental measurement of detector resolution is a simple operation requiring no more than experimental determination of the pulse amplitude distribution for X-ray photons of a known energy, and it provides a sensitive method for monitoring the condition of a detector and determining when it requires overhaul or replacement. This is because deteriorating detector performance (due, for example, to pitting of the anode wire in a gas flow detector) is reflected in loss of proportionality, which in turn produces wider pulse amplitude distributions and increased values of R. (Dirty or pitted anodes produce asymmetric electric fields, with consequent irregularities in the pulse amplitude distributions – if the detector field is asymmetric, the amplitude of the pulse produced by an X-ray photon will then depend on which part of the detector field was actually traversed by the X-ray.)

The resolution of a detector can also be expressed in absolute rather than relative terms by calculating the energy equivalent, in electron volts, of the width of the pulse amplitude distribution at half of the distribution maximum (i.e. the energy equivalent of W). This is often termed the full width at half maximum, or 'FWHM'. (It is also sometimes symbolized as 'ΔE', which must not be confused with the use of the same symbol to represent window width in some PHA systems.) It is calculated simply as the product of W and E, where E is the photon energy (in electron volts):

$$\text{FWHM} (= \Delta E) = W \times E$$
$$= 2.36\sigma \times E$$
$$= 2.36 \times \sqrt{(e / E)} \times E$$
$$= 2.36 \times \sqrt{(e\, E)}$$

If the Fano factor is incorporated,

$$\text{FWHM} = 2.36 \times \sqrt{(e\, F\, E)}$$

Thus for an argon-filled gas detector and a Fano factor of 0.8, for Fe Kα (E = 6.398 keV),

$$\text{FWHM} = 2.36 \times \sqrt{(26.4 \times 0.8 \times 6398)}$$
$$= 868 \text{ eV}$$

For a Gaussian distribution, the full width at tenth maximum ('FWTM') is equal to 1.823 times the FWHM, or, in the above example, 1581 eV. Hence an argon-filled gas detector can only completely resolve Fe Kα photons from those whose energies are more than about 2 keV (equivalent to about 0.6 Å) higher or lower. This is not good enough to separate commonly used analytical lines without the assistance of a dispersing crystal.

Detector resolution relationships can be used to calculate in advance the approximate window and gain settings appropriate for specific analytical circumstances. Thus for Si Kα and an argon-filled detector, $R = 26.0$ per cent (theoretical), and a window width of $\pm W$ about the mean pulse amplitude would be equivalent to $\pm 2.36\sigma$, and would transmit 98.2 per cent of the Si Kα pulses.

If the detector and amplifier gains are set so that the mean pulse amplitude V is 2 V, as in the example above (Fig. 4.8), and if $R = 26.0$ per cent,

$$R = (W / V) \times 100\%$$
$$W = (2.0 \times 26.0) / 100$$
$$= 0.52 \text{ V}$$

Thus for an efficient system, a window of 1.04 V (= $2W$) on a baseline of 1.48 V ($V - W$) would transmit 98.2 per cent of the desired Si Kα pulses. In practice, a window setting of 2 V on a baseline of 1 V would probably be preferred; it allows for small departures from detector perfection (i.e. a modest loss in resolution) and for the possible effects of pulse amplitude depression at high count rates (see Section 4.3.5). Calculations similar to the above will show that the 2 V window will still effectively reject the Fe Kβ pulses while transmitting virtually all the Si Kα pulses.

In setting up the amplifier/analyser system, it is necessary to adjust a number of variables, and again some degree of compromise is often involved. Although the exact procedures to be followed vary somewhat according to the design of the instrument and the preferences of individual operators, the objective is always to adjust detector voltage (gas gain) and electronic amplification to produce pulse voltages suitable for both discrimination and counting, and to set a discriminator window which will pass all the desired pulses to the counter and reject all others.

In some cases it is sufficient to use only threshold discrimination to reject low-amplitude electronic noise, i.e. to use a window of infinite width on a baseline of, say, 1 V. This mode of operation, where the window has no upper limit, is sometimes called *integral mode* operation; it contrasts with *differential mode*, in which an additional upper window discriminant level is also set to reject unwanted higher amplitude pulses (Fig. 4.10).

4.3.5 Pulse amplitude depression

Suitable amplification can be achieved by using high detector voltages (high gas gain) or high electronic amplifier gains, or by a compromise between the two. Other things being equal, high amplifier gains tend to

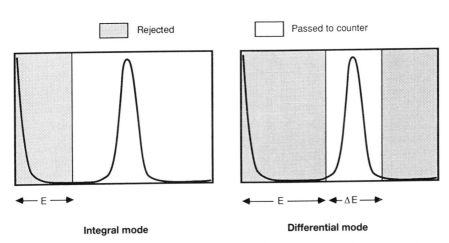

Figure 4.10 Interfering pulses, which increase the background beneath measured X-ray intensities and thus reduce sensitivities, may be removed by electronic filtering (or discrimination), provided that their amplitudes are significantly lower or higher than the distribution of amplitudes produced by the desired X-ray photons. In *integral* mode discrimination, all pulses with amplitudes below a 'threshold' value E are rejected, and all other pulses are accepted. This eliminates electronic noise and possibly other low-amplitude interferences. In *differential* mode, both low- and high-amplitude pulses are rejected, and the only pulses passed to the counting circuit are those whose amplitudes lie between the threshold voltage E and an upper limiting voltage $E + \Delta E$. ΔE is usually termed the 'window' or 'window width'.

result in high noise levels, although this effect is much less serious with modern solid-state, low-noise amplifiers than it used to be with vacuum tube equipment. On the other hand, high detector gains are also accompanied by a disturbing effect – a shift to lower mean pulse amplitudes at high count rates. This effect is always present, but becomes much more serious at high detector voltages. It arises from the overlap of successive pulses, which produces a spurious baseline voltage in the AC amplification circuitry and causes an apparent reduction, or depression, of the pulse amplitudes (Fig. 4.11).

This effect arises from the slow drift of the positive ions towards the cathode. While in the vicinity of the anode, they have a screening effect and reduce the intensity of the electric field. At high count rates this effectively reduces the gas gain and hence the amplitude of the output

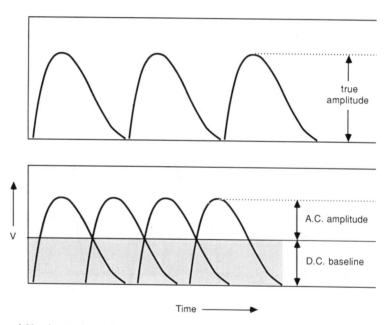

Figure 4.11 An apparent depression of pulse amplitudes arises at high count rates, when successive pulses begin to overlap each other. The output voltage of the detector/ preamplifier system then has insufficient time to return to its true baseline between pulses, and an artificially high constant-voltage baseline takes its place. This component is ignored by the AC amplifier, to which it appears as a DC signal. The amplifier processes only that component of each pulse which appears as a varying voltage, and its output pulses are thus reduced in amplitude, or 'depressed', below their true voltages. The effect becomes progressively more serious at higher count rates, as the extent of pulse overlap increases. Its magnitude also depends on pulse shape, which is a function of such variables as X-ray photon energy, operating conditions (detector voltage, amplifier gain etc.), and detector/ amplifier design and maintenance.

pulses. The effect is more pronounced at high detector voltages, at which each incident photon produces more positive ions (Fig. 4.12), and it is also more severe with the more energetic X-ray photons because these also produce more positive ions. Pulse depression is also relatively more severe if high-Z filling gases are used because of the decreasing mobility of heavier ions; in this respect argon is a better filling gas than xenon (and neon is better still).

Analytical errors can arise if a window set up at, say, a low count rate is not set sufficiently wide to allow for pulse amplitude depression at high count rates (or vice versa), since sample and standard count rates often vary by several orders of magnitude.

Pulse amplitude depression is a more serious problem in XRF than it is in EPA because XRF count rates are usually much higher. XRF operators therefore tend to operate their electronic amplifiers at high gain, in order to use only sufficient detector voltage as is necessary to obtain satisfactory total amplification. Microprobe operators, usually working with much lower count rates, are less disturbed by pulse amplitude depression; many of them tend to operate at lower amplifier

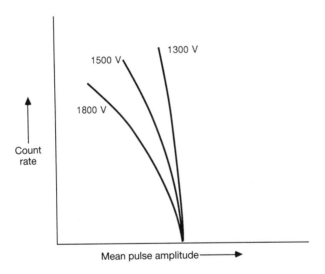

Figure 4.12 Dependence of pulse amplitude depression on detector anode voltage. The apparent mean pulse amplitude is always depressed at high count rates, but the effect is more pronounced if high detector anode voltages are used. It can be compensated by reducing the detector voltage, but higher amplifier gains are then required and electronic noise levels may be significantly increased. Alternatively, the discriminator window width can be increased to allow for the depression, at the expense of some loss in discriminator resolution. The analyst is responsible for investigating the magnitude of depression effects and for selecting the optimum strategies to cope with them.

gains (to minimize electronic noise) and at correspondingly higher detector voltages.

Pulse amplitude depression can be minimized by careful detector maintenance, by increasing the diameter of the anode wire (which, however, then requires higher detector voltages to maintain a constant level of gas gain), or by keeping the gas gain as low as possible. In rare circumstances where the effect is unusually severe, gas gains of as low as 10^2 may be used. In general, however, advances in electronic design have greatly reduced the magnitude of this problem, at least for systems in good operating condition, and although it is good practice to check periodically for possible errors attributable to this effect, it is now seldom found to be a source of major analytical difficulty.

4.3.6 Escape peaks

If the entering X-ray photons have sufficient energy, they are capable of producing characteristic fluorescent radiation from atoms of the detector gas (e.g. Ar K emissions). This typically consumes much, but not all, of their energy; the residual energy is still available for the production of ion pairs in the normal fashion, and the resulting avalanches will result in the production of a detector pulse. The amplitude of this pulse, however, will be reduced below that of a normal pulse by an amount corresponding to the X-ray photon energy that was absorbed in the generation of the gas characteristic emission.

Under these circumstances, the pulse amplitude distribution will show a second, smaller peak at a lower voltage than the main peak (Fig. 4.8), with the voltage difference between the two being equivalent to the (amplified) excitation energy of the gas radiation. Thus the absorption edge for Ar K radiation is at 3.20 keV. An Fe Kβ photon entering an argon-filled detector (Fe Kβ = 7.057 keV) may lose 3.20 keV, or 45.3 per cent, of its energy in exciting an Ar K photon. The remaining 3.857 keV can then produce ion pairs in the normal way, but will produce only 54.7 per cent (3.857 / 7.057) of the normal average number. If the system total gain is adjusted so that the 'normal' Fe Kβ pulses have a mean amplitude of 8 V (Fig. 4.8), the *escape peak* pulses will be distributed around a mean amplitude of 8 × 0.547 = 4.4 V.

Clearly an escape pulse will only be produced if the Ar K photon escapes from the detector without producing its own chain of ion pairs, otherwise the total number of ion pairs produced will be the same as that otherwise generated by the incoming photon (Fe Kβ in the above example) and the escape phenomenon will not be observed. Since the filling gas will be relatively transparent to its own radiation (its K photons, for example, will always have less energy than their own K absorption edge), there will be a relatively high probability of escape and hence of the generation of a significant escape pulse.

Generation of escape pulses is another statistical process, and the integrated intensity of the escape peak pulses will be proportional to the intensity of the detected X-ray photons to which they are related by the escape mechanism. The escape pulses could in fact be used for analytical purposes. Cases are known in which the discriminator window has been set up accidentally on the escape peak rather than the true peak and has still yielded reasonably good analyses; statistical precision is sacrificed at the lower escape peak intensities, but accuracy is not affected. It is necessary to use care, however, to ensure that the discriminant window is set up so that it either *always includes* or *always excludes* the escape peak if the latter is present.

Escape peaks will be produced in all cases where the incoming X-ray photons are more energetic than an absorption edge of the filling gas. The phenomenon is particularly pronounced in cases where the photon energy is just higher than an absorption edge of the filling gas; under these conditions the fluorescence excitation efficiency is high. The Ar K absorption edge is at 3.20 keV; hence K Kα (3.31 keV) and Ca Kα (3.69 keV) produce characteristically strong escape peaks in argon-filled detectors.

Occasionally cases arise where pulse amplitude discrimination can be used successfully to eliminate harmonic overlap interference (e.g. the Fe Kβ / Si Kα interference discussed earlier), but interference due to an escape peak remains. A classic example is that of the determination of phosphorus (P Kα = 6.16 Å) in the presence of calcium (Ca Kβ(2) = 2 × 3.09 = 6.18 Å). If the detector and amplifier gains are adjusted so that the mean P Kα pulse amplitude is 2.0 V, then the two distributions are sufficiently well resolved in a good system that the 'normal' Ca Kβ pulses can be entirely removed by amplitude discrimination (Fig. 4.13).

However, the Ca Kβ escape peak (4.012 − 3.202 = 0.81 keV) will then be centred slightly below 1 V, and its 'tail' will overlap that of the P Kα pulses (Fig. 4.13). It is then impossible to set a discriminant window that will pass all the P pulses and reject all the Ca pulses. To cope with the interference it will be necessary either to use a narrow window that rejects some of the P pulses (with due attention paid to possible pulse amplitude depression effects), or to apply an independently calibrated correction for the extent of the Ca interference. The superior resolution of Ge or quartz crystals may also be used to alleviate the problem by improving the crystal-dispersion separation of the P Kα and Ca Kβ peaks.

4.3.7 Dead time

The dead time of the detection/counting system is the interval, following

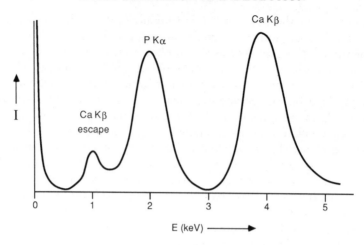

Figure 4.13 Typical pulse amplitude distributions produced by P Kα and Ca Kβ X-ray photons generated from the same sample, and not separated by the dispersing crystal because of harmonic overlap. The Ca Kβ escape peak overlaps the P Kα distribution and therefore cannot be completely separated from it by electronic discrimination. Note that this interference is not a problem in energy-dispersive spectrometers, which are not subject to harmonic effects.

the arrival of a pulse, during which the system does not respond to further pulses. As already noted, the detector is non-responsive during the period when the positive ions remain in the vicinity of the anode. However, this is not the only source of dead time in the counting circuit as a whole; in particular the pulse-shaping circuitry also has a dead time which may be comparable with or greater than that of the detector alone. Given sufficient information concerning the detector characteristics and operating conditions, and sufficient detail of the electronic circuits employed, it is possible to calculate the dead time of the whole system from first principles. In fact, however, this is seldom done – it is usually faster and more accurate to measure the system dead time experimentally.

The practical effect of dead time is that in any finite counting time some pulses will be 'lost' because the system was inactive during the period when they arrived at the detector and should have been recorded. This is equivalent to saying that the net counting time during which the detector was active and able to record pulses is somewhat shorter than the actual elapsed counting time, by an amount equivalent to the total passive time; the latter in turn will be equal to the product of the number of counts actually recorded and the dead time due to each of those pulses. If, for example, a total of 10 000 counts was recorded in an elapsed time of 1 s, and the system dead time was equal to 2 µs (2×10^{-6} s), each of the 10 000 recorded pulses would have rendered the system passive for 2 μs, and the total passive time would have been

$$10\,000 \times 2 \times 10^{-6}\,\text{s} = 2 \times 10^{-2}\,\text{s}\ (\text{or 20 ms})$$

The active counting time would therefore have been $(1 - 0.020)$ seconds, or only 0.980 seconds, and the true counting rate would have been $10\,000 / 0.980 = 10\,204$ cps. At this count rate the dead-time effect, if uncorrected, would have produced a counting error of approximately 2 per cent.

In general, if n' is the observed counting rate, in *counts per second*, and t is the system dead time, *in seconds*, then in any one second the detector will have been 'live' for only $(1 - n't)$ seconds, and the true counting rate n will be given by

$$n = n' / (1 - n't)\ \text{cps} \tag{4.5}$$

which is the expression most commonly used to calculate the extent of the dead-time effect.

In fact this expression is an approximation (as is the argument from which it was derived), since it assumes

(a) that the pulses arrive in truly random fashion rather than in bursts, which is not strictly true (particularly at low count rates), and
(b) that the dead time is *non-extendable*, i.e. that the arrival of an X-ray photon while the system is dead does not further prolong the dead period, which also is not strictly true.

At low count rates (less than about 50 000 cps) and with dead times of the order of 1–2 μs, these approximations are sufficient and the simple dead-time expression may be used without serious error. However, modern XRF systems in particular are easily capable of count rates in excess of 10^5 cps, and if such high intensities are to be used in analysis more accurate calculation of dead-time effects (which recognize that dead-time losses are not truly linear functions of count rate) will be required. The divisor on the right-hand side of the correction equation must then be replaced by a binomial expansion of the generalized form of

$$n = \frac{n'}{(1 - n't + (n')^2 t^2 / 2! - (n')^3 t^3 / 3! \ldots)} \tag{4.6}$$

Dead-time errors may be compensated either instrumentally or by calculation. In either case it is first necessary to measure the dead time, directly or indirectly, so that the appropriate corrections can be applied.

In single-spectrometer XRF systems, it is convenient to make the corrections automatically within the timing/counting systems. For example, the timing circuit which starts and stops the counting process can be modified so that, in effect, the timing clock is stopped during periods in which the counting system is dead. If the dead time is measured and found to be, say, 1.6 µs, then the clock is stopped for 1.6 µs every time a pulse is counted. Thus when the timer shows an elapsed period of 1 s, this elapsed period will be for 'live' time only and the apparent count rate will be the true count rate. If 10 000 counts were recorded the actual elapsed period will be

$$1 + 10\ 000 \times 1.6 \times 10^{-6} = 1.016\ \text{s}$$

but by showing the counting time as only 1 s the dead-time correction will automatically have been applied.

A similar approach is to add, electronically, a predetermined number of pulses to those recorded in each preset time interval. For example, suppose that the dead time has been measured and found to be 2 µs. If a count rate of 10^4 cps is observed, then the true count rate will be

$$\frac{10^4}{1 - 10^4 \times 2 \times 10^{-6}} = 10\ 200\ \text{cps}$$

Similarly, an observed count rate of 20 000 cps is equivalent to a true count rate of 20 800 cps (approximately) and an observed count rate of 40 000 cps is equivalent to a true 43 500 and so on. The proportional addition procedure thus requires the automatic electronic addition of 2 pulses for every 100 pulses observed at 10K cps, 4 for every 100 at 20K cps and so on. Devices to make this proportional addition are available and are supplied with some commercial XRF systems. Apart from periodic maintenance and recalibration they require little attention, and they effectively remove dead time as a serious source of error so long as count rates are kept below 50–100K cps. However, they cannot cope quite so simply with the effects of non-linearity encountered at higher count rates and significant errors may result if this failure is not recognized. Further, most automatic correction devices of this type assume that the actual dead time is independent of such parameters as X-ray photon energy and detector anode voltage, and these assumptions are only safe within certain limits (see below).

Direct instrumental corrections, of either of these types, are not widely used in microprobe systems, which typically have several different spectrometers counting simultaneously, usually at different count rates but controlled by a single master timing system. It is therefore not possible to make clock-type corrections without providing separate timing systems for each spectrometer; the 'added-pulse' technique is technically feasible but to date has not been widely adopted. Instead the normal procedure is to measure the dead time (separate measurements are usually required for each spectrometer) and to apply an arithmetic correction to each observed count rate in order to convert it to true intensity. At the relatively low count rates typically encountered in microprobe work the simple linear dead-time correction expression usually works quite well (indeed the dead-time error itself may be insignificantly small and it can often be neglected by comparison with other sources of error); the arithmetic correction is easily and quickly made on systems equipped with dedicated computer control and data-processing systems.

In principle, the dead time of a detector should be dependent on such factors as X-ray energy and detector operating voltage (since these determine the magnitude of the positive ion charge field and its decay time). In practice, however, the dead time of the whole detection and counting system is more likely to be determined by the electronic pulse-shaping circuitry, which is less sensitive to these parameters. It is therefore usually feasible to use a single, experimentally measured value for the dead time of a counting system spanning a reasonably wide range of X-ray energies and operating conditions. Serious errors are only likely to arise at high counting rates, but once again the X-ray analyst should always guard against a false sense of security and should make periodic experimental checks to ensure that the use of a generalized dead-time correction is warranted. This is particularly the case in XRF analysis, in which a single spectrometer is normally used to span the complete range of analytical X-ray energies, and in which count rates are usually higher and dead-time errors potentially more severe.

4.3.8 *Measurement of dead time*

Depending on the instrumentation involved, any of several procedures may be used for the direct measurement of dead time. The simplest of these are based on the assumption of proportionality between X-ray intensities and certain critical excitation conditions. In the microprobe, for example, the intensity of a measured X-ray line should be proportional to the sample current, providing all other factors are held constant. The sample current is a measure of the number of electrons absorbed by the sample per unit of time; if it is doubled then the intensity of emitted X-rays should also double. In XRF, doubling of the current in the primary tube should double the intensity of the primary X-rays and therefore should have the same effect on sample emissions. However, in both cases intensities measured at the higher current settings will be found to be lower than those calculated from direct proportionality because of higher relative dead-time losses at the higher intensities. Departures from proportionality at progressively higher count rates can be extrapolated back to zero count rates (at which, of course, there will be no dead-time losses) to determine the true proportionality constant, which can then be used to calculate the dead time.

If n is the true count rate (in cps) and i_s is the sample current (microprobe) or tube current (XRF), then

$$n / i_s = k \tag{4.7}$$

where k is a proportionality constant. If n' is the measured count rate (also in cps), then n' / i_s will not be constant for different values of n' because of the dead-time effect. Since $n' < n$ (except at $n = 0$, when $n' = n$), n' / i_s will be less than n / i_s except at zero count rate.

Therefore a series of values of n' / i_s, measured experimentally, can be plotted against n' and extrapolated to zero n', where n' / i_s is equal to n / i_s ($= k$). Such a plot is shown in Figure 4.14.

Since

$$n / i_s = k$$

and

$$n = n' / (1 - n't)$$

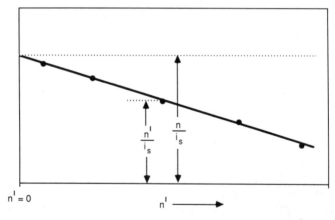

Figure 4.14 Measurement of dead time by extrapolation of a plot of measured intensity (n') against tube current (XRF) or sample current (microprobe), to zero count rate. At zero intensity there is no dead time loss, so that the true ratio n/i_s can be established and used to correct the measured intensities. Because of statistical counting errors, particularly at low count rates, the linear extrapolation should normally be fitted to at least four or five observations. Conversely, the assumption that dead-time losses are a linear function of measured intensity may be invalid at high count rates.

GAS-FILLED DETECTORS

then

$$n' / i_s = k(1 - n't)$$

Hence the experimentally determined values of n' / i_s can each be divided by the value of k found from the extrapolation, to determine a series of values of $(1 - n't)$; a value of t can be calculated from each of these. Because of statistical counting error (see Ch. 7) there will usually be some slight variation between the successive values of t calculated by this method. However, if these values are averaged over at least four or five measurements of n' / i_s (i.e. a best-fit linear extrapolation to zero n' is used) a satisfactory estimate of t will usually be obtained. Note, however, that the assumption of a linear extrapolation implies that the dead time is not extendable, and this simple procedure should not therefore be used for count rates higher than about 50K cps.

In some XRF systems it may be convenient to adopt a slight variant of this procedure which does not assume proportionality between X-ray intensity and tube current. Instead intensity measurements are made at each of a series of tube currents, in each case recording the intensities observed with coarse and fine collimators inserted successively in the primary beam. The spacing between plates is greater in the coarse collimator, so that intensities measured with this collimator are greater than those observed with the fine collimator, usually by a factor of four to five. For any given instrument, n'_{coarse} / n'_{fine} would be constant if there were no dead-time losses. Because of the latter, however, the ratio decreases at progressively higher count rates. Hence this ratio can be plotted against, say, n'_{coarse} and extrapolated to zero n' to determine the true proportionality constant, in a fashion exactly analogous to that described above. For any of the measured data sets (which again should preferably be best fit to observations made at four or five tube current settings in order to reduce experimental and statistical errors), the equation is

$$k = \frac{n_c}{n_f} = \frac{n'_c}{n'_f} \times \frac{(1 - n'_f \times t)}{(1 - n'_c \times t)} \tag{4.8}$$

All values of n and n' in these equations are expressed in *counts per second*, and the corresponding values of t are expressed in *seconds*. If other units are used (e.g. count times of 10 s, or dead times expressed in microseconds), then the equations must be appropriately modified.

4.4 Scintillation detectors

A scintillation detector contains two essential components:

(a) a *phosphor*, which converts the energy of an incident X-ray photon into a series of light pulses of about 3 eV energy, and
(b) a *photomultiplier*, which converts the light pulses into voltage pulses which can be amplified and counted in similar fashion to the voltage pulses produced by a gas-filled detector (Fig. 4.15).

The phosphor is typically an alkali halide (usually sodium iodide) doped with an impurity element such as thallium or europium. Energy from the X-ray photon is used to excite a halide valence band electron into a conduction band, leaving behind a positive 'hole' (Section 2.3.2) which migrates to the nearest impurity atom. The latter thus becomes ionized and oxidized to a higher valence state. The oxidized atom combines with an electron from the halide conduction band to enter an excited state; radiation is emitted as it then decays to the stable ground state. For Tl(thallium)-activated NaI, the decay energy is about 3 eV, so that the resulting emitted radiation consists of a light photon with a

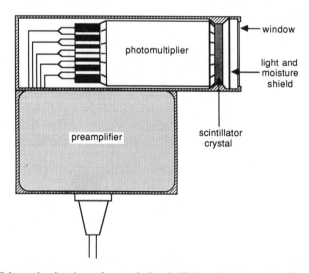

Figure 4.15 Schematic drawing of a typical scintillation detector. An X-ray photon entering the detector, through a light- and moisture-shielded window, produces a series of light pulses (each of about 3 eV energy) in a doped phosphor crystal. The light pulses are detected and amplified by a photomultiplier, which outputs a voltage pulse to a discriminator and counter. The number of light pulses produced in the phosphor, and hence the amplitude of the photomultiplier output pulse, depends on the energy of the X-ray photon, and a scintillation detector is thus basically a true proportional detector.

SCINTILLATION DETECTORS

wavelength of about 4100 Å (in the blue region of the visible spectrum).

The phosphor must be prepared in a form thick enough to absorb a high proportion of the incident X-rays, but not so thick that the light photons produced by X-ray interaction are significantly absorbed before they reach the photomultiplier. Alkali halide crystals are hygroscopic, so that the phosphor must be enclosed in an air-tight mounting which is reasonably transparent to X-rays on its outer side, but optically opaque so that the phosphor is not flooded with stray light.

Hence phosphors are usually prepared as discs 2–3 cm wide and about 2 mm thick, mounted behind windows made of beryllium about 0.2 mm thick and coated with a thin layer (about 1 μm) of evaporated aluminium. The window necessarily limits the penetrative efficiency of soft X-rays, and as a result scintillation detectors have only very limited applications for wavelengths longer than about 2.5 Å.

The rear, or light emission, surface of the phosphor is optically coupled to the glass front surface of the photomultiplier, usually with a thin smear of a low-vapour pressure oil (e.g. silicone). Alternatively, the phosphor and photomultiplier may be manufactured as a single integrated unit. Behind the front surface of the photomultiplier is a photocathode, constructed of a material such as Sb/Cs which will release 'bursts' of electrons when struck by light photons from the phosphor.

These electrons are accelerated towards and electrostatically focused on the first of a series of *dynodes*, or plates within the photomultiplier held at progressively higher positive potentials. Each electron striking the first dynode has sufficient energy to produce a number of secondary electrons. These are then accelerated towards the next dynode, where each produces further secondary electrons, and so on. The net result is an amplification analogous to the gas amplification of the gas-filled detector; if the number of secondary electrons produced at the first dynode is P, the total gain of the photomultiplier will be P^x, where x is the number of dynode stages. In commercial detectors, P typically has a value of about 4 and x is usually about 10; the total gain is therefore about 4^{10} or approximately 10^6.

The electron output from the final dynode is collected at an anode, where a small voltage pulse is produced and can be processed in exactly the same way as in a gas detector. The amplitudes of the output pulses will depend on the X-ray photon energies and on the voltages applied to the anode and the dynodes. If the applied voltages are held constant the amplitude will be proportional to the X-ray photon energy, so that a scintillation detector is also a *proportional* detector.

The density of the solid phosphor is roughly 10^3 times that of the gas in a gas-filled detector, so that the active phosphor volume required to produce a light pulse is low compared to that needed to produce an ion pair in the gas-filled detector. Further, the energy required to produce a phosphor ion pair is only about 3 eV, almost an order of magnitude less than the average ion pair production energy in a gas detector. In principle, therefore, it appears that a scintillation detector should produce a greater average number of ion pairs for a given X-ray photon energy, with reduced statistical fluctuations and correspondingly better resolution. However, the processes of light pulse generation in the phosphor and electron amplification in the photomultiplier are both relatively inefficient – approximately 20 per cent and 5 per cent efficiencies respectively – leading to a total efficiency of only about 1 per cent, so that the effective ion pair energy is thus about 300 eV.

For a scintillation counter, therefore,

$$\text{resolution } R = (236 \times 1.0 \times \sqrt{0.300}) / \sqrt{E} \quad (\text{for } F = 1.0)$$
$$= 129.3 / \sqrt{E}$$

which is appreciably poorer than that of the gas-filled detector. For Fe Kα (E = 6.40 keV), R_{scint} = 51.1 per cent (cf. 13.6 per cent for an argon-filled detector).

On the other hand, scintillation detectors are often able to handle higher count rates than gas-filled detectors before the onset of the undesirable effects of extendable dead time. Their use may be favoured in XRF situations in which count rates are high and detector resolution relatively unimportant.

The optimum wavelength sensitivity range for the scintillation detector is about 0.2–2.0 Å, which overlaps the short-wavelength end of the gas-filled detector range (c. 1.5 Å). It is therefore common practice in XRF spectrometers to arrange for interchange of the two types of detector; they are usually mounted in such a fashion that they can be operated independently or in tandem. In the latter case the flow detector is mounted in front of the scintillation detector, with an auxiliary collimator between them. The flow detector is fitted with windows at both front and rear (Fig. 4.16). Radiation which passes through the flow detector is picked up by the scintillation detector; however, the resolution of the tandem system is effectively that of the scintillation detector and is therefore poorer than that of the flow detector alone.

The dead time of a scintillation detector is usually slightly less than that of a gas-filled detector, but the difference tends to be masked by the dead time of the pulse-counting circuitry. Both systems typically exhibit net counting dead times of between 1 and 2 µs.

Scintillation detectors may also produce escape peak phenomena, caused by the incident X-ray photons ejecting K or L shell electrons from the phosphor halides. However, the iodine L absorption edge, for example, is at approximately 5 keV, outside the normal working range of

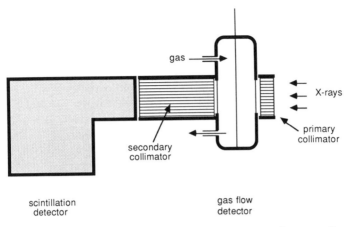

Figure 4.16 Combined, or 'tandem', gas-flow and scintillation detectors. Some X-ray photons, particularly those of high energy for which the filling-gas absorption coefficients are low, may pass through the gas-flow system without producing detectable ionizations. However, if they are allowed to leave through a second window, they can then be detected by the scintillation detector, which is more efficient at short wavelengths. The secondary collimator minimizes loss of resolution arising from divergence along the additional path length.

the detector; the K absorption edge is at 33 keV and K escape peaks are thus only produced by very high-energy X-ray photons.

4.5 Solid-state (semiconductor) detectors

Like gas-filled detectors, solid-state detectors depend on the photoelectric processes initiated when an X-ray photon enters the detection system. However, these processes take place in a solid instead of a gaseous medium, and the detection process depends on the production of 'electron–hole' pairs rather than 'electron–ion' pairs.

It has been noted earlier (Ch. 2) that the 'outer' orbital electrons of atoms in a crystalline material can only possess energies within discrete bands. Intrinsic semiconductors, such as pure crystalline silicon, have fully occupied valence bands separated by energy gaps from their higher-energy conduction bands. The energy of an X-ray photon absorbed by such materials is transferred by Auger and photoelectron interactions to the promotion of valence electrons across the band gap to the conduction band. The conduction band electrons produced in this way are relatively mobile and can act as charge carriers, thus making the semiconductor material more highly conductive. In an intrinsic semiconductor, promotion of each valence electron also generates an electron 'hole' in the valence band; though less mobile, the 'holes' can also function as positive charge carriers.

If a suitable DC bias voltage is applied across a volume of such a semiconductor, the charge carriers can be 'swept out' and collected at the DC bias electrodes, in much the same way as the ion-pair charges are collected at the bias electrodes (anode and cathode) of a gas-filled detector. Since the number of electron–hole pairs produced will be proportional to the total energy of the incident X-ray photon, the magnitude of the charges collected at the bias electrodes will also be proportional to the X-ray photon energy, and a semiconductor detector of this type is also inherently a true proportional detector.

The most widely used solid-state detector material is crystalline silicon. When pure, silicon is an intrinsic semiconductor, with a band gap of 1.1 eV. Under applied bias voltages of 500–1000 V its intrinsic conductivity (due to thermally excited electron–hole pairs) is relatively low, and it can be further reduced by maintaining the silicon at low temperatures (of the order of 100 K) with liquid nitrogen. For detector use, low intrinsic conductivity is essential; otherwise the applied DC bias produces 'leakage' currents which flow constantly, independently of the generation of charge carriers by photon absorption, and hence produce excessively high background signals.

Unfortunately, it is not technically feasible to produce silicon free of trace impurities, which significantly modify its electrical properties. The most common of these impurities is the Group III element boron, whose presence provides acceptor bands just above the silicon valence band and thus modifies the silicon to a p-type extrinsic semiconductor with increased conductivity. In practice, therefore, this undesirable effect has to be compensated by the deliberate, accurately controlled addition of another impurity chosen to provide donor bands to neutralize the boron acceptors. When properly compensated in this way silicon functions as an intrinsic semiconductor with high inherent resistivity.

The most widely used 'donor' element is lithium, which has a small atomic radius and hence can be easily diffused through the silicon, and which also has a low ionization potential (0.033 eV), making it an efficient electron donor.

A suitable solid-state detector medium can thus be prepared by adding the appropriate amount of lithium to silicon and diffusing it to a homogeneous distribution. Silicon with excess lithium will have n-type extrinsic properties, while lithium-deficient volumes will be p-type. The distribution of lithium must therefore be very accurately controlled.

In practice, lithium is applied to the surface of a wafer of silicon, and diffused through it by heating briefly to about 350 °C. This produces a gradient of diffused lithium in the silicon host (Fig. 4.17). The latter will be n-type in the high-Li region near the wafer surface, and p-type in the

SOLID-STATE (SEMICONDUCTOR) DETECTORS

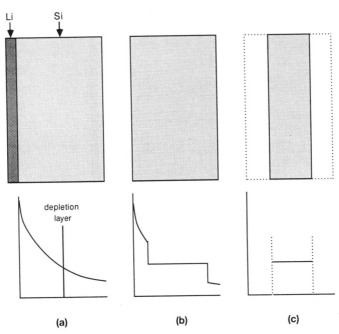

Figure 4.17 Production of a Si(Li) detector wafer by (a) controlled thermal diffusion of lithium into a silicon wafer, followed by (b) *drifting* of the lithium under the influence of an applied reverse bias voltage, and finally by (c) removal of the residual uncompensated boundary regions.

more remote low-Li region. There will be a narrow compensated zone, called a *depletion layer*, at the boundary between the n-type and p-type regions. This layer will lack free charge carriers and will thus have a high resistivity relative to the uncompensated regions on either side of it.

If a reverse bias voltage (n-type region positive with respect to the p-type region) is now applied – again under moderately increased temperatures to promote diffusion – a strong field in the resistive depletion layer will result in a migration of lithium ions from the n-region to the p-region. If properly controlled, the result of this drifting process will be an expansion of the compensated depletion layer to produce a wider region of high resistivity suitable for use as a detector (Fig. 4.17b). Excess uncompensated silicon (i.e. in the remaining n-type and p-type regions) is then removed, and the surfaces are coated with thin layers of gold to serve as electrodes. The result is a *lithium-drifted silicon detector*, or *Si(Li)* detector (Fig. 4.18).

Detectors of this type are operated at low temperatures, for two main reasons. First, as already noted, background noise due to thermally excited electron–hole pairs is minimized, and secondly, reverse diffusion

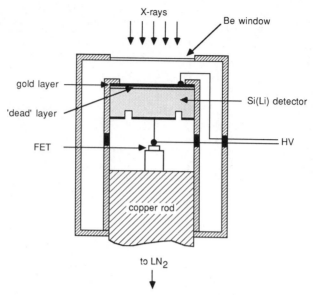

Figure 4.18 Schematic drawing of a typical *solid-state* or *semiconductor* detector, such as an Si(Li) detector. A wafer of lithium-drifted silicon is coated with gold on two opposite surfaces, to provide electrical connections for the application of a DC bias voltage, and mounted on the end of a copper rod in an evacuated, light- and moisture-shielded assembly. The other end of the copper rod is immersed in liquid nitrogen to keep the detector at low temperatures which minimize thermal interference and prevent diffusion of the lithium under the influence of the bias voltage. An X-ray photon entering the detector through a thin Be window produces a series of electron-hole pairs in the Si(Li) wafer, and these charge carriers are 'swept out' by the bias voltage to yield a small voltage pulse which is amplified by an FET preamplifier and passed to external amplification and pulse-processing circuits for energy calibration and counting.

of the lithium ions under the influence of the detector bias field is inhibited. At higher temperatures – even at room temperatures – an applied detector bias can otherwise produce a reverse drifting process and regenerate the p- and n-type regions (i.e. the detector can revert to a *p–n junction device*).

Effective cooling is achieved by mounting the detector on the end of a copper rod whose other end is kept immersed in liquid nitrogen ('LN$_2$' or 'LN2'). The cryostat mounting is carefully designed to maintain the detector itself at a constant temperature of about 100 K. Some mountings include self-regulated heat sources that act as thermostats to hold the detector at a constant temperature slightly above that of the LN2.

The basic principles of the Si(Li) detector are generally similar to those of the gas-filled detector. The electron–hole charge carriers produced by photoelectric absorption of the energy of an incoming X-ray photon are swept out by the bias field and collected at the bias electrodes to produce

SOLID-STATE (SEMICONDUCTOR) DETECTORS

a pulse whose amplitude is determined by the number of charge carriers collected and hence is proportional to the energy of the incident photon (though again subject to statistical variation, since some of the photon energy is lost in interactions such as thermal lattice vibration which do not produce charge carriers – just as some energy in the gas-filled detector is lost in non-ionizing collisions).

The pulse collected at the electrodes is then preamplified, amplified and passed to an electronic device called a *multi-channel analyser* (MCA), which fulfils the dispersive functions of a spectrometer by sorting the pulses electronically according to their amplitudes into a large number of different 'channels' (each corresponding to a different narrow range of X-ray photon energies) and counting the number of pulses collected in each channel. Further details of the operation of the energy-dispersive spectrometer (EDS) are given in Chapter 5.

Energy-dispersive spectrometers can function without need of a crystal to separate the different X-ray photon energies because their solid-state detectors have energy resolution characteristics which are superior to those of gas-filled or scintillation detectors. If an X-ray photon of energy E enters a solid-state detector, it will produce an average of N electron–hole pairs, such that

$$N = E / e$$

where e is the *average* energy required to produce one electron–hole pair. For a Si(Li) detector operated under typical conditions, e is approximately 3.8 eV; hence an Fe Kα photon ($E = 6.398$ keV) will produce an average of 6398 / 3.8, or 1683.7 electron–hole pairs. This is much higher than the equivalent charge-carrier yield produced by an Fe Kα photon in a gas-filled detector (Table 4.3). A series of Fe Kα photons will still produce a statistical spread about this average value, but, because of the larger number of charge carriers generated and collected, the *proportional* spread about the mean will be less and the overall resolution will be considerably improved.

Table 4.3 Comparative resolutions of different types of detector.

Detector	Average ionization energy	Average Fe Kα charge carrier yield	Theoretical resolution (%)*
Si(Li)	3.8	1683.7	5.8
Gas proportional (Ar)	26.4	242.3	15.2
Scintillation	300.0	21.3	51.1

* W/V for Fe Kα.

By contrast with the gas-filled detector, the Si(Li) detector has no internal gain, since it has no equivalent of the strong electron-accelerating field in the vicinity of the gas-filled detector's anode wire and thus it does not produce any 'avalanching' process. Hence the swept electron pulse collected at the anode will consist only of the (average) 1683.7 electrons, or about 10^{-16} coulombs. This small signal must be electronically amplified before sorting and counting by a system which must obviously be characterized by high-gain, linear response (to maintain proportionality) and very low noise levels.

Again a preamplifier is usually attached as closely as possible to the detector to minimize capacitance effects, and a *field-effect transistor* (FET) is almost invariably used as the preamplifier because of its low-noise characteristics. The latter are further enhanced by LN2 cooling, so that the FET is also mounted within the detector cryostat assembly. The output signals from the FET are amplified by an external device similar to the main amplifier used by gas-filled detector systems, before processing in the multichannel analyser.

Electrons and holes can move freely only through a perfect crystal. In real crystals, impurities and crystal defects may produce intermediate electron energy levels between the silicon valence and conduction bands, and these can 'trap' the moving charge carriers, at least temporarily. Trapped carriers cannot be collected at the electrodes; the output pulse is then reduced in magnitude and the energy resolution characteristics of the detector are degraded. In severe cases an asymmetrical pulse amplitude distribution results, with a significant tail on the low-energy side, leading to extended overlap of the distributions produced by photons of different energies and thus to reduced ability of the system to *resolve* or distinguish between them.

Trapping losses can be reduced by using high bias voltages to ensure rapid collection of the charge carriers, or by operating the detector at higher temperatures to enhance detrapping by thermal excitation. However, high voltages and high temperatures both result in increased noise levels and thus degrade detector performance. Again some degree of compromise is always necessary. The actual extent of trapping problems can vary considerably from one detector to another, since it is largely a consequence of the quality of detector materials and manufacture.

As shown in Figure 4.18, there is also a thin layer (up to about 0.1 mm thick) at the surface of the detector which does not function effectively as a detection medium. This 'dead layer' is transitional with the active region beneath it, and the transition zone typically contains a relatively high concentration of traps. Carrier loss effects due to trapping therefore tend to be most severe for low-energy photons which are more heavily absorbed in the surface layers.

SOLID-STATE (SEMICONDUCTOR) DETECTORS

Preamplifier noise cannot be entirely eliminated, and it always contributes a significant component to the statistical spread about the mean pulse amplitude distribution. A further term must be added in quadrature to the resolution equation to include the effect of random noise, so that

$$\Delta E \,(= \text{FWHM}) = 2.36 \sqrt{(e\,F\,E)}$$

becomes

$$\Delta E^2 = (\Delta E_{\text{noise}})^2 + (\Delta E_{\text{ionization}})^2 \tag{4.9}$$

and hence

$$\Delta E = \sqrt{\{(\Delta E_{\text{noise}})^2 + [5.57\,(e\,F\,E)]\}} \tag{4.10}$$

(Remember that use of the term 'ΔE' in the above expressions as a measure of statistical spread about the mean is quite different from use of the same expression to denote the window width in a pulse height discrimination system.)

ΔE_{noise} is typically about 100 eV and Fano factors are currently about 0.12 for good quality Si(Li) detectors. Thus the FWHM for Fe Kα, for such a detector, is given by

$$\Delta E = \sqrt{\{100^2 + 5.57 \times (3.8 \times 0.12 \times 6398)\}}$$
$$= 162 \text{ eV}$$

and

$$\text{FWTM} = 1.823 \times \text{FWHM}$$
$$= 295 \text{ eV}$$

This resolution is sufficient to separate most X-ray energies of analytical interest without the need for a dispersing crystal (except possibly in the relatively 'crowded' L and M spectral regions of samples containing several heavy elements). It should be noted that for sales purposes absolute energy resolution is sometimes quoted for lower photon energies, often 1.0 keV which (because of window absorption etc.) marks the practical low-energy limit for the current generation of Si(Li) detectors. For 1.0 keV photons,

$$\text{FWHM} = \sqrt{\{100^2 + 5.57 \times (3.8 \times 0.12 \times 1000)\}}$$
$$= 112 \text{ eV}$$

Most reputable manufacturers, however, quote their detector resolutions for a more realistic energy, usually that of Mn Kα (5.894 keV).

When these expressions are evaluated, it becomes clear that the noise term is actually the major component, and improved detectors with reduced noise characteristics can be expected to yield significantly better resolutions in the future (though it must also be noted that the rate of technical improvement in this respect has slowed considerably in recent years). It is also possible that new and superior semiconductor materials will ultimately replace lithium-drifted silicon, which at present is almost universally used in solid-state X-ray detection devices. Germanium detectors, for example, have already been used successfully for some applications.

Because of their superior energy resolution, solid-state detectors offer practical advantages over gas-filled or scintillation types for many analytical applications. Because they can be used without a dispersing crystal, they require none of the mechanical (goniometer) components needed to rotate the crystal and detector to appropriate Bragg angles; nor do they require the additional mechanical components necessary to exchange the crystals needed for different energy/wavelength ranges. In the absence of moving mechanical parts, there are no alignment problems resulting from backlash or wear.

A solid-state X-ray detector can be placed much closer to the emitting source (the sample) than the crystal in a conventional wavelength-dispersive spectrometer, and it does not need intensity-reducing collimator or slit systems to ensure Bragg relationships. Hence it can sample a much wider cone of emitted X-rays, which in itself is equivalent to a substantial increase in effective X-ray intensity under constant excitation conditions. Furthermore, since it does not need a crystal to be set at different angles for different wavelengths, it can sample the complete spectrum of emitted X-ray energies simultaneously, avoiding the time-consuming necessity for sequential scans from one energy (wavelength) to the next.

These factors combine to give the solid-state, or energy-dispersive, system its most outstanding attribute – that of speed of analysis. Multi-element analyses can often be completed, with adequate levels of precision, in total measurement times of the order of seconds rather than the minutes or even hours required for wavelength-dispersive systems. Alternatively, the efficiency of the energy-dispersive system permits substantial reduction in the excitation energy flux (e.g. much lower electron beam currents in the electron microprobe). This in turn minimizes analytical errors due to thermally induced diffusion or even, in some cases, partial volatilization of some heat-sensitive samples.

On the other hand, there are some potential problems with solid-state detectors that must be carefully controlled to ensure satisfactory operation. For example, the surface of the detector is very sensitive to contamination, so that the housing in which it is mounted must be kept permanently evacuated to low pressures (10^{-6} torr or better), and the surface must be protected by a window which in turn must be able to withstand pressure differentials of up to 1 atm. but yet have the best possible transmission of low-energy X-rays. The windows are usually made of beryllium about 8 μm thick, which gives about 50 per cent transmission of 1 keV X-rays. As with gas-filled detectors, supported thinner windows (e.g. of metallized polypropylene) may be used for special applications. The strength of the thin window material typically limits the maximum size of the window to surface diameters of 8–10 mm.

The silicon band gap energy of 1.1 eV corresponds to the infra-red region of the electromagnetic spectrum, so that electron–hole pairs can also be produced by incident infra-red or visible light. The detector must therefore be shielded by an opaque window. Since it is normally placed much closer to the analytical sample than gas-filled or scintillation detectors, it also requires additional shielding from stray electrons and/or unwanted X-rays scattered from the sample or from nearby instrumental components. This usually requires at least a simple form of collimator in front of the detector window.

Solid-state detectors must be maintained constantly at low temperatures and are usually cooled with LN2, provided from a dewar reservoir which requires regular refilling (typically every two or three days for five litre dewars). No great harm is done if the detector is occasionally brought to room temperature for short periods (e.g. for maintenance purposes), *providing the bias voltage is switched off.* Accidental warming with an applied bias voltage may result in lithium diffusion and serious degradation of detector performance.

Vibration of the detector assembly is a potential source of undesirable levels of electronic noise. This can only be controlled by careful attention to the design, construction and maintenance of the detector and preamplifier mountings.

5 X-ray spectrometers

The spectrometers used in both XRF and EPA are of two main types, those that use crystals to disperse the component wavelengths of the X-ray spectrum emitted from the sample, and those that achieve the same purpose by electronic 'sorting' of pulses of different amplitude produced by X-ray photons of different energies. These two types are known as *wavelength-dispersive spectrometers* (WDS) and *energy-dispersive spectrometers* (EDS), respectively. Of course these names are simply conveniences, since energy and wavelength are inversely equivalent to each other.

5.1 Wavelength-dispersive spectrometers

The task of the X-ray spectrometer is to *disperse* the incident radiation, i.e. to separate its constituent wavelengths, in order to allow the intensity of any selected wavelength (or, more correctly, narrow wavelength band) to be measured. Dispersion may be achieved by diffraction from a crystal of known d-spacing, appropriate to each particular application, and intensity measurement is the function of the detector(s) and its (their) associated amplification and counting circuitry.

Wavelength-dispersive spectrometers may be either

(a) *fixed*, or set up permanently to detect and measure only one selected analytical wavelength, or
(b) *scanning*, or provided with a method for systematically varying the Bragg angle θ over a reasonably wide range so that the spectrometer may be used to measure any of a correspondingly wide range of wavelengths (the latter facility being even more enhanced if the spectrometer is constructed so that several different dispersing crystals can easily be interchanged).

Fixed spectrometers, which have no moving parts other than those used for initial adjustment, are best suited to routine analytical situations involving the repetitive measurement of a relatively small number of wavelengths (e.g. metallurgical process control). Obviously at least one spectrometer is required for each element to be measured; commercial X-ray spectrographs having as many as 30 fixed spectrometers have been built. Fixed spectrometers can be optimally 'tuned' to give the best possible performance for each of the elements sought, and they have the advantage of relatively low maintenance costs, since they do not need frequent realignment. However, they provide little or no flexibility in their analytical applications.

Scanning, or *sequential*, spectrometers are better suited to environments in which it is necessary to cope with a range of different analytical problems, e.g. research laboratories. Given a sufficiently wide range of dispersing crystals and detectors, they can be used for any element or combination of elements within the capabilities of the technique, although adaptation to a range of elements often involves some compromise and it may be more difficult to tune them for optimum performance for any one particular element. With the advantage of relatively high X-ray intensities, and therefore shorter counting times, XRF spectrographs are normally provided with only one scanning spectrometer, whereas electron microprobes almost invariably have more than one (usually three or four) so that several different wavelengths can be measured simultaneously. Either XRF or microprobe systems may be set up with a range of fixed spectrometers adjusted for the most commonly determined elements (e.g. as many as eight or ten for routine silicate analysis), together with one or more scanning spectrometers to provide the flexibility necessary for other analytical work. However, good computer-control systems have helped to overcome many of the problems involved in repeated resetting of scanning spectrometers, and the relative expense of a large number of fixed spectrometers (each with its own crystal, detector and goniometer) now tends to preclude their use in most laboratories.

Many different mechanical and geometric arrangements have been employed in both fixed and scanning spectrometers, varying principally in their choice between flat and curved dispersion crystals and their use of either reflection or transmission diffraction modes. Transmission modes are not widely used, and will not be further discussed in this introductory treatment.

5.1.1 Flat-crystal spectrometers

The simplest type of spectrometer is that common to most XRF spectrometers, which use a flat crystal mounted on a central arm of a rotating goniometer so that the Bragg angle θ can be varied by simple rotation of the crystal mount (Fig. 5.1).

Divergence in the beam, leading to significant variations in θ over the relatively large surface of the crystal, is limited by mechanical collimators. Several collimator designs have been employed, but the most common type consists of a series of parallel metal plates. X-rays parallel to the plates can pass through the collimator without obstruction (except for those striking the ends of the plates), but those diverging significantly

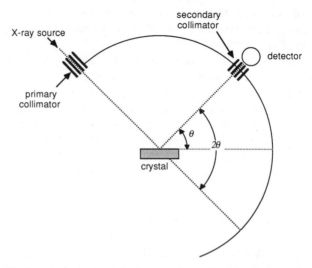

Figure 5.1 Fundamental geometry of a flat-crystal, non-focusing, wavelength-dispersive X-ray spectrometer. Beam divergence is limited by primary and secondary collimators to control the extent of divergence in the incidence and reflection Bragg angles (θ) over the relatively large surface of the dispersing crystal. The secondary collimator also shields the detector from stray X-rays scattered from various components of the spectrometer (including the primary collimator).

from the collimator axis are absorbed by the plates (Fig. 5.2). The consequent reduced divergence improves the resolution of the spectrometer, i.e. its ability to separate closely spaced wavelengths, but at the expense of some loss of intensity in both incident and diffracted beams. A collimator with closely spaced plates gives optimum resolution, at the expense of relatively severe losses in intensity. It is therefore common practice to equip flat-crystal spectrometers with interchangeable coarse- and fine-spaced primary collimators to allow a satisfactory compromise to be achieved in most analytical situations.

5.1.2 *Focusing (curved-crystal) spectrometers*

In electron microprobe analysis the X-ray intensities are usually so low that collimation losses cannot be tolerated, even if widely spaced collimator plates are used. Instead, microprobe spectrometers invariably use some form of *focusing* or at least *semi-focusing* X-ray optics. The situation is analogous to the use, in light optics, of focusing lenses or mirrors to collect low-intensity light rays subtending a finite angular aperture. In the case of X-ray optics, the flat dispersing crystal is replaced by a curved crystal which 'collects' the incident X-rays and 'focuses' the

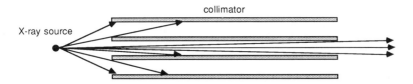

Figure 5.2 Restriction of X-ray beam divergence by a collimator consisting of a series of closely spaced, parallel metal plates. The maximum divergence of the emergent beam depends on the thickness and length of the plates and on their spacing. Long, thick and closely spaced plates result in minimum divergence, but also in maximum loss of beam intensity. X-ray spectrometers are therefore usually fitted with two or more interchangeable collimators with different plate spacings, to allow some flexibility in compromising between intensity and spectrometer resolution for different analytical applications.

diffracted X-rays on to an entrance slit in front of the detector. The slit takes the place of the collimator system, in that variation in the width of the slit controls both resolution and detected intensity – a narrow slit gives optimum resolution, but obviously reduced intensity since X-ray optics are also subject to aberrations and the focusing effect of the curved crystal is never perfect. The interchangeable collimators of flat-crystal optics are replaced by interchangeable slits (or slits whose widths can be varied mechanically) to allow compromise between the conflicting demands of resolution and intensity.

The essential geometrical features of focusing X-ray optics (which are similar to those of analogous light optics) are shown in Figure 5.3. The source of the X-rays, the surface of the dispersing crystal and the entrance slit to the detector must all lie on the circumference of a circle, called the *Rowland circle*, which can be of any mechanically suitable radius (large Rowland circles give better resolution, but the X-ray path lengths are longer and intensities are thus reduced by absorption, even in low-pressure systems; large Rowland circles also mean large spectrometers, which are more difficult to fit mechanically to the microprobe, and they require larger crystals to subtend the same angle of X-ray divergence).

It can be shown geometrically that the Bragg Equation will hold over the complete surface of the crystal and for the full attainable range of Bragg angles, provided that

(a) the crystal is *bent* to a radius equal to the *diameter* of the Rowland circle; and
(b) its front surface is *ground* to a radius equal to the *radius* of the Rowland circle.

This configuration is known as *Johannson optics*. Techniques have been devised to bend most of the commonly used dispersing crystals (or to

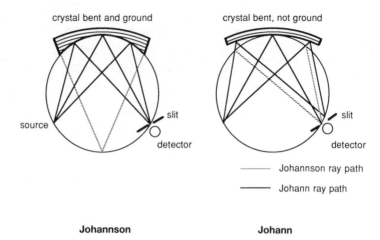

Figure 5.3 The essential geometry of fully-focusing ('Johannson') and semi-focusing ('Johann') X-ray spectrometer optics. In both cases the dispersing crystal is bent to a radius equal to the *diameter* of the focusing circle (Rowland circle). In fully-focusing optics, the front surface of the crystal is also ground to a radius equal to the *radius* of the Rowland circle; all points on the surface of the crystal then lie on the Rowland circle, and if the X-ray source lies on the same circle, diffracted X-rays are brought to a 'point' focus (a line in three dimensions) which also lies on the Rowland circle. Semi-focusing crystals are bent but not ground, resulting in some broadening and asymmetry of the focusing 'point'. These diagrams are not drawn to scale, so that they exaggerate the loss of resolution in a semi-focusing spectrometer; the effect is usually much less pronounced than they suggest, and semi-focusing spectrometers have many useful practical applications.

grow them on suitably curved backing plates). However, not all crystals are amenable to grinding of their front surfaces – which also considerably increases the cost of the crystal. Hence some spectrometers employ a simplified configuration called *Johann optics*, in which the crystals are bent, as in (a) above, but are not ground (Fig. 5.3). This is still superior to flat-crystal optics (in terms of measured intensities), and is quite adequate for many purposes. However, line profiles are broader and asymmetric, and Johann optics cannot yield optimum resolution or intensities.

Both Johannson and Johann optics take advantage of the fact that the X-ray source in the electron microprobe is effectively a point source, which is not the case in XRF. Ideally the crystals used in Johannson optics should be double-ground (toroidal), but this is mechanically very difficult to achieve and gives only a marginal improvement in performance. With single-grinding the locus of X-rays diffracted from a point source is a line rather than a point – it is for this reason that a linear slit is used.

Mechanically, the simplest form of scanning focusing spectrometer would resemble a flat-crystal spectrometer except that a curved crystal is used (Fig. 5.4). The crystal could be mounted on a 'θ arm' and the detector on a '2θ arm', as in the flat-crystal type. This arrangement is simple and would be mechanically reliable, but unfortunately it can only fulfil the basic requirements of focusing geometry at one Bragg angle (that for which $\theta = \sin^{-1}(d/R)$, where d is the interplanar spacing and R the radius of curvature of the crystal). There are no other Bragg angles at which source, crystal surface and detector (slit) can all lie on the circumference of a Rowland circle. However, since dispersing crystals are never perfect, such 'semi-focusing' spectrometers can actually work quite well over an appreciable range of Bragg angles on either side of the design angle; e.g. a Johann LiF_{200} crystal bent for $\theta = 30°$ will usually work reasonably well over a θ range of about 20–60°. Nevertheless, resolution and peak-to-background ratios are relatively poor, and this type of spectrometer is not normally used in microprobe systems intended for accurate analysis. To cover the normal working range of, say, an LiF_{200} crystal would require at least three crystals bent to different curvatures, thus detracting from the economy of the simple mechanical configuration.

The need for a series of crystals bent to different curvatures can be overcome, at least in theory, by using a flexible crystal in a Johann configuration, with the curvature being continuously varied by a cam as

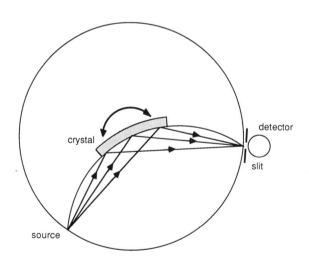

Figure 5.4 A curved-crystal (focusing) analogue of simple, flat-crystal spectrometer geometry (cf. Fig. 5.1). The detector and the crystal are mounted on a common centre of rotation, which is a mechanically simple arrangement but allows Rowland focusing geometry to be satisfied precisely at only one Bragg angle.

the crystal and detector are rotated to different wavelength settings (Fig. 5.5). The curvature radius is varied as a cosec θ, where a is the source–crystal distance (this is the basic requirement of Rowland circle geometry).

This configuration was used commercially in the 'X-ray macroprobe' marketed some years ago by the Philips group. This instrument, which was a development of the Hamos design principles (Ch. 1), utilized a flexible mica crystal which was capable of good resolution and reasonably high P : B (peak-to-background) ratios. However, a mica crystal has a relatively high d-spacing, so that high orders of diffraction must be used for the shorter wavelengths (for which such instruments are most commonly used), and these are hence diffracted with relatively weak intensities. No suitably flexible, low-d crystal has yet been developed, and the X-ray macroprobe has now been effectively replaced by the electron microprobe.

If crystal and detector are both rotated about a central bearing, so that each moves around the circumference of a fixed Rowland circle, fully focusing geometry can be maintained at all Bragg angles (Fig. 5.6). However, in this arrangement the 'take-off' angle of the X-rays (the angle between the sample surface and the direction of the detected X-rays) varies as the Bragg angle varies. The take-off angle is of critical

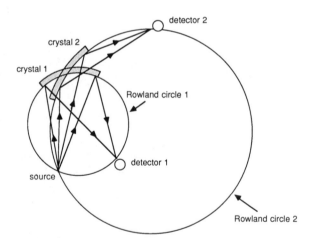

Figure 5.5 Flexible-crystal focusing geometry. The crystal is fixed at one end, and its curvature is varied by a cam rotating in synchronization with rotation of the crystal assembly to vary the Bragg angle. The source–crystal distance is constant at all Bragg angles, but the radius and location of the Rowland circle (and hence the spectrometer resolution) are variable. The detector path becomes more complex, and the detector must also be rotated as the Bragg angle is changed, to keep its window pointed towards the crystal. This design is mechanically effective, but its application is restricted by the limited availability of flexible crystals (particularly of low d-spacing).

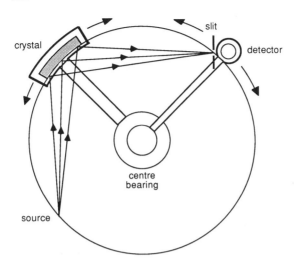

Figure 5.6 A centre-bearing, double-arm, fully-focusing spectrometer. The source–crystal distance varies with θ, and the detector must again be rotated as it is moved, to keep its window pointed towards the crystal. The 'take-off' angle (see text) also varies with θ, adding a complication to the calculation of the effects of the sample itself on X-rays generated beneath its surface (see Ch. 9).

importance because it determines the magnitude of matrix effects arising from absorption, fluorescence and other effects on X-rays generated beneath the sample surface (Chs 8 & 9). These errors are difficult to correct accurately, and the correction procedure becomes even more complex if the effects of variable take-off angles must also be considered. While this is a less serious objection than it used to be, because it can be handled fairly easily by on-line computer correction procedures, absorption effects in particular increase rapidly as take-off angle is reduced; the resulting intensity losses restrict the effective take-off angular range and hence the useful range of a spectrometer designed with this geometry.

This system is relatively simple mechanically, and it has been used in some electron microprobes. However, it has now been almost entirely replaced, in commercial systems, by the use of so-called *linear fully focusing spectrometers*. These are mechanically complex systems which maintain fixed Rowland circle geometry and a constant take-off angle by moving the crystal towards or away from the X-ray source as it is rotated to vary θ (Fig. 5.7). The Rowland circle is fixed in the sense that its radius is held constant; however, its actual position in space changes continuously as the crystal is moved and rotated. Hence the detector must move along a geometrically complex path (a segment of the 'four-leafed rose' curve), and it must be rotated during movement to keep the slit pointing towards the dispersing crystal. The mechanical difficulties are apparent,

X-RAY SPECTROMETERS

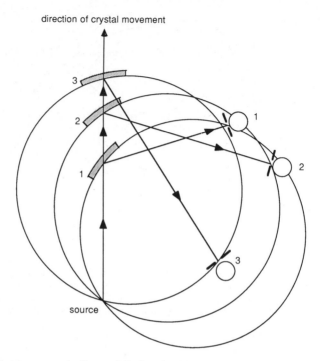

Figure 5.7 The geometry of a linear, fully-focusing X-ray spectrometer. This arrangement is mechanically complex, since a change in θ requires rotation of the crystal, adjustment of the source–crystal distance, movement of the detector along a complex path, and rotation of the detector. However, a constant Rowland circle radius is maintained, together with a constant 'take-off' angle, and the mechanical problems are not insuperable. This type of spectrometer is used in most commercial wavelength-dispersive electron microprobes, in which focusing geometry is mandatory because of relatively low X-ray intensities.

but they have been successfully overcome in several ingenious designs and the linear spectrometer is by far the most common type of WDS spectrometer currently used for electron microprobe analysis. However, it requires careful alignment and high standards of maintenance for long-term reliability.

Further complexity is introduced by the need to interchange crystals in the spectrometer to encompass a usefully wide range of wavelengths. The precision requirements of focusing geometry place much more severe constraints on the interchange mechanism than are normally encountered in flat-crystal spectrometers. Nevertheless, these difficulties have also been successfully overcome in several commercial instruments, all of which now make provision for crystal interchange (even, in some cases, under computer control during the course of an analysis).

The choice of Rowland circle radius is a fundamental design

consideration governed by such factors as the limiting radius to which the crystal can be curved, or how close it can be brought to the sample. A large radius requires a large crystal to subtend a given solid angle of focused X-rays, and large crystals are difficult and expensive to make. As noted above, they also require large spectrometers which are mechanically more difficult to place near the sample (in an area which, in the electron microprobe, is already congested with electron lenses, light microscope components, sample stage mechanisms etc.). On the other hand, crystals are difficult to bend and grind to small radii of curvature, and with small Rowland circles they must be brought very close to the sample at low Bragg angles, which is also mechanically difficult to arrange. In most cases compromise Rowland circle radii of between 10 and 25 cm are used.

A further desirable feature is the availability of continuously variable or interchangeable fixed detector slits. Wide slits (or no slits at all) give maximum intensities at the expense of relatively poor resolution. Again a compromise must usually be effected, particularly in automated systems, but it should not be forgotten that the ability to control slit width accurately can sometimes make a critical difference to the quality of an analysis.

Even with the advantage of focusing spectrometers, microprobe X-ray intensities are usually relatively low, and analysis therefore tends to require relatively long counting times to attain satisfactory levels of precision. Most microprobes therefore provide for the simultaneous operation of several spectrometers (usually between two and four). Depending on the analytical problems involved, these may be fitted with different crystals and detectors, or two or more of them may have identical configurations. Thus a three-spectrometer instrument engaged mainly in silicate analysis might be equipped with LiF, PET and TlAP crystals; the LiF and PET 'channels' might be equipped with sealed gas proportional detectors, and the TlAP spectrometer with a thin-window gas flow detector (since it will be used only for long-wavelength, low-energy emissions); the total configuration is capable of dispersing and measuring characteristic radiation from all elements heavier than fluorine. However, a similar instrument intended principally for the analysis of stainless steels (Fe, Mn, Cr, Co, Ni etc.) would be more usefully equipped with two or even three LiF spectrometers, simply to reduce overall counting times by counting two or three elements in the LiF range simultaneously. Research instruments used for a wide variety of sample compositions obviously have the most critical needs for flexible crystal interchangeability, and the operator planning an analysis procedure must often give careful thought to optimizing the spectrometer configuration.

In both XRF and electron microprobe analysis, sequential spectrometers have to be moved successively to various Bragg angles to measure the intensities of different X-ray lines. In the earlier stages of X-ray spectrometry, setting the spectrometers for each line was performed manually, and was both tedious and a major source of human error. It was not long before servo-based systems were developed to move the spectrometers rapidly and reproducibly through a series of preset angles (including backgrounds). More recently, the servo systems have been replaced by digitally controlled stepping motors, externally controlled by a computer or microprocessor and hence capable of fairly sophisticated programmed operation, e.g. peak search and location routines that have done much to reduce reproducibility problems such as those that arise from mechanical backlash in the spectrometers.

A consequence of the geometry of the linear focusing spectrometer is that the source–crystal distance D is equal to $2r \sin \theta$, where r is the radius of the Rowland circle. By substitution in the Bragg Equation (assuming $n = 1$),

$$D = r\lambda / d \tag{5.1}$$

In other words, λ is proportional to D, so that linear spectrometers can be calibrated to read directly in wavelength for any particular crystal d value, unlike flat-crystal XRF spectrometers, which can only be linearly calibrated in θ or 2θ. Hence X-ray analysts who work mostly with microprobes tend to associate spectral lines with their wavelengths (or energies, if they work primarily with energy-dispersive systems). XRF analysts, by contrast, often speak of spectral emissions in terms of their θ or 2θ equivalents. Linear spectrometers are normally calibrated by the manufacturer for LiF_{200}; if a crystal of different d-spacing is used, the apparent LiF wavelengths must be corrected arithmetically or by the use of conversion tables.

5.2 Energy-dispersive spectrometers

Because of their convenience and relative simplicity of operation and maintenance, energy-dispersive spectrometers have become widely used since the late 1960s, when advances in electronic and digital-processing technology first made them practicable for routine analysis of X-ray spectra. They are particularly well suited to applications in microprobe analysis, not the least because they provide a relatively economical way to add analytical capabilities to scanning electron microscopes. They are also useful adjuncts to wavelength-dispersive spectrometers, and they have found interesting applications in a number of compact, portable analytical systems which use probes containing radio-isotope sources to excite fluorescent X-ray spectra directly from 'field' samples (rock outcrops, underground exposures, drill core samples etc.). Energy-dispersive spectrometers are used to analyse the fluorescent spectra and provide rapid concentration estimates for one or more precalibrated elements.

An energy-dispersive spectrometer is basically an electronic system that processes the energy-proportional pulses produced by a solid-state detector, sorts them according to their amplitudes (and hence according to the energies of the X-ray photons that produced them), and counts the pulses produced in a series of amplitude ranges appropriate to the elements sought. Because of the superior energy resolution of the solid-state detector, EDS systems need no dispersing crystal and hence none of the mechanical components required to maintain Bragg relationships in wavelength-dispersive spectrometers. Since all elements of the X-ray spectrum are collected and processed simultaneously instead of sequentially, EDS systems can usually complete analyses much faster than their WDS equivalents.

Alternatively, since they work best with much lower total spectral intensities, they allow the use of lower excitation energy fluxes (such as the low electron beam current of a scanning electron microscope, or the low beam currents necessary to minimize thermal damage of some samples in the electron microprobe).

There is considerable variation in detail from one commercial EDS system to another, and EDS systems are still undergoing relatively rapid technological evolution. However, most of the units currently available commercially contain at least the following essential components (Fig. 5.8):

(a) A solid state *detector*, usually of the Si(Li) type, fitted with an FET-based preamplifier. The detector collects the charges produced when an X-ray photon interacts with the drifted silicon to produce electron–hole pairs, and the preamplifier integrates the collected charges and converts them to a low-voltage pulse whose amplitude is proportional to the number of charges collected and hence to the X-ray photon energy.

(b) A high-gain linear *amplifier*, which has two main functions:

 (i) it increases the preamplifier output pulse amplitudes to voltage levels suitable for sorting and counting, and

 (ii) it shapes the pulses to a form appropriate for processing, while accurately maintaining the proportional relationship between pulse amplitude and X-ray photon energy on which the system depends.

(c) A *single-channel analyser* (SCA), which scans the amplifier output and rejects (filters) low-voltage electronic noise and other interfering signals.

(d) An *analog-to-digital converter* (ADC), which processes the filtered

X-RAY SPECTROMETERS

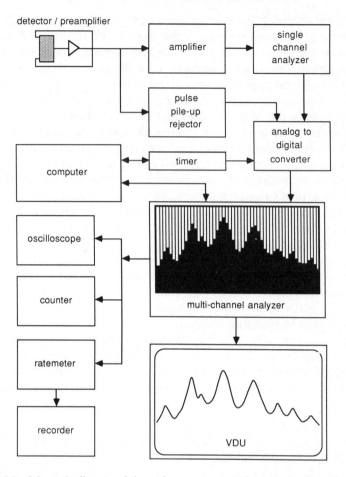

Figure 5.8 Schematic diagram of the major components of a typical energy-dispersive X-ray spectrometer. The function of each component is described in the text.

pulses from the SCA, converting each to a numerical (digital) value proportional to its original pulse amplitude and hence to the energy of the X-ray photon that produced it.

(e) A *multi-channel analyzer* (MCA), which sorts the digitized pulses into a series of computer memory registers, or 'channels'. Each channel is, in effect, a 'bin' used to store and count the pulses falling within narrow, accurately calibrated digitized energy ranges in the X-ray spectrum. A typical MCA has 1024 channels calibrated for consecutive 10 (or 20) eV increments, and thus spanning the range of X-ray energies from, say, 0 to 10.24 (or 20.48) keV. 10.24 keV corresponds to a wavelength of 1.2 Å, so that the range from 0 to 10.24 keV includes almost all of the characteristic X-ray lines routinely used in XRF or electron microprobe analysis.

(f) A *timer*, which is used to control the time intervals over which data are collected, so that the rates of pulse collection can be reduced to X-ray intensities. The timer is usually a 'live-time' clock – a device which provides automatic dead-time correction by stopping the clock during intervals when pulses are being processed and the system is temporarily insensitive to the arrival of further X-ray photons.

(g) Various types of output device which are used to display the collected data. These include scalers, ratemeters, chart recorders etc. equivalent to those used in WDS systems, but virtually all except the most compact portable systems also utilize the convenience and flexibility of some kind of visual display unit (VDU). In conjunction with a control computer (see below), a VDU can be used to display the contents of any selected channel or group of channels during or after an analysis, or to display the complete spectrum as a graphics image of all the MCA channels, side by side (Fig. 5.8). Data may also be output to a printer or to some form of storage medium (e.g. magnetic disk or tape).

(h) Processing of EDS data for quantitative analysis is complex, and not practicable for most routine analyses – the time required for calculation would greatly exceed that needed for data collection, and would detract seriously from the speed of analysis which is the greatest advantage of EDS systems. However, virtually all commercial systems now include, or can be readily interfaced to, a computer which not only handles the data processing but also controls much of the analytical procedure and the system devices. Dedicated, inexpensive microcomputers now have the power required to perform all of these functions, but it is also possible to utilize on-line larger and faster computers for unusually complex applications. This is seldom necessary for routine analysis.

(i) Provision is usually made on microprobes or scanning electron microscopes to access any specified region of the pulse spectrum (or any wavelength isolated by a dispersing crystal) and use its contents to modulate the spot brightness of an oscilloscope whose beam can be swept in synchronization with sweeping of the electron beam over a selected area of the sample surface. The resulting two-dimensional X-ray intensity images provide pictures of the distributions of selected elements over the selected sample area, in the same way that secondary electron images are used in a scanning electron microscope to display features of the sample surface topography. The oscilloscope image for any selected element or for an electron image can be photographed for a permanent record, or can be computer-processed ('digitized') for enhancement of significant features.

X-ray images produced in this way are of considerable value to the study of very small-scale compositional variations, such as mineral zoning, small inclusions and grain boundary segregations. In themselves they provide qualitative or semi-quantitative data, and they are also often used to assist in precise positioning of the static microprobe electron beam for quantitative analysis. They can be produced from the output of either WDS or EDS spectrometers, but the latter have several distinct advantages. Sample damage by the microprobe beam can be reduced by placing the EDS detector close to the sample, so that it collects a wider cone of X-rays and permits the use of lower beam currents. Constraints on the geometric relationship between sample and detector are much less severe than in focusing X-ray spectrometers, and defocusing effects at the margins of low-magnification X-ray scans are virtually eliminated. EDS spectrometers often yield much better images for samples with irregular surface topography (e.g. dust particles). Electron beam diameter, and hence small-scale detail resolution, can be better optimized at the lower beam currents, which are inadequate for WDS systems but well suited to EDS.

On the other hand, WDS spectrometers usually have better peak-to-background (P : B) ratios (see below) and may yield better images under low-sensitivity conditions (e.g. for low concentrations of the imaged element, or for light elements beyond the effective working range of the EDS detector; see below).

The major advantage of EDS spectrometers is undoubtedly the speed and convenience with which they can be used to complete complex, multi-element analyses. With computer assistance, setting up for a routine analysis is often very simple, and EDS spectrometers have proved to be particularly useful in laboratories where relatively inexperienced operators require frequent access to microprobe or similar facilities.

The lack of moving parts simplifies maintenance and also eliminates errors arising from mechanical wear and tear, e.g. the development of free play in the crystal and detector mountings and linkages of wavelength-dispersive spectrometers, which leads to problems of resettability and ultimately to poor analytical precision.

Since data for all elements are collected in the same short measurement cycle, they can nearly always be obtained from the same point or area on the sample. This is not always the case in WDS analysis. In electron microprobe analysis, for example, sequential determination of successive elements extends the total analytical time during which the target area on the sample is subject to energetic electron bombardment. Some samples are sensitive to the thermal effects of the electron beam; diffusion gradients may be established, or in extreme cases the sample may be more severely damaged. In almost all cases, prolonged exposure to the electron beam results in the slow accumulation of a 'contamination spot' on the sample surface as a consequence of thermal processes affecting pump oil vapours. The surface contamination absorbs increasing proportions of the emitted X-rays (particularly those of low energy) and can become a serious source of error in intensity measurement.

In any of these cases, the problem can be avoided only by using low beam currents (which may lead to unacceptably low X-ray intensities), or by moving the beam over the sample surface as the analysis progresses, to reduce the temperature build-up. In the latter case, sequentially determined elements may not then be measured on exactly the same sample points, and this can obviously lead to problems if the sample is not homogeneous at the scale of analysis.

On the other hand, there are also some practical disadvantages to EDS spectrometers. Although the energy resolution of an Si(Li) detector is much better than that of a gas-filled detector, by a factor of about three, it is more than an order of magnitude worse than that of a WDS spectrometer using a dispersing crystal in conjunction with a gas-filled detector. This means that overlap interference between adjacent spectral emissions (i.e. those of nearly identical energy) is much more pronounced and more difficult to correct.

Electronic noise is also a more serious problem than in WDS systems, with unfiltered noise contributing to more substantial non-peak backgrounds. Since peak-to-background ratios are lower, the ability of EDS systems to detect and measure low concentrations is significantly reduced. Light elements with low-energy characteristic spectra often cannot be measured at all, because of absorption losses in the detector window and the silicon dead layer.

The need for constant LN2 cooling of the detector and preamplifier is inconvenient, but is generally not a serious objection. However, if the LN2 supply is accidentally allowed to run out while bias voltage is maintained on the detector the resulting damage to the detector may be both extensive and expensive.

There are also some problems in the electronic processing of the detector signals. The requirements for maintaining a proportional relationship between X-ray photon energies and their corresponding pulse amplitudes are much more stringent than in WDS because EDS systems are *completely* dependent on this proportionality – it is the only basis on which they can distinguish between different X-ray energies. This in turn demands very accurate pulse shaping in the main amplifier before the pulses are digitized, and the shaping process is relatively slow. Under these conditions, system *dead times* are extendable and are almost an order of magnitude greater than those typical of WDS systems (i.e. of the order of 10–20 μs rather than 1–2 μs). Thus EDS systems perform poorly at even moderately high total count rates.

A related problem is that of *pulse pile-up*. Because of the relatively slow pulse-shaping process in the main amplifier, there is always a finite probability that the detector, which has a much shorter inherent dead time than the shaping circuitry, will register another X-ray photon and output a second pulse while the first is still being processed. (It has been noted already that this is less of a problem with gas-filled detectors, partly because pulse-processing times are much shorter and coincidence probabilities are therefore much lower, and partly because the 'screening' effect while positive ions remain in the vicinity of the anode inhibits the avalanching effect required to register the second pulse.) The probability of such a 'pile-up' occurring is approximately equal to $n\tau$, where n is the count-rate and τ is the pulse-processing time. In practice, τ amounts to the amplifier *rise time*, or the time taken to process the leading edge of the amplifier output pulse (since the ADC is relatively insensitive to irregularities in the pulse tail). For a rise time of 10 μs and a very modest *total* count rate of 1000 cps, there is thus about a 1 per cent probability of a pile-up effect being associated with the processing of any one pulse.

If uncorrected, the result would be the generation of a spurious pulse whose amplitude would correspond to the sum of the energies of the two coinciding pulses if the latter were exactly coincident in time, or something less if they were not precisely superimposed (Fig. 5.9). Pile-up may involve two or more pulses of the same or of different energies, and it may involve pulses produced by both characteristic and continuum X-ray photons. Because of the wide range of possibilities, uncorrected pile-up further enhances background intensities; sensitivities are even further reduced by the loss of peak counts due to each of the coincident pulses – each is lost to the pulse of spurious higher energy.

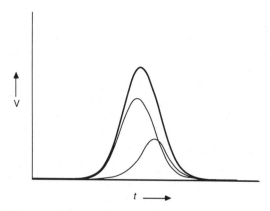

Figure 5.9 The coincidence or near-coincidence of two pulses results in the generation of a spurious pulse, the shape and amplitude of which will depend on those of the contributing pulses and on the extent of their overlap.

If the two pulses are exactly coincident in time, nothing of practical value can be done to correct the problem apart from calculation of the relevant probabilities and the application of a theoretical correction. However, the probabilities of perfect or near-perfect coincidence are very low and the errors resulting from the 'lost' counts are small.

Various ingenious techniques have been developed to cope with the potentially more serious problem of pulses arriving from about 2 to about 10 µs apart. This level of separation can be recognized electronically by a *pulse pile-up rejector*, which can automatically close a gate at the ADC input and inhibit further processing of the spurious pulse. It can also stop the live-time clock to compensate for the lost pulses.

A typical rejector consists of a fast *amplifier*, a *discriminator* and a pile-up *inspector*. The original output from the detector/preamplifier is split and fed simultaneously to the main amplifier, which begins its accurate but relatively slow amplification and shaping process, and to the fast amplifier, which produces a less critically shaped but much shorter pulse (with a rise time of about 1 ms instead of about 10 µs). The latter is immediately passed to the discriminator, which ignores it if it is of the low amplitude typical of random electronic noise. If not, the discriminator sends a 'trigger' signal to the inspector to initiate an 'inspection interval' of approximately the same duration as the rise time of the main amplifier.

If no further pulses arrive at the amplifiers during this interval, then there will be no problem with pile-up and no further action is taken. The shaped pulse output by the main amplifier is passed to the SCA and then to the ADC/MCA for measurement and counting. However, if a pile-up pulse is fed to the amplifiers during the inspection interval, the fast amplifier/discriminator sequence will recycle and another trigger signal will be sent to the inspector. Registration of this trigger within the active inspection interval causes the inspector to recognize the pile-up and send an inhibit signal to the input gate of the ADC, instructing it to ignore the next pulse coming from the main amplifier because it will be of spurious amplitude. The inspector will then initiate a new inspection interval because of the possibility of yet another pile-up pulse arriving before the amplifier has finished processing the already spurious pulse. The ADC gate remains closed until a clear inspection interval has passed.

This means, of course, that when pile-up effects become severe at high count rates the system will effectively be permanently disabled – it is then said to *lock up*. This factor limits the maximum count rates at which EDS systems can be used.

Electronic pile-up correction greatly enhances both the accuracy and the precision of EDS analysis, but at the cost of the need for careful operator attention to amplifier performance and characteristics (such as the time constants of the differential and integral circuits used for pulse-shaping). Amplifier settings optimized for one part of the X-ray spectrum, or for one total count rate, may be significantly less satisfactory for other energies or other count rates. It should also be noted that the inhibit technique does not restore the 'lost' pulses, except by the indirect method of controlling the live-time clock as well as the inhibitor.

X-ray photons having energies outside the actual recorded range will also contribute to both dead-time and pile-up counting losses. For example, if a 20 keV electron beam is used with a microprobe EDS system to analyse silicate samples, a 1024 channel MCA would typically be used with 10 eV channels to record the energy spectrum between 0 and 10.24 keV, which includes the K lines of all of the usual silicate major elements. However, the emitted (but not the recorded) spectrum will also include emissions with energies in the 10.24–20 keV range – mostly continuum, but also possibly some heavy-element characteristic lines. Although they are not recorded by the MCA because their energies are too high, these emissions will still contribute to dead-time and pile-up losses in the recorded portion of the spectrum, and the true total count rate will in fact be appreciably higher than that recorded by the MCA. A live-time clock may compensate for the dead-time losses, but not necessarily for the pile-up effects. Under these circumstances it is better not to use excitation energies substantially in excess of the maximum recorded spectral energy. In silicate analysis, for example, electron beam energies of 12–15 keV maximum are usually preferred.

Apart from the electronic problems of dead time and pulse pile-up, EDS spectra may also be complicated by undesirable effects of *peak broadening* (due principally to the contribution of noise to resolution loss; see Ch. 4), and *peak shifts*. The latter represent a phenomenon analogous to the pulse amplitude depression observed in WDS systems at high count rates, in that they result from incomplete restoration of the AC amplifier baseline between closely spaced counts. While the problem can be partially alleviated by automatic baseline restoration circuits in the main amplifier, it remains much more serious than in WDS because the EDS amplifier processes *all* of the detector pulses – not just those from one element, isolated by the dispersing crystal. This contributes further to the constraints on the maximum practical total count rates.

With lower peak-to-background ratios, accurate estimation of the background intensity beneath each analytical peak becomes more critical than in WDS (in which it can often be ignored in major-element analysis; see Section 7.3.5). It also becomes much more difficult, because of the extensive overlap between adjacent, relatively poorly resolved peaks. The simple WDS technique of extrapolating background intensities measured on either side of the peak (Ch. 8) is seldom directly applicable, since the extrapolation (if it can be made at all) must be made over broad spectral regions in which much detail (e.g. of absorption edges, low-intensity peaks etc.) is obscured by the broad tails of adjacent peaks.

Considerable attention has been given to this critical problem, and several methods have been developed to cope with it. These range from attempts to calculate background profiles from theoretical principles to the totally empirical fitting of various mathematical expressions to experimental data. No universally satisfactory solution has yet emerged, but several of the methods developed to date have been shown to yield reasonably good results in a fair range of analytical circumstances. However, even in favourable circumstances EDS sensitivities usually remain significantly poorer than their WDS equivalents.

Similarly, problems of overlap interference between adjacent peaks are nearly always more severe in EDS than in WDS systems, but progress with this problem has been rather more successful because of the availability of on-line computers which can be used for mathematical *deconvolution*, or *peak stripping*, procedures. The calculated profiles of successive peaks are systematically 'stripped' from the spectrum, beginning with the strongest; as each peak is removed, corrections are made for its overlap effects on the other remaining peaks, and weaker peaks of the same element are stripped at the same time. The final result is an estimate of the net intensity for each peak, after subtraction of the overlap effects due to all other peaks and the estimated or computed background intensity beneath the peak.

This procedure is also subject to several limitations. It is dependent on the accuracy of background estimation as well as the overlap factors. The shape of the spectral peaks is critical to any stripping procedure, but this depends on the system resolution and hence on such disparate factors as the condition of the detector or the amplifier characteristics. Where two peaks have nearly identical mean energies, or where a low-intensity analytical peak is almost entirely obscured by tail interference from an adjacent stronger peak, successful overlap correction requires a high level of accuracy that cannot always be maintained – particularly if net count rates are low and statistical precision is poor (Section 7.3.5). Under these circumstances, computer print-out of the results of hidden and – to the unskilled operator – largely incomprehensible calculations may well give the inexperienced analyst a totally unwarranted sense of analytical power and accuracy.

In terms of interference, EDS systems have one major advantage over their WDS equivalents: in the absence of a dispersing crystal, there is no problem with high-order reflections. The potential interference of fourth-order Fe Kβ with the measurement of Si Kα (quoted in a previous example) does not arise in EDS. Of course, there is also no problem with crystal fluorescence effects, since there is no dispersing crystal in the system.

Despite its problems, energy-dispersive spectrometry has already been established as a very useful technique, capable of obtaining rapid and reasonably accurate analytical results on a wide variety of sample types, including some that for one reason or another are not amenable to WDS analysis (e.g. samples susceptible to thermal damage by high electron or primary X-ray fluxes). It is particularly convenient for portable XRS systems and for adding analytical capabilities to scanning or transmission electron microscopes. Many laboratories have found it to be an economical and effective way to rejuvenate elderly electron microprobes whose wavelength-dispersive spectrometers have become worn out and are prohibitively expensive to overhaul or replace. While it is unlikely ever to replace wavelength-dispersive spectrometry completely (because it is theoretically incapable of matching the superior energy resolution of WDS), it has already been established as a powerful tool to be used in conjunction with WDS or, for many practical applications, in its own right.

6 Summary of instrumentation

6.1 X-ray fluorescence spectrographs

6.1.1 Generator

The primary tubes used in X-ray fluorescence spectrographs (Fig. 6.1) are designed to be operated at voltages up to 100 kV, typically at powers of up to 3 kW or more. These voltages (and the corresponding tube currents) require high-quality generators which must also meet stringent standards of stability over a wide range of operating conditions. Modern high-voltage systems can deliver voltages over the range 10–100 kV (usually in 1 or 5 kV increments) and tube currents of 5–100 mA. The power output of the generator (in watts) at any operational setting can be calculated by multiplying the voltage (in kV) by the tube current (in mA) – thus the power output at 60 kV and 30 mA is 1800 watts (or 1.8 kW). In theory a 100 kV/100 mA generator is capable of delivering $100 \times 100 = 10\,000$ W, which would immediately destroy a tube with a typical maximum rating of 3 or 4 kW (note, however, that many older tubes still in service are rated only at 2 kW or less). It is the operator's responsibility to know the maximum rating of the tube in use and to ensure that it is never exceeded under either manual or automatic control, and to remember that the effective tube loading is a function of *both* voltage and current.

The need for strict electrical safety precautions when using 100 kV generators is obvious.

6.1.2 Primary X-ray tubes

The function of the primary tube is to provide radiation, both characteristic and continuous, to excite the fluorescent radiation from the analysis sample. Tubes are marketed with a variety of different anode materials: Au, Pt, W, Ag, Rh, Mo, Cr, Ti and Sc are those most commonly used, but others may be appropriate for particular applications. None is suitable for all applications, for several reasons:

(a) It is usually not feasible to use a tube if its anode is the same as the element sought, particularly if the latter is present in the sample in low concentrations. Characteristic lines from the tube anode are scattered from the sample, and appear in the measured spectrum in

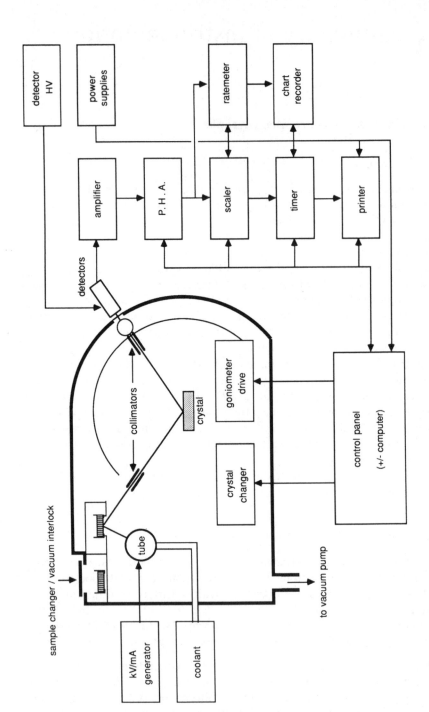

Figure 6.1 Schematic diagram of the major components of a typical wavelength-dispersive X-ray fluorescence spectrograph.

the form of high background beneath the fluorescent emissions. They cannot be eliminated by dispersion or pulse amplitude discrimination because they have essentially the same wavelength/energy as the analytical radiation (except for slight shifts due to Compton scattering), and their interference cannot be compensated by the normal off-peak background estimation procedures (Ch. 8).

(b) A tube cannot be used if its anode material yields characteristic or Compton-scattered radiation that interferes with the analytical characteristic line of the element sought. For example, Cr Kβ interferes with Mn Kα (although the magnitude of this particular interference can be greatly reduced by the use of an aluminium filter in front of the tube window, and Cr tubes can in fact be used with such a filter for the determination of all but trace concentrations of Mn). The tungsten L spectrum interferes with the Kα lines of Ni, Cu and Zn.

(c) It has been noted that the tube continuous spectrum contains much more energy than its characteristic spectrum, and in general the bulk of the fluorescent radiation from the sample is produced by the continuum. Since the total energy of the continuum increases with increasing atomic number, tubes with anodes made of heavy metals (Au, Pt, W) are particularly useful for general use. However, there are some special cases where effective use can also be made of the relatively high intensities of some tube characteristic lines, e.g. Cr Kα, with a wavelength of 2.29 Å, is an efficient excitation source for the Ba L spectrum, whose absorption edge is at 2.363 Å.

(d) While useful for general purposes, heavy anode tubes lose much of their advantage when used to excite light element spectra because the heavy anodes self-absorb a high proportion of the long-wavelength continuum. Much of the latter is also absorbed by the tube window, which needs to be fairly thick to withstand heating stresses produced by the relatively large number of electrons scattered from the heavy anode surface. Tubes with lighter anodes, such as Cr or Ti, in fact give much better results for light-element analysis since anode self-absorption of the long-wavelength continuum is much less severe; light anode tubes also require thinner windows (e.g. 0.3 mm Be instead of 1.0 mm), because electron scattering by the light anode is much less severe.

Thin window Cr tubes typically yield fluorescent intensities for light elements (e.g. Si or Al) which are five to six times higher than those obtained with a W tube operated at the same power. However, if a Cr tube is to be operated at high power (e.g. 3 kW), a thicker window may again be required and some of the advantage will then be lost.

It is therefore necessary for most XRF laboratories to be equipped with a range of tubes, selected according to the analytical problems commonly encountered. For silicate geochemistry, Cr (or Sc), Mo and Au are the most useful tubes. Rh has been claimed to be a versatile substitute for both Cr and Mo. W tubes are good all-purpose tubes, apart from the interference problems with some elements noted above.

XRF tubes are expensive and are also rather fragile; much of their construction is of glass, which can be irreparably damaged by minor shocks. Therefore they should always be handled (e.g. during tube changes) with considerable care. If tube changes, and consequent handling, are relatively infrequent, a tube should give a life of 3000 to 5000 operational hours (or even more in favourable circumstances). However, unused tubes also deteriorate fairly rapidly, so that they should not be left in storage indefinitely. Even if not in regular use, a tube should be installed in the instrument and operated for a few hours at least once every two to three months. High-power loadings should never be suddenly applied to a tube (which should be kept in mind when programming automated systems for multi-element analysis), and water-cooling of the tube should always be maintained for a few minutes at least after power has been switched off.

XRF tubes are powerful and intense sources of X-radiation, and should always be accorded the greatest respect. Modern commercial XRF systems are well shielded and, when properly operated, present virtually no radiation hazard. However, safety monitoring programs should be regarded as essential elements of laboratory procedure. All commercial instruments are fitted with safety switches to disable X-ray sources if protective shielding is removed; operation of these switches should be checked periodically, and they should *never* be bypassed except by skilled maintenance technicians. While these remarks seem self-evident there are unfortunately too many examples of accidents, some very serious, which have followed failure to observe these elementary precautions.

A generalized guide for tube selection and optimum operating conditions is given in Table 6.1. While the guide suggests useful starting points to use when setting up an analytical procedure, additional fine-tuning is almost always required.

6.1.3 Sample chamber

Provision must be made for precise positioning of the sample relative to the primary X-ray tube and to the spectrometer. The use of a collimator to limit divergence of the fluorescent X-ray beam means that the spectrometer 'views' only a limited angular range, and the sample must be correctly (and reproducibly) placed within that range. Positioning must also be precise to ensure that intensities from samples and calibration

X-RAY FLUORESCENCE SPECTROGRAPHS

Table 6.1 Generalized guide to primary tube selection and operation.

K spectra		
O(8)–Ti(22)	Cr	55 kV
V(23)–Co(27)	W	55 kV
Ni(28)–Zn(30)	Au	65 kV
Ga(31)–Y(39)	Mo	100 kV
Zr(40)–U(92)	Au	100 kV
L spectra		
Mo(42)–Cs(55)	Cr	55 kV
Ba(56)–Hf(72)	W	55 kV
Ta(73)–Re(75)	Au	65 kV
Os(76)–U(92)	Mo	100 kV

standards are measured under conditions which are held as nearly constant as possible.

Analysis samples are therefore usually inserted into special holders which are loaded into a positioning turret, which may be operated manually or under some form of automatic control. To simplify analytical procedures, provision is often made for loading at least three or four samples in a turret which can present them sequentially to the analytical position. Automated X-ray spectrographs may be equipped with additional facilities for automatic loading of 100 or more samples; such instruments are then capable of lengthy periods of unsupervised operation. 'On-stream' process control instruments may be fitted with special sampling and sample-loading devices appropriate to their specialized applications.

If the spectrograph is to be operated in vacuum (see below), it is desirable to allow sample interchange without losing the complete spectrometer vacuum, which would be wastefully time-consuming to restore. Sample chambers therefore usually include an airlock to reduce the volume that must be repumped every time a sample or group of samples is changed.

The paths of both primary and fluorescent X-ray beams are inclined to the sample surface, so that the intensity of X-ray generation is not uniform over the area analysed and minor inhomogeneities in sample composition can have a disturbing effect on reproducibility. The magnitude of this problem can be considerably reduced if the sample is rotated during analysis about an axis normal to its surface, and most sample chambers incorporate a 'spinner' to accomplish this rotation, typically at speeds of 50–100 rpm.

6.1.4 Spectrometer

XRF systems can employ either wavelength- or energy-dispersive spectrometers, but because their X-ray intensities are relatively high and their detection limits correspondingly low, the superior resolution of WDS systems tends to be favoured for trace element analysis. Crystal-dispersing, or 'Bragg Law', spectrometers are therefore still fitted to the majority of commercial instruments, although energy-dispersive spectrometers are undoubtedly useful or superior for some applications.

Many different geometries have been used for XRF wavelength-dispersive spectrometers, but the mechanically simple, flat-crystal, non-focusing type is the most common. Its essential components include dispersing crystals (usually with provision for easy interchange of up to five different crystals), detectors (usually gas flow and scintillation types which can be operated separately or in tandem) and a pair of collimators. The primary collimator is fixed in position between the sample and the dispersing crystal, while the secondary collimator is attached to the front of the detector. The collimators are usually of the Soller type, consisting of an array of closely spaced parallel metal plates.

The angular resolution of the collimator depends on the length of the plates and the spacing between them. Long plates and/or narrow spacings result in high resolution, but at the expense of lower intensities; it is therefore common practice to equip spectrometers with at least two interchangeable collimators: a coarse collimator giving an angular divergence of about 0.60°, and a fine collimator with an angular divergence of about 0.20°. For secondary collimation, flow counters are normally fitted with relatively low-resolution systems, and scintillation counters with high-resolution collimators.

6.1.5 Vacuum/helium X-ray path system

For wavelengths up to about 3 Å, the spectrometer can be operated in air, although there will be appreciable losses in intensity, as a consequence of air scattering and absorption, for wavelengths longer than about 1.5 Å. Above 3 Å these losses become so severe that the air path must be replaced, either with a lower scattering medium such as helium or by evacuation. The latter is generally to be preferred, but cannot be used for samples of high volatility (e.g. aqueous solutions, which would boil at low pressures), for which a helium path usually gives satisfactory results.

The spectrometer and sample chamber must therefore be equipped with an efficient, reliable vacuum pumping system, capable of fast pumping to pressures of the order of 0.1 torr and stable maintenance of these pressures for very long periods. Providing the system is well designed and properly maintained (with particular attention to cleanliness), pressures of this magnitude are not difficult to achieve and maintain with a single-stage mechanical rotary pump, and modern vacuum pumps can run continuously for very long periods with minimal (but not negligible) maintenance.

The vacuum is monitored with a gauge, normally of the thermal conductivity type, e.g. a Pirani gauge. For automated spectrographs, a vacuum-sensing interface to the control system is necessary.

6.1.6 Detector power supplies

Both flow-proportional and scintillation detectors require high-quality, stable DC power supplies, typically capable of producing adjustable anode voltages of up to 2 or 2.5 kV. A single power supply may serve for both detectors, but provision for independent simultaneous operation is highly desirable. Adjustment of the detector voltage is a critical factor in successful analysis, so the power supplies should include a sensitive potentiometer and an accurate voltmeter.

6.1.7 Pulse-processing equipment

Pulses output from the detector(s) must be amplified and shaped to a form suitable for counting. Each detector is usually fitted with a preamplifier whose output is further processed by a linear AC amplifier incorporating the necessary pulse-shaping circuitry. The preamplifiers normally have fixed gains of about $10\times$, and the amplifiers have adjustable gains of up to $100-150\times$. The signal-processing equipment also includes provision for pulse amplitude discrimination (variously called 'PHA', 'PHS', 'PAD' etc.), which permits rejection of unwanted high-order harmonics, scattered radiation etc., and thus improves peak-to-background ratios and enhances sensitivity in the low-concentration ranges.

6.1.8 Pulse counters

The purpose of the detector is to provide an estimate of intensity, or the rate of emission, of X-ray photons of a selected wavelength. To provide the analyst with this information, the rate at which detector pulses are received by the detection system can be established in either (or both) of two ways:

(a) A *ratemeter* can be used to monitor continuously the rate at which pulses are being produced. This is essentially an integrating, or 'averaging', device with an analogue (voltage) output that varies in proportion to the count rate. Since it operates by averaging the count rate over a finite time interval, it has a smoothing effect on short-term fluctuations, which can be varied by changing the averaging interval, or *time constant*, of the system – time constants can typically be selected in the range of about 0.1 to about 10 seconds. A short time constant is equivalent to a short averaging interval, so that the ratemeter responds to short-term fluctuations. With a longer time constant, short-term fluctuations are smoothed, or *damped* (but significant short-term signals can then be lost in the damping).

Output of the ratemeter can be either to an adjustable-range voltmeter or to a chart recorder, and is particularly convenient for qualitative or semi-quantitative applications (e.g. wavelength scans to locate peak positions or for qualitative analysis). However, analogue output is no longer much used for quantitative analysis, since it does not easily provide information needed for statistical evaluation of the results (Section 7.3.5).

(b) The amplified detector pulses can be displayed, as they are processed, on a digital device (a *scaler*), and electronically counted over finite intervals of time established by a *timer*. The interval over which counts are accumulated may be selected by the analyst either by fixing the time interval and measuring the number of counts accumulated in that period (*preset time* mode) or by fixing the total number of counts to be accumulated and measuring the time interval that it takes to collect them (*preset count* mode).

Digital processing has the advantage that the data can easily be passed to a teletype or similar device for printing, or to an on-line computer for immediate processing of the analytical data. It is now almost universally used for quantitative analysis.

6.1.9 Power supplies

Although not always apparent to the user, power supplies are required for all of the electronic sub-systems, and for various electric motors, cooling fans etc. The overall reliability of the total system is very much a function of the reliability of these supplies.

6.1.10 Automatic control systems

Most modern XRF spectrographs are equipped with some level of automatic control, ranging from simple devices for automatic scanning of a Bragg spectrometer to sophisticated 'dedicated' computer systems which are capable of assuming control of the complete analytical procedure and of immediate 'on-line' processing of the results. By eliminating many of the more tedious aspects of routine X-ray spectrographic analysis, these systems have done much to ensure widespread acceptance of XRF (and electron microprobe) analysis; in particular, they permit much of the work to be done on a routine basis by relatively unskilled operators. However, it should never be forgotten that the results that they produce are determined by such factors as the quality of the instrument to which they have been fitted (and of its maintenance), the standards of sample preparation, and the validity of the procedures that they have been programmed to follow. As noted elsewhere, the analyst using highly automated equipment must be particularly aware of the need for properly critical standards of appraisal, and of the dangers of excessive reliance on the computer for protection against errors over which it has, in fact, little or no control.

6.2 The electron microprobe

Since Castaing first demonstrated the practical feasibility of electron microprobe analysis in 1951, about a dozen instrument designs have been marketed commercially, and most of them have since passed through a very rapid process of technical evolution, particularly during the 1970s. Some aspects of instrumentation reflect differences in design philosophy or patent protection, and some are the result of technological advances over the past three decades. Nevertheless most of the basic features of Castaing's original design are still to be found in current instruments. They include the essential components shown in Figure 6.2.

6.2.1 Electron gun

The function of the electron gun is to provide a source of electrons, of appropriate energy and current density, to form the electron beam used to excite X-rays from the analytical sample. In order to select optimum conditions for excitation of the sample X-ray spectra, provision must be made for variation of both energy and current, and each must be regulated within narrow tolerances. For maximum spatial resolution (i.e. production of an electron beam of minimum diameter at the sample surface), the electrons must be produced at an effective point source, or as near to it as practicably possible. The electron optics (see below) produce a demagnified image of this source at the sample surface.

In most microprobes these requirements are met by the use of a self-biasing triode type of electron gun, in which the electrons are produced by thermal emission from a pointed tungsten filament, held at high negative potential, and accelerated towards an anode plate at ground

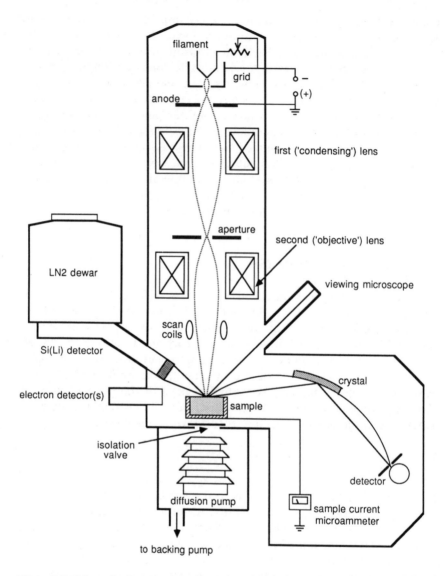

Figure 6.2 Schematic diagram of the electron beam-forming column and spectrometers of an electron microprobe (only one wavelength-dispersive spectrometer is shown, but most microprobes have two or more, disposed radially around the beam column and sample chamber, in addition to an energy-dispersive spectrometer). Electronic components for processing the X-ray signals are generally similar to those shown in Figures 5.8 and 6.1. Electron microprobes also usually have additional facilities for processing signals from one or more electron detectors (e.g. secondary and/or backscattered electrons) and for the display of X-ray or electron images produced by raster-scanning of the electron beam over the sample surface.

potential. A hole in the anode plate allows the accelerated electrons to leave the gun and enter the beam-forming electron optical system.

The filament emission current is controlled by a grid, usually in the form of a cylindrical cap ('Wehnelt cylinder') fitted over the filament. A small negative bias voltage on the grid is controlled with a variable resistor to determine the gun emission current; low bias voltages provide little opposition to the passage of the accelerated electrons and hence result in relatively high emission currents, and vice versa. The emitted electrons pass through a small hole in the centre of the grid cap, which itself has a focusing effect on the electron beam and produces a crossover whose demagnified image at the sample surface ultimately forms the electron 'probe'.

Other types of electron guns, including lanthanum hexaboride and field emission guns, have been investigated experimentally and found to offer some advantages. They have not yet come into general use in electron microprobes, but they are already widely used for certain types of electron microscope (e.g. scanning transmission electron microscopes, or 'STEMs'), some of which also include X-ray spectrometers to give them some degree of chemical analytical capability.

Many of the day-to-day problems encountered in routine electron microprobe analysis arise from instability in the electron beam, which can often be traced to the gun; filaments, for example, have limited lives (particularly if the vacuum system is not maintained at peak efficiency), and they almost invariably fail at the most inconvenient moment. This is one aspect of microprobe instrumentation which can be expected to improve considerably as the new types of electron gun come into wider use.

6.2.2 *Electron beam-forming column ('electron optics')*

The gun crossover image is demagnified by a system of electron lenses (usually electromagnetic lenses, although there is no reason in principle for not using electrostatic types). Most beam columns contain two such lens systems, which are commonly described as 'condensing' and 'objective' lenses respectively because of a loose analogy with light microscope optics. The focal lengths of the lenses are controlled by means of variable, stabilized lens current supplies; adjustment of the condensing lens current serves effectively to control the probe current incident on the sample, and adjustment of the objective lens current controls the diameter of the electron beam at the sample surface (this is the basis of the analogy with light optics – a microscope condenser controls the intensity of light illuminating the object, and the objective aperture determines the resolution of the optical system; the analogy should not be pursued much further).

One or more apertures placed in the electron optical column (usually between the lenses) are used for further control of the probe current and also, indirectly, for its measurement. They also serve to minimize some of the aberrations inherent in electron optics, in a fashion analogous to the use of limiting apertures to control aberrations in light optical systems. Other aberrations in the electron beam are controlled by auxiliary electromagnetic or electrostatic *stigmators*, which allow the operator deliberately to distort the beam cross-section so as to oppose the aberration effects.

The very small effective beam diameter which can be attained in the electron microprobe means that X-ray spectrometric analyses can be made on extremely small sample volumes, and this in turn gives the microprobe its most powerful attribute – that of very high spatial resolution. To obtain maximum benefit from this capability, it must be possible to position the electron beam very accurately on the sample surface. This is achieved partly by precise mechanical positioning of the sample under the beam (see below), and partly by the use of scanning coils or similar systems, similar in principle to those used in an ordinary television receiver to control the raster scanning of the image-producing electron beam. These devices can be used in the microprobe to produce scanning X-ray or electron images and also to allow controlled deflection of the static beam in order to position it on the sample with high precision.

If the scanning coils can be 'driven' in similar fashion to those of an oscilloscope or a TV set, the beam can be systematically raster-scanned over a selected portion of the sample surface; analytical information (e.g. X-ray or electron emissions) obtained during such scans can be processed digitally or displayed on a coupled oscilloscope to produce two-dimensional images of surface topography, elemental distributions etc. The scanning coils can also be used for non-mechanical 'blanking' of the electron beam between analyses or during sample interchange, by deflecting the beam away from the sample at all times when no measurements are actually being made. This minimizes thermal damage to the sample, and also reduces the rate at which 'contamination spots' build up on the sample surface as a result of thermal effects in the course of an analysis.

In addition to the electron optics, the microprobe also normally includes a light optical microscope, to permit location of analytical areas and viewing during analysis (although some earlier systems made no provision for the latter feature). Depending on the designer's approach, the light optical axis may or may not be coaxial with the electron beam. If it is, the electron beam must pass through holes in the centre of the aperture of the microscope objective, necessitating the use of partly reflecting, Cassegrainian objectives which have poorer resolution than comparable fully refracting systems (though they are usually adequate for the positioning purpose for which they are primarily intended). In either case, resolution of the light optics at the high magnifications necessary for reasonably accurate location of the analytical area is limited by the need for long working distances and hence low numerical apertures – otherwise the light objective would interfere mechanically with the electron objective lens and with the emitted X-rays. In fact the physical relationship between the electron lenses, the light optics, the sample and the X-ray spectrometers is one of the areas of major design differences between the various commercial microprobes, since it also necessarily represents an area of compromise.

6.2.3 Sample chamber

The basic functions of the sample chamber are common to all microprobes, but the methods by which they are achieved vary considerably from one design to another. The principal requirements of a sample chamber include the following:

(a) It must provide for simple and rapid interchange of samples and standards, preferably by an airlock system so that it is not necessary to break the entire instrumental vacuum or to shut down the electron beam during interchange. The vacuum constraint is more critical than in XRF systems, whose sample changers must meet the same demands, because the operating pressures of the microprobe are several orders of magnitude lower than those of normal XRF systems and pumping times are correspondingly longer.

(b) In order to position the samples for analysis, provision must be made for orthogonal (X, Y and Z) and preferably also for rotational mechanical translations of the samples, with high standards of reproducibility – especially in automated systems in which sample movements are performed under computer control.

(c) To allow at least limited interchange of samples and/or standards without breaking any vacuum at all, the sample chamber should provide for simultaneous inclusion within its stage of at least three or four sample mounts. Some allow up to 12, but with improved vacuum technology this requirement is no longer as critical as it used to be, and smaller, more compact stages have become more common.

(d) In conjunction with the light optics, the sample chamber must provide for simple, rapid and reproducible positioning of the analysis sample with respect to the electron beam and the X-ray spectrometers.

(e) On occasions when the vacuum must be broken to interchange samples, it must be possible to restore analytical conditions in a few minutes at most. Longer periods waste valuable instrument time and may introduce problems with time-dependent machine drift.

(f) The sample holder must be grounded ('earthed') to minimize or prevent charging of the sample during analysis by a build-up of absorbed electrons from the beam – which would result in erratic deflection of the beam. The grounding circuit normally includes a sensitive microammeter to measure that portion of the beam current that has been absorbed by the sample and hence has contributed to X-ray excitation (variously called the 'absorbed current', 'sample current' or 'specimen current').

6.2.4 Vacuum system

The electron gun, beam column, sample chamber and X-ray spectrometers must all be maintained at low pressures during operation in order:

(a) to maintain satisfactory filament life by minimizing processes of corrosion (e.g. oxidation) and ionic bombardment;
(b) to minimize loss of electrons or of electron energy in the beam as a consequence of collisions with gas molecules;
(c) to minimize absorption of the emitted X-rays, particularly those of relatively low energy.

These requirements are met by maintaining internal pressures of less than 10^{-4} torr (1 torr = 1 mm Hg); the figure of 10^{-5} torr is frequently quoted, but this usually refers to readings on a vacuum gauge whose sensing head is placed for convenience close to the vacuum diffusion pump. Actual pressures in more remote parts of the system, particularly in the electron gun and in the beam column (which is often pumped through the lens bores and even in some cases through the apertures), may well be as much as two orders of magnitude higher.

The vacuum must also be relatively 'clean' to minimize the deleterious effects of contamination deposits, produced on the surface of the sample during analysis by decomposition of hydrocarbon molecules derived from the vacuum pumps or from careless handling of the internal parts of the system (including samples and sample mounts). Once formed, these contamination deposits absorb both incident electrons and emitted X-rays and can have serious effects on analytical accuracy.

Mechanical rotary pumps such as those used in XRF systems are fast-pumping at relatively high pressures and are therefore effective in microprobe systems as 'roughing' pumps, i.e. pumps used to attain the first stages of a working vacuum. However, they are ineffective at pressures below 10^{-1}–10^{-2} torr, so that microprobes require an additional pump, effective at lower pressures, to achieve a working vacuum. This is usually an *oil diffusion pump* – a type that has no mechanical components and is therefore simple, effective and (with proper care) very reliable.

THE ELECTRON MICROPROBE

Diffusion pumps are ineffective at high pressures (above 10^{-1} torr), and indeed will emit copious quantities of pump oil and produce very undesirable contamination effects if operated in a high-pressure regime. It is therefore necessary to *rough pump* the system down to about 10^{-1} torr with a rotary pump before the diffusion pump is switched in, and the rotary pump is then kept in operation to 'back' the diffusion pump, i.e. to maintain low pressures in the exhaust of the diffusion pump. The diffusion pump must *never* be operated at pressures higher than 10^{-1} torr, and *must never be operated without a backing pump*. Most microprobes include automatic valving to isolate the diffusion pump in the event of power loss or of unexpected failure of the backing pump.

The same, or a second, rotary pump is also used to rough pump the sample chamber after sample interchange. The pumps therefore usually require a moderately complex system of valving to allow them to be switched in or out of various parts of the system as required. Early microprobes required manual operation of the valves and permitted some disastrous combinations to be selected by an inexperienced operator. On later models the vacuum system is usually automated to the push-button level, but it is still highly desirable for the operator to have a good understanding of the operating principles of the vacuum system.

Rotary and diffusion pumps both use oil as a pumping medium (though in very different ways), and the vacuum can therefore be contaminated by backstreaming of pump oil vapours. Backstreaming can be reduced (but not entirely eliminated) by water- or LN2-cooled baffles above the diffusion pump and by adsorption traps in the backing line. Backstreaming can also be reduced by using narrow-bore pumping hoses, but at the expense of slower pumping speeds.

In some critical cases the diffusion pump may be replaced by a more powerful and more sophisticated device, such as a turbo-molecular pump or an ion pump. For most purposes, however, the simple diffusion pump will give satisfactory results if it is properly operated and maintained.

6.2.5 X-ray spectrometers

X-rays emitted by the sample during analysis can be analysed with either wavelength- or energy-dispersive spectrometers. Each type has its advantages and limitations, as discussed in Chapter 5. WDS types have better total resolution, but normally require much longer total counting times, even when fully focusing geometry is used (as is almost invariably the case). EDS spectrometers are better suited in many respects to the low count rates commonly encountered in microprobe analysis and, by contrast with XRF, are now probably more widely used in microprobes than their WDS equivalents. To provide maximum flexibility, a microprobe should incorporate at least two and preferably three WDS

SUMMARY OF INSTRUMENTATION

and one EDS spectrometers (some instruments have been built with as many as 12 spectrometers altogether).

The need for multiple WDS spectrometers arises from the relative difficulty of routine crystal interchange, imposed by the more rigid constraints of focusing geometry, and also from the difficulty of detector interchange. With longer counting times consequent upon lower X-ray intensities, multiple spectrometers also contribute usefully to lower overall analytical times by permitting the simultaneous measurement of several different elements.

There is some general advantage in keeping the *take-off angle* (the angle, measured at the sample surface, between the surface and those X-rays that enter the spectrometers) as high as possible – at least higher than about 35°. This reduces the magnitude and hence the probable error of corrections that have to be applied for adsorption effects (Section 9.3.2), but it also imposes major design constraints.

The range of crystals available for microprobe WDS spectrometers is much the same as that for XRF spectrometers, except that some crystals are difficult or impossible to prepare in the bent and ground configuration required for focusing X-ray optics. Selection of crystals for specific analytical applications follows the same rules as in XRF analysis, but may be more constrained by limitations on the number of crystals available or needed for interchange and the number of spectrometers with which the microprobe is equipped. XRF spectrometers are commonly equipped with four or five interchangeable crystals, since most XRF systems have only one spectrometer, which has to handle the entire analytical wavelength range, and the geometrical constraints on the interchange mechanism are not severe. Microprobe spectrometers seldom provide for more than two crystals, or three at most; microprobes usually have at least three spectrometers, each of which can be set up to handle a portion of the complete wavelength range. Many microprobe analysts prefer to avoid crystal interchange altogether, if possible, because of the difficulty of maintaining the precise alignment demanded by focusing X-ray optics.

Few, if any, microprobe spectrometers use scintillation detectors, which are relatively large and unwieldy. Instead, sealed gas detectors are used for the shorter wavelengths and thin-window gas flow detectors for the longer. A typical three-spectrometer microprobe configuration will have an LiF crystal with sealed detector on the first spectrometer, a PET (or ADP) crystal with a sealed or a gas flow detector on the second, and a TlAP (or RbAP) crystal with a thin-window gas flow detector on the third. This configuration allows the detection of all elements heavier than oxygen, without need of any crystal or detector interchange.

6.2.6 Electron detectors

The sample can be viewed during analysis or scanning by one or more electron detectors, which are used principally for secondary and/or backscattered electron current imaging (see Ch. 9 for a brief discussion of the formation of secondary and backscattered electrons; for practical purposes, secondary electrons may be regarded as photoelectrons emitted from the sample surface with very low energies, and backscattered electrons as beam electrons scattered back from the sample surface or shallow depths, with significant residual energy).

The generation of secondary and backscattered electrons depends on factors such as sample composition and surface topography, so that the microprobe can be operated as a scanning electron microscope (SEM) by imaging the secondary electron or the backscattered electron distribution produced by a scanning probe beam. However, in the interests of improved beam stability, microprobes are usually designed to operate at lower electron gun brightnesses and higher beam currents, so that they cannot attain quite the same standards of resolution as a dedicated SEM unless they are equipped with versatile guns that provide for convenient control of gun brightness by simple methods such as adjustment of the filament-to-grid spacing. Similarly SEMs fitted with auxiliary spectrometers will often not yield microprobe standards of stability in the analytical mode.

6.2.7 Power supplies and support electronics

These components include AC and DC amplifiers for X-ray and electron detector signals, discriminators used to reject harmonic and other unwanted signals, timing and counting systems for measurement of X-ray intensities, imaging systems (including display oscilloscopes), and an extensive range of power supplies. With the exception of the imaging systems and the electron detectors, these components are broadly similar to those used in XRF instruments.

6.2.8 Control and data reduction systems

The earliest microprobes recorded X-ray intensities by means of relatively simple analogue devices, such as ratemeters or chart recorders. While very useful for some purposes, these devices are generally inadequate for accurate quantitative analysis, and digital scaler–timer systems are now universally employed. Digital data can easily be written to a teletype or similar writer, or they may be passed directly to a card or tape punch or to a magnetic recording device for storage before subsequent computer reduction (thus eliminating the possibility of transcription errors). It is

SUMMARY OF INSTRUMENTATION

even more convenient, and now more or less standard practice, to pass them directly to an on-line digital computing system, either for disk or tape storage or for immediate processing.

Selection between these alternatives is usually a matter of economics and of other facilities available within the laboratory.

If the microprobe is indeed fitted with a dedicated computer (and the advent of microprocessors has made this a relatively low-cost option), it is also particularly useful to assign responsibility for many of the repetitive functions of instrument control to the computer; indeed this, rather than data processing, was the first area to which small, microprobe-dedicated computer systems were applied. Computer control, to a greater or lesser degree, is now almost universal. Hardware-, software- and hybrid-controlled systems are all available commercially, with many of them having been designed for retrofit to older instruments. However, it is fair to note that there is still considerable variation in their capabilities.

7 Qualitative and quantitative X-ray spectrometric analysis

7.1 Qualitative analysis

Electron microprobe and XRF techniques are both ideally suited to many applications in qualitative analysis, for elements within their detectable ranges. The methods employed in each case are very similar: sample spectra are recorded over the wavelength (energy) range of interest and elements emitting lines in the spectra are identified from their wavelengths (energies) and relative intensities. Comprehensive tables of characteristic spectra are available for this purpose. The procedure is relatively fast, particularly if the spectral search is computer controlled, and/or if an energy-dispersive spectrometer is used (either technique permits recording of the complete emission spectrum in a matter of minutes). Multiple spectrometers on an electron microprobe enhance the speed of qualitative analysis by allowing simultaneous recording of different portions of the spectrum.

Microprobe and XRF procedures are essentially non-destructive, although some kind of sample preparation is often required and this may modify the original physical and/or chemical state of the sample (particularly in XRF). Adequate XRF spectra can be obtained from samples of only a few milligrams, and the microprobe is capable of analysing samples of dust particle size.

X-ray spectra are relatively simple to interpret, providing a little common sense is used. Thus, for example, if a line is identified as the Kα line of a particular element, it is an elementary precaution to check that the Kβ line of the same element occurs, with appropriate intensity, at the correct wavelength; if it should be obscured by interference from another element, confirmation can be sought in the L spectrum or by careful examination of higher-order reflections. It is always potentially dangerous to base the identification of an element on the presence of a single line, particularly in the relatively 'crowded' short-wavelength region (unless the element is present in such low concentrations that only its strongest line is evident). A common source of confusion is the occasional satellite (non-diagram) line; many wavelength/energy tables do not include these lines.

Detection limits in qualitative analysis are strongly dependent on sample composition and on the conditions of analysis. They are commonly of the order of a few tenths of 1 per cent in electron probe analysis, and significantly less in XRF providing the sample is large enough and the spectrum is not scanned too fast. Norrish and Chappell (1977) quote the following approximate detection limits in silicate samples (e.g. rocks and soils):

C, O	1%
F	0.2%
Na, Mg	100 ppm
Al	30 ppm
S	8 ppm
Sc–Mo	1–3 ppm
I–U	2–6 ppm

If the spectrum is recorded, as is often the case, on a ratemeter/chart recorder system, use of too high a time constant on the ratemeter may obscure the presence of weak emissions from elements present in low concentrations. Elements present in concentrations close to their detection limits cannot normally be identified on ratemeter traces, even using short time constants, since their emission 'peaks' cannot be resolved from the statistical fluctuation of the background.

Energy-dispersive spectrometers are particularly well suited to qualitative analysis because of the speed with which the total emitted spectrum can be accumulated. However, for wavelengths greater than about 0.8 Å they are inferior to WDS spectrometers in their overall resolution, so that weak emissions tend to be obscured by relatively high backgrounds and/or overlaps from adjacent stronger peaks, and sensitivities are reduced. Computer-based spectral stripping techniques can be used to enhance weak emissions and reduce the effects of overlapping spectra, but these are better suited to quantitative rather than rapid qualitative analysis.

If low concentrations of a specific element are sought, then sensitivity may be considerably improved by carefully optimizing excitation and detection conditions for that particular element and making detailed examinations of relevant portions of the spectrum. Quick manual scans often serve to check a sample for specific major elements. If computer-control facilities are available, the speed of qualitative WDS analysis may be considerably enhanced by fast sequential checking of one or two selected emission lines for each element sought, rather than by systematically scanning the entire spectrum.

If optimum sensitivities are sought, it is important to keep backgrounds as low as possible, both by careful control of excitation conditions and by effective use of pulse amplitude discrimination. Since pulse amplitude is proportional to X-ray energy, this means that either the discriminator window baseline or (preferably) the amplifier gain must be continuously adjusted during a scan to compensate for the progressive change in X-ray energies. Most XRF systems make provision for automatic compensation of amplifier gain by means of a sin θ potentiometer coupled to the goniometer drive. As the goniometer scans from low to high Bragg angles (i.e. from shorter to longer wavelengths, or higher to lower X-ray energies), the potentiometer constantly increases the electronic amplifier gain as a linear function of sin θ and thus automatically keeps the mean amplitude distribution centred in the discriminator window. Similarly, fixed attenuators are used to select the gain range appropriate to each dispersing crystal. The appropriate attenuator is automatically selected whenever the crystal is changed.

One or more additional attenuators may also be switched in if the analyst wishes to examine second- or higher-order reflections from the dispersing crystal.

It is possible to equip microprobe spectrometers with similar devices (with the controlling potentiometer on a linear focusing spectrometer being driven as a linear function of θ rather than sin θ), but in fact few microprobes possess this facility and it is common practice instead to set very wide discriminator windows or even to operate only in 'integral' mode (low-amplitude rejection only) during qualitative scans. This is not good practice, since background rejection is not optimized and there will be a significant loss in sensitivity. It is better to maintain a relatively narrow discriminator window and vary the amplifier gain manually, if necessary, throughout the scan to maintain appropriate settings. These can be determined in advance by calibration against several elements within the range of each spectrometer. With a little experience it is then not difficult to make the necessary adjustments. If computer-control facilities are available, these can often be used to maintain optimum discriminator settings.

In XRF analysis, characteristic lines from the tube anode and from tube contaminants (e.g. W, Cu) will be scattered from the sample surface and will always be present in the recorded spectrum. Clearly, then, a gold tube is not well suited to the detection of trace amounts of gold. The scattered tube spectrum will also include Compton-scattered lines which will be of relatively high intensity if the tube lines are of short wavelength.

A particularly convenient feature of many EDS spectrometers is the incorporation in the data-processing system of a facility for automatic identification of lines ('peaks') present in the spectrum. The latter may be displayed on a video monitor; when a cursor on the video image is moved to a peak centroid, possible transitions producing the appropriate pulse energy are automatically displayed on the screen. This facility further enhances the speed of qualitative EDS analysis; it is particularly useful for the examination of otherwise unidentifiable small inclusions in microscope sections, dust-size particles etc.

7.2 Quantitative analysis

The basic premise of X-ray spectrometric analysis is that of a simple relationship between concentration c_A of an element A and its emitted X-ray intensity I_A, under controlled analytical conditions. If the relationship is linear, then

$$c_A / I_A = k_A \qquad (7.1)$$

where k_A is a proportionality constant and I_A is the (background-corrected) intensity of a characteristic emission line of element A. Quantitative analysis for element A would then require no more than preliminary establishment of k_A for that particular element, which could be done simply by measuring the X-ray intensity I_A emitted from a calibration standard of known c_A and substituting in the above equation. Once the constant is known, concentrations of the element in a series of samples could be determined, using the same equation, by measuring appropriate X-ray intensities emitted from the samples under identical conditions.

Unfortunately, theory predicts and experience soon confirms that this simple linear relationship is no more than an approximation that will yield accurate analyses only in exceptional circumstances. In particular, because much of the measured X-ray emission is generated at depth within the sample and hence interacts with the sample itself on the way to the detector, then variations in chemical composition and/or physical form between the samples and the calibration standards will produce departures from linearity that affect the accuracy of such simple comparative analyses.

The resulting errors may be severe, but in fact they are often only of the order of a few per cent (relative), or even less; for many practical applications, errors of this magnitude can easily be tolerated. It should not be forgotten that although a good analyst will always strive for the highest reasonable standards of accuracy, additional effort will not be productive beyond a point determined by independent errors associated with the collection and preparation of the analysis samples. In other words, there is little point in wasting time and effort in the pursuit of levels of analytical accuracy that are several orders of magnitude better than sampling error. Many analytical problems are in fact best approached on a semi-quantitative basis, with due allowance being made for the limitations of both sampling and analytical data.

Certainly the beginner in X-ray spectrometric analysis is well advised to gain as much experience as possible in semi-quantitative work before becoming too involved in the additional complexities of quantitative analysis. This experience should be used to ensure familiarity with the instrumentation, facility in planning analytical strategy and reliability in making and interpreting the necessary measurements. This is particularly important for those who are ultimately going to make use of sophisticated, highly automated instruments to produce quantitative data; without sufficient fundamental 'hands-on' experience they will be unable to appreciate the processes by which their data are derived and they will tend not to be properly critical in the appraisal of their analytical results.

7.3 Sources of error in quantitative analysis

Of the many possible sources of error inherent in X-ray spectrometric analysis, the following are the most significant:

(a) operator errors;
(b) sampling and sample preparation errors;
(c) instrumental errors;
(d) statistical errors;
(e) interelement ('matrix') errors.

These groups of errors are common to both XRF and electron microprobe analysis, although there are some differences in detail and in the methods adopted to cope with some of them. They are to some extent interrelated (e.g. sample preparation errors may result from failure to appreciate the consequences of some interelement effects), but for convenience they are treated separately in the following brief discussion, which is oriented towards XRF analysis but is in most respects equally applicable to the electron microprobe.

7.3.1 Operator errors

The most serious potential source of error in X-ray spectrometric analysis is the operator himself/herself. He/she is responsible for defining the analytical technique (including selection of instrumental conditions), for preparation of the standards and samples, for operation of the instrument, and for processing and interpretation of the analytical data. Errors in any of these phases of analysis can have disastrous effects on the quality of the final results.

It is certainly true that in many modern laboratories routine analytical procedures are standardized and performed on highly automated instruments; opportunities for casual error are then considerably reduced, though never entirely eliminated. Even the best automated equipment still needs to be programmed for the functions it is to perform, and this requires competent and thorough evaluation of the analytical task and a good understanding of the operation of each component of the instrument. Good maintenance and periodic performance checks are also essential, but are often overlooked by poor operators.

There is not much to say about operator errors that is not self-evident. The most important point is that, although often potentially serious, they can mostly be avoided, which sets them apart from most of the others discussed below. The responsibility is entirely in the operator's hands.

In the general case, the analyst must first give attention to the definition of what might be called the *analytical strategy*, which involves decisions on the nature, conditions and sequence of experimental observations and calculations that have to be made to complete the analysis. This phase of the work typically requires some preliminary experimental work to determine the optimum conditions of analysis – precise location of peak positions, preliminary estimation of statistical parameters etc. – but even this work should not be commenced before the potential analytical problems have been properly evaluated.

The first step is to ensure that the analytical task has been adequately defined. This involves a review of the expected compositional range of the samples, as far as it is known in advance, and a systematic listing of the elements to be determined, together with a realistic assessment of the levels of accuracy and precision required for each. In geochemical work, for example, this assessment is normally based on a thorough mineralogical and petrographic examination of the samples prior to analysis, which also allows a preliminary estimate of the compositional range of the samples to be made. The physical nature of the samples is considered, techniques for sample preparation are selected, and the general analytical procedures are decided. Appropriate calibration standards are chosen; as far as possible these should be generally similar in composition to the samples (for reasons that will be discussed in more detail in Chs 8 & 9), and they should preferably span the full compositional range of the samples, or very close to it. Of course the compositional range of the samples may not be well known prior to analysis, in which case it is often advantageous to perform some preliminary semi-quantitative work to obtain at least rough estimates.

Analytical conditions must then be selected for each element to be determined. These include the following:

ANALYTICAL LINE

Maximum intensity of emission is usually a desirable criterion, so that the line chosen is normally the strongest in that part of the element X-ray spectrum within the available range of spectrometer settings. The Kα line is normally used for the light elements (Z up to about 33), the Lα line for the intermediate group (Z in the range 34 to about 80), and the Mα for heavy elements (Z greater than about 80). However, the possible effects

SOURCES OF ERROR IN QUANTITATIVE ANALYSIS

of spectral interference due to emissions from other elements present in samples and/or standards must be carefully evaluated, particularly if the element sought occurs in low concentrations and/or the potential interfering element is a major constituent. Possible interferences due to scattered tube anode or contamination lines must also be considered.

Evaluation of interference effects requires careful study of a good set of wavelength tables (i.e. one which includes satellite lines and at least low-order harmonics which may prove difficult to exclude by pulse amplitude discrimination). If there is any question about the extent of a possible interference, it should be investigated experimentally, using the techniques described below in Section 8.1. This is almost mandatory for trace-element analysis, in which otherwise trivial interferences can loom as major sources of potential error.

Thus, for example, the strongest Ba L line (Ba Lα at 2.776 Å) is normally avoided in trace-element geochemical analysis because it is close to Ti Kα (2.750 Å). Since Ti is often much more abundant than Ba (frequently by several orders of magnitude), even the weak 'tail' of the Ti Kα peak overlapping the Ba Lα line at 2.776 Å may be significantly more intense than the characteristic Ba emission and it will be difficult or impossible to correct adequately for the interference. Similarly, Ba Lβ_1 (2.567 Å) is subject to slight interference from Ti Kβ (2.514 Å) and also from the Ce Lα line at 2.561 Å. If either of these elements is a potential source of interference then Ba Lβ_2 (2.404 Å) can be used; it is less intense than Ba Lβ_1, but is efficiently excited by Cr Kα (because it is part of the Ba L$_{III}$ spectrum) and can give reasonably good results if a Cr tube is used for primary excitation.

DISPERSION CONDITIONS

The operator must choose an appropriate dispersing crystal for each element (at least for WDS spectrometers), the order in which the reflection is to be measured, and the primary collimator spacing (or slit width in focusing spectrometers). The approximate Bragg angle at which the reflection is to be measured is calculated or determined from tables, and precise spectrometer calibration is then established by using a calibration standard to make a slow scan across the peak to determine its true centroid. This is best done not by attempting to position the spectrometer precisely at the position of maximum intensity (which, because of statistical fluctuations in the count rate, is virtually impossible to locate), but by scanning across the peak at a uniform rate and noting the spectrometer readings at, say, 80 per cent of maximum intensity on either side of the peak. The peak centroid can then be taken as the spectrometer position half way between these two readings. It is advisable to use at least 80 per cent of the maximum intensity for this procedure; lower values may lead to errors because of peak asymmetry arising from incomplete resolution of, say, $K\alpha_1$ and $K\alpha_2$ emissions, or from interferences due to adjacent spectral lines.

On computer-automated instruments it may be convenient to use 'peak-seeking' procedures which make a series of intensity measurements across the approximate position of the peak and compute the position of the true peak centroid. To avoid problems arising from mechanical backlash the measurements must always be made consecutively in the same direction across the peak, and the spectrometer must be backed-off and reapproached from the same direction when set to the peak. 'Peak-seeks' can be made in a few seconds and in some procedures are used routinely whenever the spectrometer setting is changed. However, they should not be used to determine the position of the peak centroid for low-intensity emissions – in such cases poor statistical precision can lead to erroneous spectrometer settings.

Selection of the crystal is usually straightforward, except for the cases of those elements which can be determined with equal facility in the low-angle range of one crystal or the overlapping high-angle range of another (e.g. in the overlapping ranges of LiF/PET, PET/RAP or PET/TlAP). In such cases the high-angle ranges are generally preferred because of superior resolution, but this consideration may be outweighed by questions of relative reflection efficiency of the crystals concerned or, in multiple-spectrometer instruments, by the need to optimize total analytical time.

It may also be necessary on occasion to select a less efficient crystal and sacrifice some intensity in order to improve resolution and hence reduce the magnitude of interference effects. Thus the interference of Ti $K\beta$ with Ba $L\beta_1$, described above, can be reduced significantly by using a LiF_{220} crystal instead of LiF_{200}.

Because of its higher intensities, the coarse collimator (or wide slit) is normally used for all applications unless the better resolution of the fine collimator (narrow slit) offers material advantage – which is often the case in trace-element analysis. Similarly, reflections are normally measured in the first order, but there may be cases where the better resolution attainable in higher orders compensates for the necessarily lower intensities. It should be noted that this latter stratagem may result in difficulties, however, on instruments that are equipped with $\sin \theta$ potentiometers and crystal attenuators to provide automatic compensation for the effect of variable X-ray photon energy, unless additional attentuation for high orders is available.

CHOICE OF DETECTOR

In XRF analysis, most spectrometers provide for the use of either a gas flow proportional detector or a scintillation detector (or both, operated in tandem). The former may be used for wavelengths greater than about

SOURCES OF ERROR IN QUANTITATIVE ANALYSIS

1.4 Å and the latter for wavelengths less than about 3 Å. Choice in the 1.4–3 Å range depends primarily on performance and may require experimental calibration; the flow counter is usually superior over much of this range because of its better pulse amplitude resolution. In some cases, particularly in the lower part of the range, there may be advantages in improving total intensity by using the two detectors in tandem, though in such cases it must be remembered that the overall pulse amplitude resolution will be that of the inferior (scintillation) detector.

If light elements (e.g. Na or Mg) are to be determined, better results will be obtained from a thin-window gas flow detector which, however, will require more frequent maintenance (and hence more frequent performance checks).

Microprobes do not normally use scintillation detectors, but some choice between sealed gas, gas flow or thin-window gas flow detectors is usually available, with the latter again offering advantages for light-element analysis – or indeed being mandatory for the low-Z range of the microprobe, which is more extensive than that for XRF. However, the choice of detectors available for a specific application may be constrained by the spectrometer configuration, since detector interchange is more difficult and often cannot be effected during the course of an analysis.

EXCITATION CONDITIONS

Electron beam (voltage and current) or primary tube (anode, tube voltage and current) conditions must be selected to yield efficient excitation of the analysis element and optimum emission intensities. Low count rates are generally undesirable, but of course are often unavoidable, particularly in trace-element analysis.

Excessively high count rates should also be avoided, both because of dead time and pulse amplitude shift problems and because of excessive load on the primary tube (XRF) or the risk of unnecessary sample damage (EPA). In determining XRF tube settings, it should be recalled that most of the primary beam energy is contained in the continuum, and tube voltage should therefore be adjusted, within reasonably broad limits, to ensure that maximum continuum intensity lies just short of the appropriate absorption edge. This factor is often not highly critical, and a single compromise setting can usually be chosen to serve for several different elements in order to avoid unnecessarily frequent adjustments to the tube power supplies.

Proper choice of tube anode has considerable effect on XRF sensitivities, but tube interchange during routine analysis is usually not practical (multiple-anode tubes have been developed and are available commercially for some instruments; for various reasons, however, they are not widely used). Selection of a single anode to be used for the determination of several different elements again often involves some compromise. For best results it is common practice to group the elements best determined with each of several different anodes and to run the samples in batches at times when different tubes are installed. Thus in geochemical work, major-element analyses (normally involving only elements lighter than Fe) may be run with a Cr-anode tube. Trace elements are subsequently run in groups appropriate to each of two or more additional tubes, e.g. Mo and Au anodes. Since tube life is materially enhanced if tube changes are kept to a minimum, this demands good organization of the XRF laboratory if the final analyses are to be produced in reasonable overall time.

PULSE AMPLITUDE DISCRIMINATOR SETTINGS

For each analytical wavelength, appropriate settings of detector voltage, amplifier gain and discriminator baseline and window width have to be established. In general these must be determined experimentally, making due allowance for the possible effects of the anticipated variation in X-ray intensity on the pulse amplitude distribution. Proper adjustment of the discrimination system is particularly critical in trace-element analysis, when it is essential to keep background intensities as low as possible.

Most analysts prefer to maintain detector voltage at a constant level yielding optimum amplitude resolution coupled with minimum count rate-dependent amplitude shifts. In principle, either amplifier gain or window baseline/width can then be adjusted for each characteristic emission to bring the desired amplitude distribution into the range of the discriminator window.

This is achieved very simply on those XRF instruments that use goniometer-controlled θ or $\sin \theta$ potentiometers to make automatic adjustments to the amplifier gain as the spectrometer moves through its X-ray energy range, thus maintaining the pulse amplitude distribution within a fixed discriminator window. Such systems need only to be set up for one element and, at least in principle, the emissions from all other elements will be automatically maintained within the fixed window (providing the latter is set wide enough to allow for poorer detector resolution at long wavelengths/low energies). Even so, provision should be made for 'fine-tuning' the amplifier gain and/or window settings for each individual element – for example if Mg Kα intensities are measured using a TlAP crystal, an asymmetric window is generally used to discriminate against the TlM spectrum pulses produced by crystal fluorescence; the same window is seldom appropriate for other elements measured in the same sample.

Whether automatic compensation facilities are available or not, it is usually necessary to undertake a thorough experimental study, prior to analysis, to determine the settings to be used or to check that the automatic settings are appropriate. In particular, it is important to ensure

SOURCES OF ERROR IN QUANTITATIVE ANALYSIS

that the settings used are sufficiently flexible to cope with count rate-dependent amplitude shifts over the complete range of sample and standard compositions likely to be encountered. This may be critical in trace-element analysis, in which calibration standards are likely to yield much higher intensities than many of the analytical samples.

7.3.2 Sampling and sample preparation errors

It is obviously essential that the sample analysed should be truly typical of the material which it purports to represent, at least within properly understood statistical limits. XRF analyses are commonly performed on small samples, r̃   ̃ in size from a few milligrams to several grams, and microprobe ana   e actually made on even smaller samples – possibly as little as abo ^{15} g. The problem of ensuring that such minute quantities are representative is a critical one, but is beyond the scope of this discussion. It is, however, necessary to consider whether that portion of an analytical sample that actually emits measured X-rays is itself representative of the whole analytical sample. It is easy to overlook this problem, particularly in XRF analysis in which the sample appears to be relatively large: the sample area irradiated by the primary beam (and from which emitted X-rays are measured) is typically of the order of 2–10 cm^2 and it is not so apparent that the depth of penetration of the primary beam is normally only a few micrometres, so that the total analysis volume is usually less than 0.1 cm^3.

Since it is clearly essential that the calibration standards and the analytical samples be presented to the spectrometer in identical fashion, then physical or chemical heterogeneity (of either samples or standards) is a potential source of error which the analyst must consider in deciding on appropriate methods of sample preparation. One of the major advantages of XRF methods is their ability to handle samples prepared in any of many different ways; the analyst must be careful to choose the technique most appropriate to each problem.

If the sample is inherently homogeneous at the scale of analysis, then there is clearly no problem in this respect. Bulk single-phase metal castings can simply be machined to discs of suitable size, with flat analytical surfaces prepared by milling or surface grinding. Filings or powdered samples (e.g. pure mineral concentrates) can be compressed into durable pellets (but be careful – 'durable' pellets may disintegrate all too easily, particularly when loaded into a vacuum chamber, and cleaning the fragments out of the spectrometer can be both difficult and time-consuming). Liquid samples usually give no problem provided that they are stable under primary beam irradiation and that helium-path facilities are available if they are to be analysed for light elements. 'Thin-film' samples (e.g. those collected and directly analysed on filter paper or similar substrates, or perhaps samples thinly plated on a metal substrate) also give little trouble in this respect because the volume analysed is more or less equal to the total volume of the sample.

The greatest difficulties are encountered in the analysis of powdered multiphase samples, e.g. of rocks or soils. Experiment with such samples will soon show that emitted X-ray intensities depend not only on composition but also on particle size for all except very fine or very coarse grained samples. The effect will generally be found to be most severe when analysing light elements such as Na or Mg – it is instructive to crush a multi-element sample, screen it into a series of size fractions, and measure intensities emitted by a single light element from samples of each size fraction. (If the sample is inhomogeneous some of the differences observed may be attributable to differences in the crushing behaviour of minerals of different composition – but extensions to the experiment can easily be designed to eliminate this complication.)

These effects are known as *particle-size* or *microabsorption* effects, and they arise in the following way. A sample will be effectively homogeneous only if each ray in both primary and fluorescent spectra traverses an 'average' section of the sample. If the various phases present in the sample differ significantly in their physical effects on the X-rays passing through them (most particularly in their absorption effects), an 'average' path implies a path that traverses many different grains so that the physical effects are averaged out. Since effective primary-beam penetration depths in XRF are so shallow (typically of the order of 1–10 μm), it follows that multiphase samples must be finely ground to micrometre or sub-micrometre particles if the effects of heterogeneity are to be avoided. The critical parameter D is equal to dl_{diff}, where d is the average linear dimension of the particles and l_{diff} is the difference between the *linear* absorption coefficient of any one particular phase and the average linear absorption coefficient of all particles in the sample. The sample will be effectively homogeneous only if the absolute value of D is very much less than unity for all phases present in the sample.

It is evident that as the linear absorption coefficient for a particular type of particle approaches the average linear absorption coefficient for all particles, then larger values of d can be tolerated. In other words, if there are substantial differences in linear absorption coefficient between the various phases present in a sample, then the microabsorption effect will be severe and the sample will need to be very finely ground to eliminate it. Conversely, if all phases present in the sample have linear absorption coefficients (NB *not* mass absorption coefficients) close to the average for the sample as a whole – as in single-phase samples – then the microabsorption effect disappears.

Because linear absorption coefficients vary approximately as the cube of wavelength (between absorption edges), measurement of the long wavelengths characteristic of light elements demands much finer particle sizes than are needed for shorter wavelengths. For the determination of elements such as Na and Mg in rock samples, for example, sub-micron particle sizes are required, and these are difficult or impossible to achieve with conventional grinding equipment.

Since the particle-size effect cannot then be eliminated by fine grinding, it is common practice instead to homogenize the sample. This could be achieved by dissolving the sample chemically and analysing it in, say, aqueous solution. However, this procedure is time-consuming and often of little practical value, since the same light elements require vacuum conditions for XRF analysis. Instead a rock sample is usually dissolved by fusing it with an appropriate flux and casting the melt into a homogeneous glass disc ('button'), which is then used as the analysis sample; it follows that the calibration standards are usually prepared in the same way.

This procedure also helps to cope with the problem of inter-element effects (see Ch. 8), so that major-element rock and mineral analysis involving light elements is usually performed on glass-fused samples, with sodium or lithium metaborate or tetraborate being the most commonly used flux materials. However, this procedure necessarily involves dilution of the sample and hence loss of sensitivity for trace elements; it is also time-consuming and presents opportunities for weighing errors and for contamination which can be quite serious at the trace- or minor-element level. Pressed powder pellets are thus preferred for the determination of trace elements having characteristic emission wavelengths of less than about 3 Å; for such wavelengths the necessary particle sizes (1–5 μm) can usually be achieved by fine grinding. For longer wavelengths there is no simple solution; the analyst must consider the trade-off of particle-size effects on one hand against loss of sensitivity and possible contamination effects on the other, in the context of each specific analytical problem.

If the pressed powder sample requires dilution (e.g. if use of a binder is found to be necessary to ensure a durable pellet), it is best to choose a diluent whose absorption coefficients are as close as possible to the average coefficients of the sample in order not to introduce additional microabsorption effects. Powdered quartz is a suitable diluent for most silicate samples, and its presence also assists the fine-grinding process because of its hardness. Low atomic number organic diluents, on the other hand, may accentuate microabsorption effects and may also inhibit fine grinding.

If potential microabsorption effects are anticipated, it is advisable to test their magnitude by measuring the effects of repeated regrinding on emission intensities. The latter will usually be found to increase or decrease as particle size is reduced, until effective homogeneity is established (depending, in a particular instance, on whether the offending particles have absorption coefficients greater or less than the average absorption coefficients). It should be noted, however, that stabilization of emission intensities with progressive regrinding is not necessarily indicative of homogeneity. It may in fact be due to cessation of effective grinding because of compaction of the sample in the grinding mill.

Fusion procedures will be further discussed in the section dealing with inter-element effects in XRF analysis (Ch. 8).

In electron microprobe analysis, sample heterogeneity is often the subject of the study rather than just a potential source of error (except in cases where long analysis times and the possibility of sample damage force the analyst to move the beam position during the analysis, so that successive elements are not determined in the same sample volume). If necessary, microprobe samples can also be homogenized prior to analysis by fusion procedures analogous to those used in XRF, but even with this procedure homogeneity at the micron scale is very difficult to achieve and XRF or other methods are often superior under these conditions.

Sample preparation for the microprobe is instead largely confined to preparing a suitable surface for analysis and ensuring that a grounding path is provided to prevent the sample from becoming charged by absorbed electrons and thus deflecting the beam.

Surface preparation usually involves grinding and polishing a flat, relief-free surface to ensure uniform within-sample path lengths for X-rays generated below the sample surface. Irregular surface topography results in irregular absorption and fluorescence effects (see Ch. 9), particularly for long-wavelength emissions, and must be avoided as far as possible. The range of materials frequently analysed with the microprobe is substantial – minerals, alloys, bones, teeth, dental amalgams, glasses, ceramics, chemicals, dust particles and many others – and there is no standard technique for sample preparation in even one of these groups. Instead it remains the analyst's responsibility to ensure that the finished quality of the sample surface is as high as available experience and facilities permit, and particularly to guard against alteration or contamination (e.g. by particles of fine-grained abrasives used for grinding and polishing; alumina and chromium oxide can be particularly troublesome in this respect).

Unless the sample is a good conductor of both heat and electricity, a conductive coating is usually applied to its surface to prevent or minimize charging and/or thermal damage. The usual procedure is to 'flash' a thin layer of a suitable conducting material on to the sample surface, using a vacuum evaporator. The most widely used coating medium is carbon, although metals such as copper, aluminium and even gold have also been used. A carbon layer about 200 Å thick provides adequate thermal and electrical conduction unless electron beam currents and/or energies are unusually high. It will necessarily absorb some energy from the incident electron beam (about 5 per cent for 10 keV electrons), and also attenuate the emitted X-rays (about 0.3 per cent for Al Kα) – neither of these effects is severe, providing samples and standards are all coated *with uniform carbon thicknesses*. If variable coating thicknesses are used, absorption differences and differences in the effectiveness of electrical conduction may produce calibration errors of several per cent.

This effect can often be suspected whenever groups of analyses made on samples coated in a single batch show a tendency for consistently high or low analysis totals. It can be avoided by coating samples and standards in the same batch, but this is often not convenient – particularly when 'library' standards are used and have to be repolished whenever they are to be recoated. It is better for the operator to spend some time and effort on learning to apply coatings of consistent quality and thickness.

7.3.3 *Instrumental errors*

There are two major groups of potential errors associated with the instrumentation used in X-ray spectrometric analysis. The first of these includes those errors that are necessary consequences of the operating principles of various components of the spectrometer, e.g. detector/counter dead time and pulse amplitude shift. These cannot be eliminated, but they can be compensated or the extent of their effects can be measured and corrected if the analyst is prepared to be sufficiently thorough.

The most serious potential source of error in this group is probably that of detector/counter dead time. It will be particularly severe if excessively high count rates are used, or if there is considerable variation in count rates between the various standards and samples. Dead times must be measured periodically, and correction procedures must be established and tested. These may involve automatic compensation in the counting circuitry, or they may require correction of the raw count rate data by calculation (or on-line computation); either way the correction procedure depends, directly or indirectly, on experimental measurement of the extent of the dead-time effect. It should be noted that most correction procedures assume that dead time is a non-extendable phenomenon and hence that a simple linear correction function can be applied. However, this assumption may not be valid at high count rates, and is certainly not justified in the case of the long dead times of energy-dispersive systems. The correction procedure adopted must always be tested over the full range of count rates encountered in analytical situations.

In principle, dead time of a detector is dependent on X-ray photon energy and will therefore vary from one element to another. In practice, the system dead time is as much or more a function of the amplifying, counting and pulse-shaping circuitry as it is of the detector itself. The variation with X-ray energy is relatively small (except in solid-state systems, in which the energy range of X-rays registered by a single detector may be very wide), and a single value for the dead time can often be used successfully over an appreciable range of wavelengths. It is wise, however, to check this assumption from time to time by experimental dead-time measurements made over the full analytical wavelength range.

Dead time is also sensitive to other parameters such as detector voltage or amplifier gain (which may affect pulse shape). This emphasizes the need for periodic checking of the correction procedure, and for complete recalibration following any major change to the instrumental operating conditions (e.g. overhaul of the detector, or even the fitting of a new flow gas cylinder, which may result in significant changes in the flow gas composition and density).

Despite these words of caution, dead-time effects are not normally troublesome unless unusually extreme variations in count rate are encountered. Simple correction procedures are usually adequate for count rates up to about 50 000 cps for scintillation detectors, and 25–35 000 cps for gas-filled proportional detectors. At higher intensities, precautionary checks are always desirable. It is more difficult to quote a general figure for energy-dispersive systems, in which dead-time effects are more strongly dependent on X-ray photon energies and on amplifier characteristics, but careful checks are always advisable whenever *total* count rates approach or exceed 10 000 cps.

The second major group of instrumental errors includes those arising from the fact that XRS analysis is a comparative technique, in which intensities measured on the samples are compared with those measured on calibration standards. The comparative measurements are normally made at different times, often with intervening adjustments to various components of the analytical instrument, so that they cannot be made under absolutely identical conditions. No electronic system is entirely free of short- or long-term instability, so that there will always be a time-dependent drift in the measurement system. Mechanical components of a wavelength-dispersive spectrometer, no matter how carefully manufactured and maintained, will always be subject to backlash, which means that a goniometer, once moved from a peak, cannot be reset to the same peak with absolute precision. As might be expected, such difficulties tend to increase with the age of the instrument.

These problems also cannot be eliminated, though their magnitude can be controlled by using electronic and mechanical equipment of the highest quality, and by careful maintenance. With good, well maintained equipment they should not result in aggregate errors higher than about 0.1–0.2 per cent; for example, modern X-ray generators have their voltage and current outputs stabilized to better than 0.05 per cent

SOURCES OF ERROR IN QUANTITATIVE ANALYSIS

fluctuations over periods of an hour or more. However, it is still necessary to keep a constant watch to ensure that the *aggregate* effects are contained within acceptable limits.

The most reliable check procedure to adopt is to return one or more *monitor* standards to the system from time to time and observe any significant differences in measured X-ray intensities (a variant often used in microprobe analysis is to insert a Faraday cage in place of the sample for periodic measurement of effective beam current, although this provides no check on the mechanical reproducibilities of the spectrometers). Intensities measured on intervening samples can then be corrected, if necessary, by interpolation from a *drift curve* – a curve of monitor intensities plotted against time. If excessive drift is observed, its cause must obviously be located and corrected.

If drift is monitored in terms of X-ray intensities, it should not be forgotten that they are subject to statistical error (Sec. 7.3.4), and they may give a misleading impression if not properly evaluated.

7.3.4 Measurement and statistical errors

Since X-ray emission is a quasi-random process, any measurement of X-ray intensity made over a less than infinite time period represents a *sample* and will have an associated statistical error. However, the probable extent of this statistical error can readily be evaluated and, to at least a certain extent, conditions of analysis can be adjusted to ensure that the error is contained within selected limits. This evaluation also requires preliminary experimental investigation to determine the approximate range of intensities (peak and background) that will be encountered during the analysis – not forgetting that analysis involves both samples and standards, and the net contribution of statistical error ('counting error') to the final analysis involves errors in the measurement of both sample and standard intensities.

Quantitative XRF or EPA analysis involves making a series of X-ray intensity measurements to collect data from which analytical information is derived. It is necessary, therefore, to consider the magnitudes and kinds of error that may be associated with the actual measurements themselves in order to assess the quality of the final result.

Broadly speaking, errors may be considered as being of two main types, viz. *random errors* and *systematic errors*. To these may be added a third group, *'wild' errors* (including, for example, mistakes – which are not quite the same as errors); this group is less amenable to systematic evaluation and will not be further considered in this discussion, even though it may well contribute to significant error and should not be overlooked by the analyst.

(a) *Random errors* are always involved when natural physical phenomena are measured, though they may arise collectively from many different causes. Their practical effect is that a series of repeated measurements of the same phenomenon will not, in general, agree exactly but instead will be distributed about some mean value (Fig. 7.1). The distribution is usually assumed to be a *normal*, or *Gaussian*, distribution but may not necessarily be so. The mean of a series of results with random errors will only correspond to the true result if the distribution is not *skewed* and in the absence of *systematic errors* (see below). These possibilities must be evaluated before it is assumed that measurement errors can be treated entirely as random errors by the methods described below. Random errors arise in X-ray spectrometric analysis from such diverse sources as operator inconsistency (e.g. in repeated 'peaking' of a spectrometer), instrumental instability of certain kinds (mechanical imprecision, short-term fluctuations in power supply outputs etc.) and, most importantly *but not exclusively*, the random nature of X-ray generation and emission. With proper operation of a well designed and well maintained instrument, the first two of these can be minimized – in most cases to a level of the order of 0.1 per cent relative or better over analytical periods of an hour or so (but it should never be assumed that they are no worse than this). Errors associated with X-ray generation cannot be controlled in the same way; instead their magnitude must be

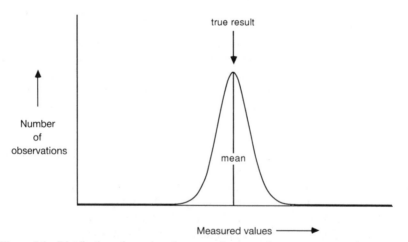

Figure 7.1 Distribution of a series of measured values of some parameter about a *mean*, which in the absence of any contrary indications is usually assumed to be the *true* value of the parameter (the 'true result'). The distribution may be symmetrical, as in this example, or it may be *skewed*. The probability that a single, random measurement will coincide with the mean depends on the form of the distribution, which is often assumed to be Gaussian, or 'normal', but need not be so (see Appendix B).

SOURCES OF ERROR IN QUANTITATIVE ANALYSIS

evaluated and an analytical strategy must be devised to contain them within acceptable limits. Because they cannot be physically controlled, 'counting' errors are usually the most severe source of random error, and will be the subject of most of the following discussion.

(b) *Systematic errors* collectively cause the mean of a series of observations, regardless of their random errors, to be systematically different from the true result (Fig. 7.2). Many factors can contribute to systematic error, but in X-ray spectrometry the most significant are usually consequences of the physical forms of samples and standards, or of variations in inter-element or matrix effects arising from differences in composition between or within samples and standards. Methods for treating these errors are considered in Chapters 8 and 9.

COUNTING STATISTICS

The following discussion will concentrate on the treatment of random errors associated with the generation and detection of X-rays, but it is again emphasized that comprehensive investigation of random errors in XRS should also include evaluations of other sources of error, such as sampling error, machine and operator reproducibility etc. Such evaluations are common to the assessment of error in any analytical procedure.

The basic problem with counting statistics is that X-ray generation is a random, not a uniform, process. If one examines the time-dependent output of an X-ray detector (Fig. 7.3), it is immediately apparent

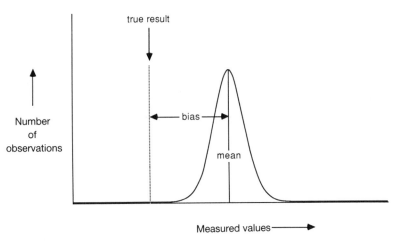

Figure 7.2 Systematic errors result in an offset, or *bias*, between the true result and the mean of the random-measurement distribution. Estimation of the probability that a single measurement will coincide with the true result then requires evaluation of *both* systematic *and* random errors.

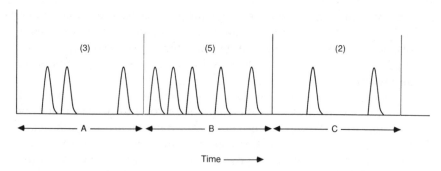

Figure 7.3 Since X-ray generation is a random rather than a uniform time-dependent process, the result of any intensity measurement will depend on the actual time interval over which it is made. In this example, measurements made over the equal time intervals A, B and C would be 3, 5 and 2 counts respectively. The extent of random error can be controlled by extending the counting time to reduce statistical fluctuations to a practical level, but it cannot be totally eliminated in less-than-infinite counting intervals.

(particularly at low count rates) that the pulses are not produced at a uniform rate. This means that a finite-time measurement of the count rate (i.e. the number of counts detected in a specified time interval) will give a result that depends on the time interval actually sampled; in the extreme example of Figure 7.3, the equal time intervals A, B and C would each yield a different measured count rate. Hence the true intensity of an X-ray line emitted from a sample under analytical conditions can only be precisely established by measurements made over an infinite time. Since this is not practicable, intensity must be estimated from a finite-time sample, and the possible extent of sampling error must then be evaluated by statistical techniques. Alternatively, given some preliminary information on expected count rates, an operator can use the same techniques to determine, in advance, the counting conditions required to yield a chosen level of precision (at least to the extent that precision is a function of counting error).

Although one feels intuitively (and is supported to some extent by theory) that long counting times will improve the 'reliability' of the results, it is also usually important to keep counting times as short as possible, for two main reasons:

(a) Although most modern instruments are far more stable then their earlier counterparts, none is perfectly so. Analytical conditions may change appreciably with time (the phenomenon of 'drift'), and the probability of significant changes between one counting interval and the next increases with increasing counting time. Reduction in the magnitude of counting errors is thus ultimately outweighed by increases in other time-dependent sources of random error.

(b) Microprobes and X-ray spectrographs are expensive to operate and maintain, and there is usually considerable demand for their facilities. It is inefficient and inconvenient to waste machine time on the collection of superfluous data. Not only should counting times be kept to the minimum necessary for a chosen level of precision, but that level of precision should also be carefully evaluated as an essential element of analytical strategy. There is no point in pursuing high standards of counting precision if systematic and sampling errors cannot be evaluated to the same standards.

A brief summary of fundamental statistical concepts relevant to the estimation of random counting error is presented in Appendix B. Further details are given by Jenkins and de Vries (1969), Norrish and Chappell (1977) and Tertian and Claisse (1982).

ANALYTICAL STRATEGY – STATISTICAL ERRORS

It is apparent from these relationships that the analyst can readily assess the magnitudes of probable statistical error involved in a series of X-ray intensity measurements. Since these errors are functions of count rates and counting times, it is also possible to calculate optimizing parameters for increased efficiency in analysis – particularly when a long series of more or less repetitive analyses is to be undertaken. It is also possible to calculate, from information derived from standards of known composition, the statistical lower limits of detection for samples of generally similar composition; i.e. in the case of each element, that concentration that yields the lowest measurable count rate that is statistically distinguishable from the background.

Determination of analytical conditions for a chosen precision Calculation of counting times required to yield a chosen level of random-error precision is relatively simple if preliminary experimental measurements of the appropriate count rates can be made (and if it is established that random errors can indeed be attributed wholly or almost wholly to counting errors). In the simplest case of a single measurement it is necessary only to calculate the number of counts required for the desired precision, and then to determine from the experimental data the counting time necessary to accumulate that number of counts. Even simpler, the analysis can be performed in the *fixed-count* mode (see below), in which the time taken to collect a preset number of counts is measured and used to calculate the count rate.

Suppose that a particular sample is found to yield, under the conditions of analysis, a count rate of approximately 500 cps, and it is desired to measure the count rate with sufficient precision to ensure that the relative standard deviation of the random counting error is less than 2 per cent.

Since

$$\varepsilon = 100 / \sqrt{N} \% \qquad \text{(Eqn B.9)}$$

then

$$N = (100 / 2)^2$$
$$= 2500 \text{ counts}$$

Thus the count time required to yield a relative standard deviation of 2 per cent would be (2500/500) = 5 seconds. The desired precision can be obtained in fixed-count mode, without any preliminary intensity measurements, by determining the time required to accumulate 2500 counts.

A similar calculation shows, for this example, that the count time required for a precision of 1 per cent (rsd) would be (10 000/500) = 25 seconds. In general, an n-fold improvement in precision of a single measurement requires the counting time to be extended by a factor of n^2, which establishes the important point that precision cannot be indefinitely improved simply by extending the counting time. At some level the count times become impracticably long, and/or other sources of error such as machine drift become predominant.

The calculation procedures can readily be extended to cases entailing two or more measurements; these usually require optimization of the effects of different count rates.

Counting strategy The simple example quoted above suggests that, since counting error depends on the total number of counts collected, it would be advantageous just to terminate counting when a sufficient number of counts has been collected to yield a chosen level of precision (*fixed-count* analysis), rather than at the end of a preset time interval (*fixed-time* analysis). For single measurements this may well be the case, at least in purely statistical terms, but it may be outweighed in other cases by various practical considerations. For example, electron microprobes are usually equipped with multiple spectrometers controlled by a single master timer, so that it is often not practicable to specify appropriate fixed counts for each of several measurements being made simultaneously (though it is by no means impossible, particularly with computer-controlled systems).

Intensity parameters such as net peak intensity (peak − background) or an intensity ratio (sample : standard) typically require more than one measurement, each with an associated random counting error. It is then necessary to consider the *propagation of errors* in order to derive an appropriate counting strategy, either to minimize the aggregate error obtained in a specified counting time or to determine the counting times

SOURCES OF ERROR IN QUANTITATIVE ANALYSIS

required to attain a specified error level. (It should never be forgotten, however, that random counting errors are not the only potential sources of error.)

The principles of random error propagation are briefly reviewed in Appendix B, in which it is shown that

(a) if C_p and C_b are the count rates measured on an analytical peak and its background, with counting times of t_p and t_b respectively, then the relative standard deviation of the estimated net count rate $(C_p - C_b)$ is given by

$$\varepsilon_{net} = \frac{\sqrt{\left[\left(\frac{C_p}{t_p}\right) + \left(\frac{C_b}{t_b}\right)\right]}}{C_p - C_b} \quad \text{(Eqn B.15)}$$

(b) if C_{sam} and C_{std} are count rates measured on a sample and a standard, with counting times of t_{sam} and t_{std} respectively, then the relative standard deviation of the intensity ratio $(C_{sam} : C_{std})$ is given by

$$\varepsilon_{ratio} = \sqrt{\left(\frac{1}{C_{sam} t_{sam}} + \frac{1}{C_{std} t_{std}}\right)} \quad \text{(Eqn B.16)}$$

In both cases the relative standard deviation of the net estimate depends on the count *rates* and the counting *times* for each component, as would be expected. It is appropriate to examine the implications of Equations B.15 and B.16 for three commonly used counting strategies, in terms of the component count rates C_1 ($=C_p$ or C_{sam}) and C_2 ($=C_b$ or C_{std}) and count times t_1 ($=t_p$ or t_{sam}) and t_2 ($=t_b$ or t_{std}). In each case the total counting time T is the sum of t_1 and t_2.

(a) *Fixed time*: in this case $t_1 = t_2 = T/2$. For a *difference* estimate, such as $(C_p - C_b)$,

$$\varepsilon_{net} = \frac{\sqrt{\left(\frac{C_p + C_b}{t_p}\right)}}{C_p - C_b}$$

$$= \frac{\sqrt{\left(\frac{2(C_p + C_b)}{T}\right)}}{C_p - C_b}$$

$$= \frac{\sqrt{2}}{\sqrt{T}} \frac{\sqrt{(C_p + C_b)}}{C_p - C_b} \quad (7.2)$$

For a *ratio* estimate, such as $(C_{sam} : C_{std})$,

$$\varepsilon_{ratio} = \frac{1}{t_{sam}} \sqrt{\left(\frac{1}{C_{sam}} + \frac{1}{C_{std}}\right)}$$

$$= \frac{\sqrt{2}}{\sqrt{T}} \sqrt{\left(\frac{1}{C_{sam}} + \frac{1}{C_{std}}\right)} \tag{7.3}$$

(b) *Fixed count*: in this case, $N_1 = N_2$ (where N is the total number of counts accumulated in each estimate). Since $N_1 = C_1 \times t_1$ and $N_2 = C_2 \times t_2$, then

$$t_1 / t_2 = C_2 / C_1$$

For a *difference* estimate, substitution in Equation B.15 then yields

$$\varepsilon_{net} = \frac{1}{\sqrt{T}} \cdot \frac{\sqrt{\left[(C_p + C_b) \cdot \left(\frac{C_p}{C_b} + \frac{C_b}{C_p}\right)\right]}}{C_p - C_b} \tag{7.4}$$

For a ratio estimate, such as $C_{sam} : C_{std}$, substitution in Equation B.16 yields

$$\varepsilon_{ratio} = \frac{\sqrt{2}}{\sqrt{T}} \cdot \sqrt{\left[\frac{(C_p + C_b)}{C_p \times C_b}\right]} \tag{7.5}$$

(c) *Optimized time*: Optimal precision (i.e. a minimum relative standard deviation for a given total counting time T) requires that T be correctly apportioned between the two measurements. It can be shown that the optimal condition is met when

$$t_1 / t_2 = \sqrt{(C_1 / C_2)}$$

for a *difference* estimate, when

$$\varepsilon_{net} = \frac{1}{\sqrt{T}} \cdot \frac{1}{\sqrt{C_p} - \sqrt{C_b}} \tag{7.6}$$

or

$$t_1 / t_2 = \sqrt{(C_2 / C_1)}$$

SOURCES OF ERROR IN QUANTITATIVE ANALYSIS

for a *ratio* estimate, when

$$\varepsilon_{ratio} = \frac{1}{\sqrt{T}} \cdot \frac{\sqrt{C_{sam}} + \sqrt{C_{std}}}{\sqrt{C_{sam}} \cdot \sqrt{C_{std}}} \qquad (7.7)$$

Examples

(a) *Difference*: suppose that it is necessary to determine the difference between two intensities C_1 and C_2, of approximately 10 000 cps and 500 cps respectively, with a relative standard deviation of 0.2 per cent (i.e. $\varepsilon = 0.002$). The total counting time T may be determined by transposition and substitution in Equations 7.2, 7.4 and 7.6 for fixed-time, fixed-count and optimized-time strategies respectively; its apportionment between the counting times for measurement of C_1 and C_2 is then determined from the appropriate relationship between t_1 and t_2.

For fixed-time strategy:

$$T = \frac{2}{\varepsilon^2} \cdot \frac{C_1 + C_2}{(C_1 - C_2)^2} \qquad \text{(from Eqn 7.2)}$$

$$= \frac{2}{0.002^2} \cdot \frac{10\ 500}{9500^2}$$

$$= 58 \text{ seconds}$$

$$t_1 = t_2 = T/2 = 29 \text{ seconds}$$

For fixed-count strategy:

$$T = \frac{1}{\varepsilon^2} \cdot \frac{(C_1 + C_2) \cdot \left(\dfrac{C_1}{C_2} + \dfrac{C_2}{C_1}\right)}{(C_1 - C_2)^2} \qquad \text{(Eqn 7.4)}$$

$$= \frac{1}{0.002^2} \cdot \frac{10\ 500 \times (10\ 000\ /\ 500 + 500\ /\ 10\ 000)}{9500^2}$$

$$= 583 \text{ seconds}$$

$$t_1/t_2 = C_2/C_1 = 500\ /\ 10\ 000 = 0.05$$

$$t_1 = 28 \text{ seconds}; \; t_2 = 555 \text{ seconds}$$

For optimized time strategy:

$$T = \frac{1}{\varepsilon^2} \cdot \frac{1}{(\sqrt{C_1} - \sqrt{C_2})^2} \qquad \text{(Eqn 7.6)}$$

$$= \frac{1}{0.002^2} \cdot \frac{1}{(\sqrt{10\,000} - \sqrt{500})^2}$$

$$= 41 \text{ seconds}$$

$$t_1 / t_2 = \sqrt{(C_1 / C_2)} = 4.5$$

$$t_1 = 34 \text{ seconds}; \; t_2 = 7 \text{ seconds}$$

These examples show clearly that if there is a substantial difference between C_1 and C_2 (that is, if one is less than about 0.4 to 0.5 of the other), fixed-count strategies consume excessive time in accumulating the necessary counts at the lower count rate. If $C_1 = C_2$ they become equivalent to fixed-time strategies; otherwise they are always inappropriate for difference estimates. Optimized-time techniques are more efficient than fixed-time, particularly if there is a substantial difference between C_1 and C_2. However, the improvement is often not sufficient to warrant the additional experimental complexities. Thus fixed-time strategies are almost always chosen for routine difference estimates. With multiple-spectrometer instruments, they are usually the only practical choice.

(b) *Ratio estimates*: suppose that it is necessary to determine the ratio of two intensities C_1 and C_2, of approximately 5000 cps and 10 000 cps respectively, with a relative standard deviation of 0.2 per cent ($\varepsilon = 0.002$). From Equations 7.3, 7.5 and 7.7,

For fixed-time strategy:

$$T = \frac{2}{\varepsilon^2} \cdot \left(\frac{1}{C_1} + \frac{1}{C_2} \right)$$

$$= \frac{2}{0.002^2} \cdot \left(\frac{1}{5000} + \frac{1}{10\,000} \right)$$

$$= 150 \text{ seconds}$$

$$t_1 = t_2 = 75 \text{ seconds}$$

For fixed-count strategy:

$T = 150$ seconds (Eqn 7.5 is the same as Eqn 7.3)

SOURCES OF ERROR IN QUANTITATIVE ANALYSIS

$t_1 / t_2 = C_2 / C_1 = 2$

$t_1 = 100$ seconds; $t_2 = 50$ seconds

For optimized-time strategy:

$$T = \frac{1}{\varepsilon^2} \cdot \frac{(\sqrt{C_1} + \sqrt{C_2})^2}{C_1 \cdot C_2}$$

$$= \frac{1}{0.002^2} \cdot \frac{(\sqrt{5000} + \sqrt{10\,000})^2}{5000 \times 10\,000}$$

$= 145$ seconds

$t_1 / t_2 = \sqrt{(C_2 / C_1)} = 1.4$

$t_1 = 80$ seconds; $t_2 = 65$ seconds

In this case, fixed-time and fixed-count strategies require the same total counting time, but the fixed-count method may be less convenient for routine applications. Optimized time again gives only a marginal improvement unless one of the count rates is substantially lower than the other (C_1 / C_2 – or vice versa – is less than about 0.2). The reader should verify that if the count rates in this example are changed to 1000 cps and 10 000 cps the saving in total counting time is more significant.

Trace-element analysis At low concentrations, C_p will approach C_b and counting time should then obviously be apportioned equally between peak and background. Under these conditions the net counting error will increase rapidly as ($C_p - C_b$) approaches zero.

The lower limit of detection (LLD) is then often defined as the lowest concentration yielding C_p *significantly higher* than C_b, with the level of significance being defined in terms of standard deviation of the background count rate. 2σ or 3σ confidence levels are commonly adopted.

At the 3σ confidence level, the lower limit of detection is the case in which

$$N_p - N_b = 3\sigma_{N_b} \tag{7.8}$$

In terms of net count rate (cps),

$$C_{net} = 3\sigma_{N_b} / t_b$$

$$= 3 \sqrt{N_b} / t_b$$

$$= 3 \sqrt{(C_b \times t_b)} / t_b$$

$$= 3 \sqrt{(C_b / t_b)} \tag{7.9}$$

In practice it is necessary to make at least two measurements to determine peak and background intensities. Since

$$(\varepsilon_{net})^2 = (\varepsilon_p)^2 + (\varepsilon_b)^2 \qquad \text{(Eqn B.14)}$$
$$= 2(\varepsilon_b)^2$$

then $\qquad \varepsilon_{net} = \sqrt{2} \times \varepsilon_b$

and the random error is thus increased by $\sqrt{2}$.
Hence at the 3σ limit of detection,

$$C_{net} = 3 \times \sqrt{2} \times \sqrt{(C_b / t_b)}$$
$$= 4.24 \times \sqrt{(C_b / t_b)} \qquad (7.10)$$

In terms of concentrations,

$$\text{LLD} = \frac{4.24}{m} \times \sqrt{(C_b / t_b)} \qquad (7.11)$$

where m is the sensitivity (counts per second per unit concentration) for the element determined, under the analytical conditions employed.
Alternatively, since $T = 2t_b$, the expression may be given as

$$\text{LLD} = \frac{6}{m} \times \sqrt{(C_b / T_b)} \qquad (7.12)$$

It must be emphasized that the LLD calculated from these expressions is a *statistical lower limit of detection*, and not necessarily a *practical lower limit of determination* – in fact the error at the statistical lower limit is infinite. For this reason the practical lower limit of determination is usually taken to be at least twice the statistical lower limit of detection, i.e. a concentration equivalent to at least six standard deviations of the background count.

Sensitivities are found in practice to vary in response to several factors: excitation efficiencies, crystal efficiencies, detector response, matrix compositions etc. As a very broad generalization, they tend to be better in the intermediate wavelength region (1–3 Å approximately). At shorter wavelengths excitation is relatively inefficient because of the problems of attaining sufficient overvoltage (microprobe) or short-wavelength continuum intensities (XRF), and at longer wavelengths absorption effects in the X-ray path (including detector window absorption) reduce the measurable intensities. Sensitivities also tend to be poorer in the low Bragg angle region of a particular crystal spectrometer, where

SOURCES OF ERROR IN QUANTITATIVE ANALYSIS

background is considerably enhanced by scattered tube X-rays (XRF) or electrons (microprobe).

It should also be noted that the term *sensitivity* may be applied either to the ability of a technique to detect low concentrations or to the related concept of slope of the calibration curve (Fig. 7.4). In the latter sense, sensitivity may vary as a function of composition if the calibration curve is not linear. In such cases, values of m (cps/%) determined at high concentrations may not be appropriate for calculation of the lower limit of detection.

It must also be emphasized again that lower limits of detection or determination estimated in this way refer only to random statistical error, and do not take systematic errors into account. In particular, non-statistical errors in background measurement may be of much greater magnitude, particularly in cases such as:

(a) non-linear background profiles, where the simple measurement of off-peak background(s) is inadequate, or
(b) interference by overlapping spectral lines due to other elements (even very weak emissions from major elements may have severe effects on the accuracy of measurement of trace-element emissions).

These problems are further considered in Chapters 8 and 9.

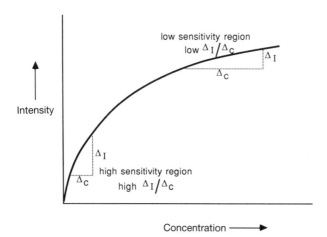

Figure 7.4 The sensitivity of a comparative analytical technique such as X-ray spectrometry depends on the slope of the calibration curve for the chosen analytical conditions. If the curve is not linear, sensitivity will vary with concentration, and neither the accuracy at low concentrations nor the absolute detection limits can be reliably estimated from emission intensities measured on high-concentration calibration standards. It is important to establish the form of a calibration curve over the complete range of concentrations for which it will be used.

7.3.5 Composition-dependent ('matrix') effects

It has already been noted that the assumption of a simple proportional relationship between the intensity of a characteristic spectral line and the concentration of the element emitting it serves only as a first approximation. As long as the analysed sample resembles the calibration standard(s) fairly closely in composition, the approximation is often sufficient for practical purposes; however, this constraint is a severe limitation on the potential versatility of X-ray spectrometric techniques, and it is necessary to explore the principal composition-dependent sources of error, and some of the methods that can be used to compensate for them.

Some of these matrix effects are common to both XRF and EPA, but there are also some fundamental differences warranting independent treatment. Matrix errors in XRF are considered in more detail in Chapter 8, and those encountered in EPA are discussed in Chapter 9.

8 Composition-dependent errors in X-ray fluorescence analysis

Although they are to some extent interrelated, the major composition-dependent sources of error encountered in XRF analysis can be conveniently grouped for discussion as follows:
(a) interference effects
   (i) background interference
   (ii) interference from characteristic spectra
(b) composition-dependent attenuation/enhancement effects

8.1 Interference

8.1.1 Background interference

The intensity of a characteristic line measured for analytical purposes always includes at least some contribution from the non-characteristic background on which it is superimposed. This effect is usually less severe in XRF than in EPA, and in major-element analysis it is often a negligible source of error. However, it must always be properly assessed, and in minor- or trace-element XRF analysis it will nearly always require compensation.

'Background' is a general term encompassing all radiation which is measured in the same wavelength/energy band as the analytical line but which does not originate as a characteristic emission from the analysed element. Interference of this type may arise from any or all of several possible sources: continuum from the primary tube scattered by the sample into the spectrometer; X-rays of other wavelengths scattered from the dispersing crystal rather than diffracted by it; fluorescent X-rays emitted from the dispersing crystal; background gamma-radiation etc. Some of these have the same wavelength/energy as the analytical line, and therefore cannot be separated from it by the dispersing system (either WDS or EDS). Others have different energies but are not isolated by the dispersing system because they do not follow the normal path through it, e.g. γ-rays, which may enter the detector directly, or X-rays of any wavelength scattered from the surface of the dispersing crystal directly into the detector.

The interference effect of background photons that differ appreciably in energy from the analytical line can be reduced considerably by pulse amplitude discrimination. Proper adjustment of the discrimination system is essential if maximum sensitivity is to be obtained in trace-element analysis; it will be recalled that the lower limit of either detection or determination is defined as the minimum concentration whose effective emission intensity can be significantly distinguished from the background at a specified confidence level, and that the 'significant distinction' is defined in terms of the standard deviation of the background intensity. It follows that the lower the background intensity, the lower will be the detection limit. Under some circumstances, careful adjustment of the pulse amplitude discrimination ('PAD', 'PHD' or 'PHA') conditions can improve sensitivity by an order of magnitude or more.

Of course, these conditions also imply that if P : B ratios are sufficiently high (e.g. in major-element analysis) the effect of background interference becomes insignificant and can often be ignored. However, this situation can never be safely assumed, and it must always be evaluated when determining analytical strategy.

In most other cases, and particularly in trace-element analysis, the background intensity must be established and subtracted from the apparent 'peak' (i.e. line) intensity to determine the true emitted characteristic intensity.

Since the peak and the background lie at the same apparent wavelength, neither can be measured directly and independently of the other – the only direct measurement that can be made is the intensity of peak plus background (P + B). The normal procedure, then, is to estimate the background intensity from one or more indirect measurements, and to subtract this estimate from the measured (P + B) to obtain an estimate of the true peak intensity.

The simplest and most widely used method is to measure the background intensity at each of two 'off-peak' positions, at equal intervals on either side of the peak, and average them. The averaged background intensity can then be subtracted from the measured peak intensity to yield an estimate of the true emitted intensity. The 2θ positions chosen for the background measurements must be far enough from the peak position to avoid serious interference from the peak itself (Fig. 8.1), but also close enough to provide an accurate estimate. They should ideally be free of spectral interference from other elements (see below); as with the peak itself, the background measurement positions should be checked in the wavelength tables for potential interferences, and it is also good practice to make some experimental scans over both peak and background positions to verify appropriate profiles. However, the analyst should be aware that even slow scans may not reveal slight interferences which can nonetheless be serious at trace-element intensity levels.

If it is known, or can be safely assumed, that the background profile under the analytical peak is linear and has a negligible slope, a single

INTERFERENCE

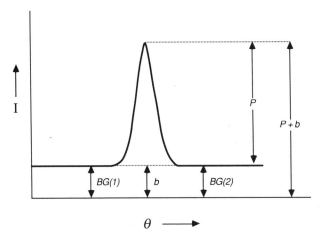

Figure 8.1 Estimation of background by linear interpolation between two off-peak measurements BG(1) and BG(2). The interpolated estimate b can then be subtracted from the measured peak intensity $(P + b)$ to yield the true peak intensity P. In this case the background is linear and has negligible slope, so that interpolation is unnecessary and either BG(1) or BG(2) would suffice for an estimate of b.

background measurement on either side of the peak will suffice (Fig. 8.1). This approach may be necessary if, for example, spectral interferences from other elements in the sample are present on one side of the analytical peak but not the other (Fig. 8.2), or if the background profile is discontinuous for some reason, e.g. there is a discontinuity due to the presence of an absorption edge of another element (Fig. 8.3). In the latter case, the absorbing element may be present in the sample, in which case it will result in more effective absorption of shorter wavelengths in the sample–scattered tube continuum, or it may be a constituent of the dispersing crystal, when it will have a similar effect on continuous background scattered from the upper layers of the crystal.

An alternative approach to background estimation is to replace the analytical sample with another of very similar composition except that it contains none of the element sought. In the absence of the analytical peak, it is then possible to determine the background profile very accurately. This technique is often used in XRF analysis because it is possible to synthesize samples of the desired composition, using reagents of high purity, and because background intensities are not highly sensitive to small changes in gross chemical composition. In silicate analysis, for example, a calibration sample of pure SiO_2 has a mean atomic number sufficiently close to those of a wide variety of silicate samples (and therefore has very nearly the same scattering characteristics) to serve for background calibration. For critical trace-element work, at or near

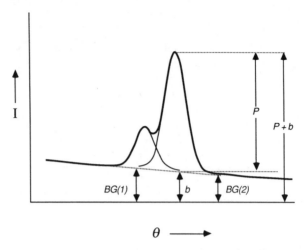

Figure 8.2 Off-peak background measurement positions must be free from spectral interference, which would clearly yield an erroneous estimate for BG(1). If a single off-peak estimator, such as BG(2), is used it may be necessary to establish and apply a correction factor to compensate for a sloping and/or a non-linear background profile.

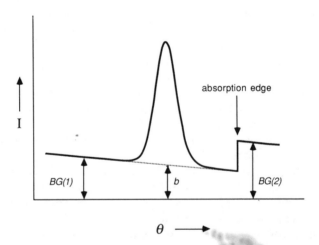

Figure 8.3 An erroneous estimate of background intensity will result if it is derived by interpolation between two off-peak measurement positions which span a significant discontinuity in the background profile, such as that due to an absorption edge of an element present in the sample in significant concentrations. Such an element will strongly absorb background continuous radiation generated within the sample and having wavelengths just short of the absorption edge, but will be a less effective absorber of slightly longer wavelengths.

INTERFERENCE

maximum levels of sensitivity, the approximation is not sufficient to provide a sufficiently accurate estimate of absolute background intensity, but it will still serve to establish the form of the background profile.

When profiles are determined in this way, it is often found that they are not linear, and certainly that they do not usually have negligibly small slope. Instead the more general situation is that shown in Figure 8.4; the profile is both curved and sloping, and an estimate of background made by averaging off-peak measurements (b') will clearly be either greater or less than the true background intensity (b), depending on the direction of profile curvature.

Correction for this non-linear background (NLB) requires determination of the profile curvature, either by the sample substitution technique described above, or by simple extrapolation of the background profiles on either side of the peak. The latter is clearly less satisfactory, but may be necessary in cases where it proves impossible to prepare background calibration standards free of the analytical element. From the profile so established, two suitable off-peak BG measurement positions (b_1 and b_2) can be selected and a correction parameter q can be established, such that

$$b = q \ (b_1 + b_2)$$

This parameter can then be used to correct off-peak backgrounds measured at the same positions on the analytical samples.

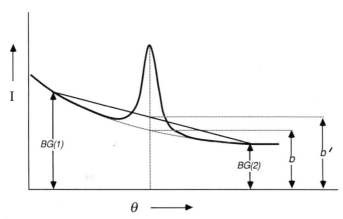

Figure 8.4 Linear interpolation between two off-peak background measurement positions yields an erroneous background estimate if the background profile is non-linear. An additional correction factor must be established from experimental measurements of the profile curvature.

8.1.2 Spectral interference

X-ray spectra are relatively simple (in the sense that they contain relatively few emission lines compared with, say, optical spectra). In major-element analysis, one does not often encounter severe difficulties arising from spectral *interference* – the presence of another characteristic emission line in the sample spectrum which cannot be effectively resolved from an analytical line. The presence of such an interfering line in effect increases the background beneath the analytical peak, but most X-ray emissions are weak in intensity and hence most potential interferences have very minor or negligible effects. There are, however, some cases in which the interference is sufficiently severe as to require correction, particularly in minor- and trace-element analysis, when the intensities of even weak interfering lines emitted by major elements can be comparable to or even greater than those of the analytical peaks of the trace constituents. A potentially troublesome interference in major-element analysis can often be avoided by choosing an alternative but less intense analytical line; however, this option may not be acceptable in trace-element work, for which maximum emission intensity is almost mandatory.

In determining analytical strategy, the probable incidence and magnitude of spectral interference with a chosen analytical line can be estimated by close study of a good set of wavelength tables, reinforced if necessary by experimental scans across the analytical peak position on several appropriate samples (preferably including a sample, synthesized if necessary, that contains all of the potential interfering elements but none of the element sought – see above). Such scans, which should be small-increment step scans with sufficient counting time at each step to minimize statistical background fluctuations, are usually also necessary to establish background measurement positions.

It should be remembered that many wavelength tables do not include satellite lines, and many do not include high-order harmonics of low intensity. However, even a very weak interference can become serious under certain circumstances, e.g. in the determination of trace cobalt in steel samples. The most sensitive Co line is Co Kα, which has minor interference from the very weak Fe Kβ_0 satellite line; this interference is normally negligible, but will present problems when very low concentrations of Co are determined in samples containing major Fe.

While it is not common for an analytical line and an interfering line to have exactly the same wavelengths/energies, it is not at all unusual to find some 'overlap', so that a portion of the 'tail' of the interfering peak coincides with the centroid of the analytical peak (Fig. 8.5) – or, possibly, with an off-peak position that might otherwise have been chosen for background measurement. Under these circumstances, the measured

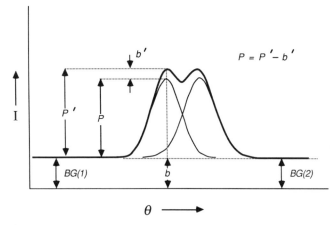

Figure 8.5 Spectral interference due to the 'tail' of an adjacent, overlapping peak adds an additional component b' to the normal background b. The magnitude of b' will be proportional to the (background-corrected) peak intensity of the interfering line itself, and the proportionality constant can be established experimentally and used to correct analytical intensities. b is estimated separately, usually by interpolation between off-peak measurements made at positions free from spectral interference.

intensity will include a component due to the interfering peak (b' in Fig. 8.5) as well as the normal background (b). Determination of the true peak intensity requires estimation of both b and b', and their subtraction from the measured peak intensity.

Correction for the effects of spectral interference is most commonly based on the fact that the extent of the interference (b') will be proportional to the (background-corrected) intensity of any peak of the interfering element – including the interfering peak itself. Corrections can be easily applied to the measured peak intensity if the proportionality constant is first determined experimentally, using suitable samples (synthesized if necessary) that contain

(a) none of either the analytical or the interfering element, and
(b) some of the interfering element, but none of the analytical element.

The first sample is used to establish the true background intensity at the peak measurement position. The second provides an additional background component at the peak position, due to the spectral interference, and this can be compared to the measured (background-corrected) intensity of the same or another peak of the interfering element to yield a correction factor that can be applied to all subsequent analytical peak measurements.

Norrish and Chappell (1977, p. 236) quote a simple example of the interference of Sr in the determination of low concentrations of Zr.

Zr Kα (0.788 Å) is measured at a 2θ angle of 22.56°, using a LiF_{200} crystal; a peak due to Sr Kβ (0.783 Å) is found at 22.42° 2θ. If a synthetic high-purity silica sample containing neither Zr nor Sr is first examined to determine background intensities at the two peak positions, and a second similar sample containing, say, 1000 ppm Sr (but still no Zr) is then analysed, it will be found that the apparent Zr background is increased by an amount equivalent to approximately 15 per cent of the intensity of the Sr Kα peak (0.877 Å, measured at 25.10° 2θ). The actual proportion will of course depend on the spectrometer resolution, which in turn is a function of collimator spacing, crystal rocking curve etc. Subsequent Zr analyses can then be corrected for Sr Kβ interference by measuring the Sr Kα peak intensity, and subtracting 15 per cent of it from the measured Zr Kα intensity.

Sr Kα is preferred as the basis for the correction, since it is proportional in intensity to Sr Kβ and, unlike the latter, it is not subject to mutual interference by Zr Kα.

If suitable calibration standards cannot be obtained or synthesized, it will again be necessary to resort to extrapolation of profiles determined by wavelength scans across the analytical and interfering peaks. This method can give quite good results as long as the interference is not too severe and the intensity of the interfering peak is at most not too much greater than that of the analytical peak. Otherwise both accuracy and sensitivity may be adversely affected.

Table 8.1 Typical spectral interferences in silicate analysis.

	Analysis line	Interfering line
Major elements	Al Kα	Cr Kβ (4)
	Mn Kα	Cr Kβ (tube)
	P Kα	Ca Kβ (2)
	P Kα	Cu Kα (4)
	Si Kα	Fe Kβ (4)
	Ti Kα	Ba Lα$_1$
Minor and trace elements	Ba Lα$_1$	Ti Kα
	Ce Lα$_1$	Ba Lβ$_1$
	Cl Kα	Cr Kα (2)
	Co Kα	Fe Kβ$_0$
	Cr Kα	V Kβ
	Cu Kα	W Lα (tube)
	Hf Lα	Zr Kα (2)
	Rb Kα	U Lα$_2$
	V Kα	Ti Kβ
	Y Kα	Rb Kβ
	Zr Kα	Sr Kβ

INTERFERENCE

Examples of spectral interference commonly encountered in XRF silicate analysis are given in Table 8.1.

Norrish and Chappell also report a more comprehensive example of the use of experimental measurements to calibrate corrections for background and spectral interference in XRF; that of the mutual interference of Rb Kα (0.927 Å) and Sr Kα (0.877 Å), which are not completely separated from each other by spectrometers of moderate resolution. In measuring background intensities for Sr Kα, Rb Kα interferes with the high-angle off-peak measurement and Rb Kβ with the low-angle measurement. Conversely, Sr Kα interferes with the low-angle off-peak background measurement position for Rb Kα. 'Tail' effects also produce enhanced backgrounds at the two peak positions (Fig. 8.6).

The extent of background and spectral interference with the measurement of Sr Kα was investigated by Norrish and Chappell by means of a series of measurements on a sample of pure quartz, and on a sample of pure quartz with 500 ppm Rb added. Peak and off-peak background (BG) positions were measured for both Rb Kα and Sr Kα positions, as follows:

2θ:	23.75 (Rb Kβ)	24.36 (BG)	25.10 (Sr Kα)	25.84 (BG)	26.58 (Rb Kα)	27.32 (BG)
I_{Qtz}	—	2 658	2 487	2 122	1 951	1 786
$I_{Qtz + Rb}$	—	2 805	2 551	2 483	22 497	2 107

Averaging of the backgrounds and determination of the net peak intensities then yields:

	Sr			Rb		
	Av. BG	(P + B)	Net P	Av. BG	(P + B)	Net P
Qtz	2 390	2 487	97	1 954	1 951	−3
Qtz + Rb	2 644	2 551	−93	2 295	22 497	20 202

It is immediately evident from these data that

(a) background beneath the Sr Kα peak is non-linear, since the intensity measured at the peak position is 97 counts higher than the average of the two off-peak backgrounds; and

(b) the addition of Rb has increased the apparent intensity at all positions, including peak *and* background positions for Sr Kα (because the Sr peak lies between Rb Kα and Rb Kβ).

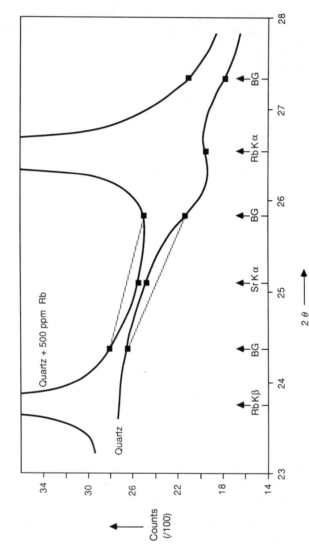

Figure 8.6 Profiles of portions of the spectra of two samples, one consisting of pure quartz and the other of quartz with 500 ppm of Rb added. As expected, addition of the Rb slightly increases the overall background levels, but it apparently *depresses* the net measured Sr Kα peak intensity, since each of the two off-peak background measurement positions is enhanced by spectral interference from the tails of the Rb Kα and Rb Kβ lines respectively, and subtraction of the enhanced background results in depression of the net peak intensity. The magnitude of this effect must be established experimentally and used to calculate an additional correction factor, based on measurement of the net Rb Kα or Rb Kβ peak intensity (data from Norrish & Chappell 1977).

Addition of 500 ppm of Rb has increased the net peak intensity of Rb Kα by 20 205 counts but, after background corrections have been applied, has *depressed* the apparent intensity at the Sr Kα position by 190 counts (by increasing background at both off-peak background measurement positions). The Sr depression of 190 counts is equal to 0.94 per cent of the net Rb peak intensity. In making subsequent Sr analyses it will therefore be necessary to compensate for this effect by reducing the apparent Sr background intensity (or increasing the apparent peak intensity) by 0.94 per cent of the Rb Kα intensity.

A similar correction will also have to be applied to compensate for the effect of Sr on the estimation of Rb Kα intensity. This will, however, be less severe because Sr interferes with only the low-angle Rb background measurement (and, to a lesser extent, with the peak). Sr Kβ, at 23.75° 2θ, does not interfere with Rb Kα peak or backgrounds.

Similar correction procedures can be followed in other cases of spectral interference, providing appropriate calibration standards can be synthesized. In this respect XRF has a big advantage over EPA because the irradiated sample area is relatively large and minor inhomogeneities in the standards are often of only minor or negligible consequence (apart from particle-size effects; see Ch. 7). It is difficult – indeed in most cases impossible – to prepare equivalent microprobe calibration standards which are chemically homogeneous at the sub-micron level, and hence the microprobe analyst is forced to use extrapolation techniques of lesser accuracy. Because of the inevitable presence of the electron-excited continuum in microprobe spectra, background intensities are also normally substantially higher than in XRF, which further exacerbates the problem. EPA trace-element sensitivities are thus typically several orders of magnitude poorer than XRF.

In either EPA or XRF, it is again emphasized that the analyst can expect difficulties whenever the level of potential error in the correction estimate is of the same order of magnitude as the net corrected intensity. Under these conditions the effective lower limit of detection will be substantially higher than that predicted from count rate statistics alone. The same remark applies to each of the other possible sources of analytical error.

8.2 Interelement ('matrix') effects

It has already been noted that quantitative X-ray spectrometry, whether by XRF or EPA, is based on an assumption of proportionality between concentration of an element present in a sample and the intensity of emission of its characteristic spectrum, i.e.

COMPOSITION-DEPENDENT ERRORS IN XRF

$$c_A = k_A I_{\lambda(A)} \qquad \text{(Eqn 7.1)}$$

For any group of standards and samples, this expression implies a linear relationship which requires only the experimental determination of the proportionality constant k_A from one or more standards of known composition, and then measurement of $I_{\lambda(A)}$ for each sample for analysis. If two or more calibration standards are used, the linear relationship can be expressed in the form of a *working curve*, or *calibration curve*, which will plot as a straight line (Fig. 8.7) if Equation 7.1 holds.

The equation of a linear working curve can be expressed in either of two forms, viz.

$$I_{\lambda(A)} = Bc_A + A \qquad (8.1a)$$

or (more conveniently, since c_A is to be determined),

$$c_A = D + E I_{\lambda(A)} \qquad (8.1b)$$

However, as previously noted it is found in practice that linearity is only approached in cases where the samples and standards show little variation in composition and physical form, and observation errors are maintained at very low levels. In other cases, marked departures from linearity are observed, and they can lead to serious analytical error if they are not corrected.

The sources of error which have been reviewed in Chapter 7 include those cases in which intensities of X-ray emission are, for one reason or another, incorrectly estimated. If unrecognized they can be serious, but they can all be controlled, or, if not, their magnitudes and hence the

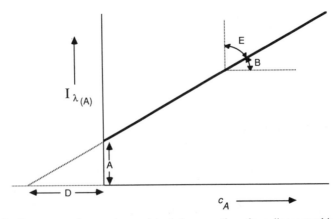

Figure 8.7 Parameters that can be used to define equations for a linear working curve.

effects that they have on the quality of the analysis can be readily evaluated.

A more difficult group of errors are those arising from non-linearity of measured intensity as a function of sample composition. Even if operator, instrumental and statistical errors are reduced to a minimum, analytical error remains because the intensity of the emitted X-rays is *not in fact strictly proportional* to the concentrations of the emitting elements.

This situation arises because, in both XRF and EPA, the emitted X-radiation is generated not only at the sample surface but also to some finite depth within the sample. Unless the sample is effectively infinitely thin, it is thus necessary to take account of

(a) interaction between the sample and the exciting radiation (the 'primary' beam in XRF or the electron beam in EPA), and
(b) interaction between the sample and the generated X-radiation, most of which must pass through a finite thickness of the sample before reaching the spectrometer (Fig. 8.8).

In each case, the interaction is composition-dependent, and if samples and standards vary in composition then the interaction will also vary from one to another and linearity can no longer be assumed.

Consider the situation in XRF (which is analogous to, but not identical with, that in EPA). Fluorescent radiation is generated in a series of thin layers dl, each at a depth l beneath the surface (Fig. 8.9). The total measured fluorescent intensity will be the integrated *emergent* intensity I_c

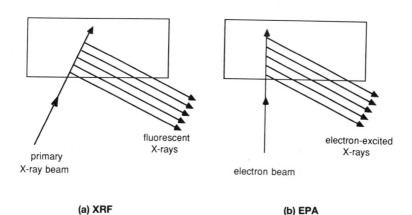

(a) XRF (b) EPA

Figure 8.8 In both XRF (a) and EPA (b), some analytical emissions are generated beneath the sample surface and pass through part of the sample itself before dispersion and detection. In so doing, their intensities may be either attenuated or enhanced by photon interaction with atoms of the sample, resulting in composition-dependent 'matrix' errors in the measured intensities (see also Fig. 8.9).

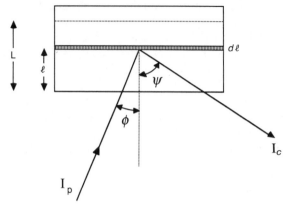

Figure 8.9 The basic geometry of fluorescent X-ray excitation. The primary beam, of intensity I_p, strikes the sample surface at an incidence angle ϕ and penetrates to an effective 'infinite' depth L. X-rays are generated in a series of thin layers, each of thickness dl at a depth l below the surface. They leave the sample at a departure angle ψ and with intensity I_c, which depends not only on the concentrations of emitting elements in the layer dl and on the efficiency of excitation, but also on sample attenuation of I_p (a function of ϕ, l and sample composition) and attenuation or enhancement of I_c (a function of ψ, l and sample composition). The net intensity of an emission measured in the spectrometer will be the sum of all intensities I_c, integrated over the depth range to 'infinite' thickness L.

for all thin layers dl over the depth range from zero to 'infinite thickness' L, which is the depth at which either

(a) the primary beam has been so attenuated that it no longer has sufficient energy to excite the measured characteristic radiation, or
(b) the fluorescent radiation generated at this depth is fully absorbed, by the sample itself, on its way towards the spectrometer; it 'emerges' at the sample surface with zero intensity and hence is not measured.

At greater depths within the sample, either the primary radiation generates no fluorescent radiation (case (a)), and/or any generated fluorescent radiation will be totally absorbed by the sample because of its longer path length through the sample (case (b)). For any depth, the extent of absorption of either the primary or the fluorescent X-rays will be a function of sample composition (which determines the relevant mass absorption coefficients) and of the incident angle θ (primary beam) or the emergent ('take-off') angle ψ, which determines the path lengths within the sample.

8.3 Correction procedures

Complete analysis of the generation and emission of any XRF analytical wavelength involves consideration of the effects of the sample composition on each of the following:

(a) *Excitation*:

  (i) Excitation of the analytical wavelength λ_x by the primary beam continuum.
  (ii) Excitation of λ_x by characteristic primary radiation.
  (iii) Excitation of λ_x by energetic electrons (EPA only).

(b) *Absorption*:

  (i) Photoelectric absorption of the primary X-ray beam as it passes through the sample.
  (ii) Energy loss in the primary X-ray beam as a consequence of scattering as it passes through the sample.
  (iii) Absorption of the analytical fluorescent radiation λ_x, as it passes through the sample before emerging at the surface.

(c) *Enhancement*:

  (i) Enhancement of λ_x by characteristic radiation λ_y, generated from another matrix element Y, or from two or more such elements ('secondary fluorescence'); this can happen whenever λ_y has a wavelength shorter than the absorption edge for λ_x.
  (ii) 'Third element' or 'tertiary' fluorescence, in which λ_x is excited by λ_y, which in turn has been excited by λ_z generated from a third matrix element Z.
  (iii) Enhancement of λ_x by continuous radiation generated within the sample (EPA only, since incident X-radiation does not generate a continuum from the sample but the primary electron beam of EPA does).

It is theoretically possible to derive a set of equations to describe each of these physically complex processes, and to combine the equations into a rigorous mathematical description of the generation of the emergent fluorescent X-rays (see, e.g., Tertian & Claisse 1982, Ch. 4). This mathematical model, when complete, could be used to calculate corrections to apply to the measured intensities to compensate for composition-dependent errors.

However, though scientifically desirable this approach is not easy to apply in practice, except in ideal circumstances. The descriptive equations are predictably complex, requiring considerable computing power and relatively long computing times for their solution. Although access to powerful but relatively inexpensive computing facilities has done much in recent years to alleviate this problem, it nevertheless detracts somewhat from the inherent simplicity of XRS methods and has encouraged analysts to seek simpler solutions.

Evaluation of some of the correction parameters requires knowledge of the sample composition, which of course is usually the object of the analysis; it is then necessary to begin with an approximate composition, calculate correction parameters, compute a corrected composition, calculate new correction parameters, and repeat the process through several cycles until the calculated compositions converge to more or less constant values. *Iterative* procedures of this type are frequently used in both XRF and EPA correction procedures.

A significant drawback to this technique is that many of the essential correction parameters are still not known with sufficient accuracy, e.g. mass absorption coefficients for low-energy, long-wavelength radiation, which cannot be measured directly but must be derived by extrapolation, using empirical expressions which are sometimes of doubtful accuracy.

More seriously, the X-ray generation, absorption and enhancement calculations must be integrated over the finite thickness of the sample and over the complete range of X-ray energies involved (e.g. in the primary beam continuum). The limits of integration are difficult to define, and the integrals themselves are rather complex.

These factors combine to yield uncertainties that may approach or even in some cases exceed the magnitudes of the matrix errors themselves. Although there can be no doubt that this is fundamentally the best approach to the problem, and that it will ultimately yield to continued research, XRS analysts are still largely dependent on more pragmatic solutions. Many have been proposed, and *within reasonable limits* many of them work quite well – hence the undoubted popularity of both XRF and EPA.

A comprehensive review of the methods adopted to cope with various problems is beyond the scope of this introductory treatment, but it can be noted that most of them can be assigned to one of the following philosophies:

(a) The most obvious approach is simply to tolerate the errors. In fact they are often not of major consequence, particularly if the samples and calibration standards do not vary greatly in composition. For the purposes for which the analyses are required, modest errors may not be troublesome, e.g. in process control applications, where detection of changes in sample composition is often more important than absolute accuracy. Indeed it can be argued that time, effort and resources are often wasted in the pursuit of excessive analytical accuracy which is not warranted by the nature of the samples

CORRECTION PROCEDURES

themselves or of the problem for which they are being analysed.

(b) Since matrix errors are primarily consequences of compositional differences between the samples and the reference standards, they can be eliminated by using standards of the same or very nearly the same compositions as the samples. Clearly this approach is of limited value to a laboratory which must process a wide range of unpredictable sample compositions, but it is very useful for repetitive work such as process control.

(c) The samples and standards may be physically and/or chemically modified (e.g. by dilution in liquid or solid solutions) in order to reduce the *magnitude* of compositional differences and hence the magnitude of the matrix errors. While this technique normally cannot eliminate the errors altogether, it may permit them to be contained within acceptable limits.

(d) The theory of X-ray generation and emission may be simplified for certain specific applications, in order to yield a less complicated mathematical correction procedure. For example, if it can be shown that secondary and tertiary fluorescence effects are trivial in samples of a particular range of compositions, then correction for those effects can be eliminated from the calculations. Many simplified procedures, for example, are based on the assumption that absorption is the predominant source of matrix error; secondary and tertiary enhancement corrections are treated in simplified form or omitted altogether. This is often not an unreasonable assumption, but the analyst must be careful not to extend the simplified procedure to other cases in which it is invalid.

(e) The correction parameters, and particularly the absorption coefficients, may be measured directly or indirectly on the samples (and standards) themselves. Although this involves an additional series of measurements and can be very time-consuming, such an approach is essential in cases where the complete major-element composition is unknown and cannot be estimated or calculated with sufficient accuracy (e.g. in the determination of minor or trace elements in samples which for some reason have not been analysed for some or all of their major constituents).

(f) Several ingenious methods have been developed to compensate for matrix effects without necessarily calculating, measuring or even estimating them. These methods are also useful in cases where the bulk sample compositions are unknown.

(g) If it is practicable to prepare and analyse the sample as a very thin film (of the order of a few hundred to a few thousand Å in thickness), matrix effects consequent upon depth generation, absorption and enhancement are effectively eliminated and matrix errors are of negligible magnitude.

8.3.1 The 'simplified theory' approach

Neglecting depth considerations for the moment, let us consider the excitation of characteristic radiation $\lambda(i)$ from element i in the thin layer dl of Figure 8.9.

The total excitation efficiency of the primary beam, for element i in a matrix of j elements, may be expressed as

$$I_{\lambda(i)} = K_i \times c_i \times \int_{\lambda_{min}}^{\lambda_{edge(i)}} J(\lambda) \times \frac{\mu_i(\lambda)}{\Sigma\, \alpha_j\, c_j} \qquad (8.2)$$

where $I_{\lambda(i)}$ = intensity of $\lambda(i)$

c_i = weight fraction of element i

K_i = proportionality constant for element i for fixed excitation conditions

$\int J(\lambda)$ = *effective* excitation energy of the tube spectrum, i.e. the total energy of that portion of the tube spectrum with λ shorter than the absorption edge of $\lambda(i)$

$\Sigma \alpha_j\, c_j$ = 'total sample absorption', in which α_j is defined by

$\alpha_j = [\mu_j(\lambda) + A\mu_j(\lambda(i))]$

in which $\mu_j(\lambda)$ = mass absorption coefficient for element j for a wavelength λ in the primary beam, and

$\mu_j(\lambda(i))$ = mass absorption coefficient for element j for the fluorescent wavelength $\lambda(i)$, and

A = a geometric constant ($= \sin \phi\, /\, \sin\Psi$)

Although it is written above in a simplified form, it is immediately apparent that Equation 8.2 is difficult to evaluate. It must be separately calculated for each analytical wavelength $\lambda(i)$ in a multi-element sample, and it must be further evaluated for each layer dl from the surface of the sample to infinite thickness L (Fig. 8.9). Further – and this is a major source of difficulty – it must be evaluated over all wavelengths in the primary beam between λ_{min} and the absorption edge for each of the analytical wavelengths, including the tube continuum and any tube characteristic radiation in this range. There are thus two major problems for which simplified solutions are desirable – that of the *effective primary beam energy*, and that of the *depth generation factor*.

CORRECTION PROCEDURES

EVALUATION OF THE EFFECTIVE PRIMARY BEAM ENERGY

Several practical solutions to this problem have been proposed, and they can be illustrated by two examples of approaches that have been applied successfully to reasonably wide ranges of sample types.

Experimental measurement of the effective tube energy Criss and Birks (1968) suggested the 'fundamental parameters' method as an alternative to other empirical correction procedures which were then in common use, but which they had found to yield some inconsistencies. Briefly, the 'fundamental parameters' method is based on

(1) assumption of an approximate composition for the sample, based on the raw intensity data (relative to one or more calibration standards);
(2) calculation of the fluorescent intensities which theoretically ought to be emitted from a sample of this composition under the experimental conditions employed;
(3) comparison of the theoretical and observed intensities;
(4) adjustment of the approximate composition, in the light of (3);
(5) iteration of steps (2) to (4) until satisfactory convergence is achieved.

The theoretical intensities of step (2) are calculated from fundamental equations for primary and secondary fluorescence (tertiary fluorescence is assumed to be insignificant and is not calculated). The problem of integration of the effective tube energy is overcome by actually *measuring* the tube output experimentally, in successive small increments (0.02 Å) over the effective wavelength range from λ_{min} to the longest wavelength absorption edge involved in the multi-element analysis. The experimental data are tabulated, and in the correction calculation the integral of $J(\lambda)$ is obtained by summation of the incremental energies (expressed as measured intensities) over the wavelength range appropriate to each analyte element.

Calculation of the theoretical intensities, using a more rigorous version of Equation 8.2, still requires knowledge of the composition to yield the terms in c_j, and this is the reason for the iterative procedure of steps (2)–(4). Satisfactory convergence (i.e. agreement between observed and calculated intensities) is typically reached after three to five cycles.

The major advantages of the 'fundamental parameters' method are its applicability to a wide range of sample types and the fact that, at least in principle, it does not require a large library of calibration standards. However, it is a relatively complex procedure demanding good computing facilities. The experimental measurement of the tube output at each of the voltage/current settings used in a multi-element analysis can be tedious and time-consuming, and it also requires a special experimental configuration which may be difficult or impossible to set up in a standard commercial instrument. Although it has been shown to give good results with some difficult analytical problems, this approach does not at present appear to be widely used for routine analysis.

The 'effective wavelength' approach The integrated primary beam energy may be replaced in calculations by the energy of a hypothetical *monochromatic* primary beam having the same 'effectiveness' in exciting the analytical emission as the integrated tube energy. Although this may seem at first sight to be an oversimplification, it is not really an unreasonable assumption if it is recalled that the most effective excitation wavelengths are those just short of the analyte absorption edge. Experiment shows that

$$\lambda_{\text{effective}} \approx (2/3) \times \lambda_{\text{absorption edge}}$$

provided that

(a) strong tube *characteristic* lines are not substantial contributors to the excitation process; and
(b) the sample has no major absorption edges, due to other elements present in significant concentrations, between $\lambda_{\text{effective}}$ and $\lambda_{\text{absorption edge}}$ for each analyte emission.

If these conditions are met, this assumption provides a simple solution to the problem of integration of the primary beam energy (Fig. 8.10). Otherwise this approach is clearly invalid. It is important to note that many of the simplified methods used for quantitative XRF analysis, including several of the methods described in this chapter, tacitly assume the 'effective wavelength' concept in one form or another. The analyst must ensure that the assumption is warranted.

THE FINITE DEPTH FACTOR

Based on the concept of effective wavelength, let us now consider the depth factor, firstly for the primary beam. The intensity I_p of the primary beam at depth l will be given by

$$I_p = I_0 \times \exp(-\mu_1 \, l \, \sec \phi) \qquad (8.3)$$

where μ_1 is the *linear* absorption coefficient of the sample for the effective primary wavelength, and I_0 is the initial intensity of the effective primary wavelength.

The intensity $I_{\lambda(i)}$ of the characteristic radiation emitted per unit area of the thin layer will be proportional to I_p and to $c_i \varrho \cdot dl$, where ϱ is the density of the sample, i.e.

$$I_{\lambda(i)} = k \, I_p \, c_i \, \varrho \cdot dl \qquad (8.4)$$

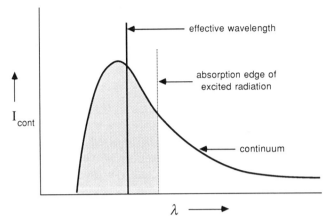

Figure 8.10 The primary continuum that contributes to the excitation of an analytical emission includes all wavelengths short of the appropriate absorption edge (the shaded area under the continuum curve), with those wavelengths just short of the absorption edge being more effective than those of higher energy. For convenience in calculating the integrated effect of all continuum wavelengths, a useful simplification is to consider them to be represented by a hypothetical single 'effective wavelength' at about 2/3 of the wavelength of the absorption edge.

The characteristic radiation will be attenuated (i.e. reduced in intensity) as it passes through the sample towards the spectrometer, by a factor of

$$\exp(-\mu_2 \, l \, \sec\Psi)$$

where μ_2 is the *linear* absorption coefficient of the sample for the analytical wavelength $\lambda(i)$.

Hence the intensity of the fluorescent radiation generated per unit area of the thin layer dl and *emerging at the sample surface* will be

$$I_{\lambda(i)} = \{k \times c_{(i)} \times \varrho \times \exp[-\mu_1 \, l \, \sec\phi - \mu_2 \, l \, \sec\psi]\} \cdot dl \quad (8.5)$$

The total emitted intensity per unit area from a sample of finite thickness L will then be

$$I_{\lambda(i)} = \int_0^L I_{\lambda(i)} \cdot dl$$

$$= k \, \frac{I_0 \times c_i \times \varrho \times \{l - \exp[-L(\mu_1 \sec\phi + \mu_2 \sec\psi)]\}}{\mu_1 \sec\phi + \mu_2 \sec\psi} \quad (8.6)$$

If the sample is effectively infinitely thick, then

$$\exp[-L(\mu_1 \sec \phi + \mu_2 \sec \psi)]$$

will be very small, so that

$$I_{\lambda(i)} = \frac{k \times c_i \times \varrho \times I_0}{(\mu_1 \sec \phi + \mu_2 \sec \psi)} \tag{8.7}$$

If we now write $A_1 = (\mu_1 / \varrho)$ and $A_2 = (\mu_2 / \varrho)$ (i.e. the *mass absorption coefficients* of the primary and fluorescent radiations respectively), then

$$I_{\lambda(i)} = \frac{k \times c_i \times I_0}{A_1 \sec \phi + A_2 \sec \psi} \tag{8.8}$$

If there are no major absorption edges between λ_{eff} and λ_{edge}, then A_1 and A_2 will be related such that

$$A_1 = k\, A_2$$

Further, if the spectrometer is designed so that $\phi \approx 30°$ and $\psi \approx 60°$, then $\sec \phi \approx 1$, and $\sec \psi \approx 2$.

The expression $[A_1 \sec \phi + A_2 \sec \psi]$ may then be replaced by the equivalent expression $[(2 + k)A_2]$: i.e. not only is A_1 related to A_2, but *the effect of* A_2 *will predominate* (given the specified spectrometer geometry) and, providing the assumptions made are valid, the absorption problem can be considered in terms of A_2 alone. If the constants of Equation 8.8 are combined, it reduces to the much simpler expression

$$I_{\lambda(i)} = kc_i\, /\, A_2(i) \tag{8.9}$$

A good reason for asymmetric angles of primary beam incidence and fluorescent beam emergence in spectrometer designs such as those of the Philips series of XRF instruments is thus apparent – it ensures that the effect of A_2 will predominate, and hence allows the consequent simplification. The latter may not be appropriate for other spectrometer designs.

Now, considering both sample and calibration standard,

$$\frac{(I_i)_{\text{sample}}}{(I_i)_{\text{standard}}} = \frac{(c_i)_{\text{sample}}}{(c_i)_{\text{standard}}} \times \frac{(A_2)_{\text{standard}}}{(A_2)_{\text{sample}}} \tag{8.10}$$

CORRECTION PROCEDURES

i.e. for analysis,

$$c_{sample} = c_{standard} \times [I_{sample}/I_{standard}] \times [(A_2)_{sample} / (A_2)_{standard}] \quad (8.11)$$

If the sample and standard are of similar composition, $(A_2)_{sample} \approx (A_2$ $_{standard}$. Equation 8.11 then becomes

$$c_{sample} = c_{standard} \times [I_{sample} / I_{standard}] \quad (8.12)$$

which is the simple (first approximation) linear relationship, in general holding only for this special case. This is not necessarily an unusual analytical situation, e.g. the determination of K in K-feldspars, using one or more previously analysed K-feldspars as standards. However, since A varies approximately as Z^3, relatively minor variations in the content of heavier elements can produce quite significant variations in A. For example, the addition of 20 per cent Fe_2O_3 to SiO_2 effectively doubles the value of A for Sr Kα. Equation 8.11 must then be used instead of Equation 8.12, and the necessary values of A_2 must be either (a) calculated from known mass absorption coefficients, or (b) experimentally measured, or (c) empirically derived.

Alternatively, it may be possible to reduce the variation in A_2 between sample(s) and standard(s) by an appropriate sample preparation technique (e.g. dilution).

ENHANCEMENT (SECONDARY AND TERTIARY FLUORESCENCE)

Consider the problem of determining Fe in zinc ores, in which the Zn content is usually substantially greater than the Fe content. Zn Kα has a wavelength of 1.437 Å, and Zn Kβ = 1.296 Å. The Fe K absorption edge is at 1.744 Å, so that both Zn K emissions are capable of exciting Fe K radiation. Fe Kα emitted from the sample thus will include not only the normal component excited by the primary beam but also additional radiation produced by Zn K fluorescence. Observed Fe Kα intensities will thus be erroneously high.

Furthermore, zinc ores often contain appreciable concentrations of copper. The Cu K absorption edge is at 1.36 Å, so that Cu Kα (1.54 Å) and Cu Kβ (1.39 Å) are both excited by Zn Kβ (but not by Zn Kα). Both Cu Kα and Cu Kβ have wavelengths short of the Fe K absorption edge, so that each of them is capable of exciting additional Fe K radiation. Enhancement of Fe Kα is then due to

(a) secondary fluorescence due to Zn;
(b) secondary fluorescence due to Cu; and
(c) tertiary (Zn → Cu → Fe) fluorescence, sometimes called 'third element' fluorescence.

Regrettably, no simple, practical and universal solution to this problem has yet been forthcoming. Attempts to treat it have been included in several of the 'fundamental calculations' approaches, but the effects are rather complex and complete success has not yet been achieved. In some cases (particularly those in which tertiary fluorescence is not a major difficulty), it has proved possible to treat enhancement as a *negative absorption* and hence to correct for it by modification of the appropriate absorption coefficients – particularly if the latter are empirically determined (e.g. using the 'α-factor' method discussed below). However, this does not always give entirely successful results.

Fortunately, errors arising from enhancement effects are mostly relatively small. Norrish and Chappell (1977, p. 235) note that in the case of 1 per cent Fe in ZnO, enhancement is less than 15 per cent, and the consequent error can frequently be tolerated if the calibration standards are not too different in composition from the analytical samples. In fact, judicious choice of standards is probably the most effective method of controlling enhancement errors, pending the development of more versatile and more accurate calculation procedures.

Fluorescence errors are often more substantial in electron microprobe analysis because of additional enhancement by the electron-excited continuum. Fluorescence correction procedures are therefore discussed in greater detail in Chapter 9.

8.3.2 *Attenuation of matrix effects by sample modification*

DILUTION

If samples and standards can be sufficiently diluted with an 'inert' component – i.e. preferably one containing none of the elements to be determined – then variations in A can be effectively reduced, though never completely eliminated. Dilution can be effected by any of several methods, e.g.

(a) solution of samples and standards in a suitable liquid solvent, such as water (which, however, will usually then preclude analysis *in vacuo* and thus greatly reduce sensitivity for light elements);
(b) fine grinding with a suitable solid diluent (which, although often convenient, may result in undesirable microabsorption, or particle-size, effects);
(c) fusion of the sample, using a suitable flux that can serve effectively as a diluent and also eliminate microabsorption effects.

Any of these techniques will reduce variations in A_2, at the necessary expense of some loss in sensitivity as a result of the dilution. Generally speaking, the loss in sensitivity is more or less proportional to the suppression of matrix effects, so that once again some degree of compromise may be required.

Fusion into a borate glass is a widely used technique in major-element silicate analysis, in which it also has the desirable effect of eliminating microabsorption ('particle-size') errors which would otherwise affect the determination of light elements such as sodium or magnesium (see Ch. 7). Lithium tetraborate or metaborate are commonly used as fusion fluxes; they both fuse relatively easily, at temperatures below 1000 °C, they readily dissolve most silicates and oxides (but there are some exceptions), and they do not emit interfering characteristic X-rays in the wavelength region used for silicate analysis.

For many analytical wavelengths, the relatively low mass absorption coefficients of the borate glass are not greatly different from those of the analytical samples, so that excessive dilution may be required to achieve significant reduction of the absorption effect; the loss in sensitivity may then be unacceptable, and the risk of reagent contamination is increased. It is therefore common practice to deliberately add a *heavy absorber*, such as lanthanum oxide, to the dilution mix. Addition of a heavy absorber has a similar effect to dilution with a light absorber in that it reduces the magnitude of the differences between samples and standards, and it has the additional advantage of not increasing the scattered background as much as excessive dilution with a light absorber. Heavy absorbers also have their greatest effects on the heavier elements, which is desirable since these are generally subject to the most severe matrix effects.

Dilution in this way may, in some cases, reduce absorption (and enhancement) errors to an acceptable level. For many silicate samples, for example, it can yield accuracies adequate for many applications, particularly when viewed in the light of sampling or other non-analytical errors. Details of a widely used and effective fusion procedure are given by Norrish and Hutton (1969). However, even this technique has some major limitations, e.g.

(a) It is not well suited to the determination of trace elements because of problems with reagent contamination and with the loss of sensitivity inherent in dilution.
(b) In order to retain reasonable sensitivities, particularly for the light elements, the extent of dilution must be limited to a level that reduces, *but does not eliminate*, the matrix errors. For analyses of high quality, it is therefore still necessary to use calibration standards of similar composition to the analytical samples, and/or to apply some further correction for the residual matrix errors. In other words, dilution is used to reduce the matrix errors to the lowest feasible levels, but in general these will still require some additional correction.

In major-element silicate analysis, two simplified-theory procedures have been used with considerable success to make the additional corrections. In the method of Norrish and Hutton, the absorption coefficients of Equation 8.11 are calculated and substituted into the equation to yield corrected compositions. In an alternative approach, absorption coefficients are not calculated as such; instead empirical interelement *influence coefficients* (or '*α-factors*') are determined by multiple linear regression analysis of the data obtained from a suite of calibration standards, and these factors are then used to correct the raw intensity data obtained from the analytical samples.

Both approaches have been refined to a level where they are currently used successfully for the purposes for which they were derived and, in each case, subject to certain bounding assumptions. Each is briefly reviewed in Section 8.4; it is, however, important to note that they cannot necessarily be applied indiscriminately to other analytical problems. The Norrish and Hutton method, for example, is based on the simplified theory used to derive Equation 8.11; if the assumptions inherent in that derivation (e.g. that the effect of A_2 predominates over that of A_1, and that there are no major absorption edges between λ_{eff} and λ_{edge}) are not satisfied then the method may not yield satisfactory results.

THIN FILM METHODS

This technique is based on the simple assumption that if samples and standards can be prepared as 'infinitely thin' films there will be no generation of the analyte emission at depth and matrix errors will disappear. The emitted intensity will be proportional to the mass of analyte per unit area of the sample film. If the film thickness is L and μ_1 and μ_2 are the mass absorption coefficients of the sample for $\lambda_{\text{effective}}$ and λ_i respectively, it can be shown that the criterion for 'infinitely thin' is met if

$$L \times (\mu_1 \sec \phi + \mu_2 \sec \psi)$$

is less than about 0.1.

The relatively small emitting mass necessarily results in a substantial decrease in total generated intensity of the analyte emission – typically of about an order of magnitude. However, the total mass thickness is in fact about two orders of magnitude less than an 'infinitely thick' sample, so that there is really a significant increase of sensitivity in thin film samples, if sensitivity is defined as measured intensity per unit of analyte mass. This improvement results, of course, from the absence of absorption losses for any of the generated emissions.

Samples may be prepared by vacuum evaporation on to a suitable substrate, by evaporation of a solution on to a substrate, by impregnation of a filter paper with sample solution, by collection on ion exchange papers or membranes, by coprecipitation of the analyte from a solution followed by collection of the precipitate on a filter paper, or by various other ingenious methods developed to meet the needs of specific circumstances.

Limitations of the method are principally related to the difficulty of preparing homogeneous films of reproducible thickness (which must be kept constant within the desired precision limits, since emitted intensity is proportional to the mass of analyte per unit area of the film rather than the analyte concentration), suppression of background (including background scattered from the supporting substrate), and preparation of sufficiently thin films. For light elements emitting low-energy characteristic radiation the value of μ_2 in the thickness parameter is usually large, so that L must be small – typically of the order of 0.02 g cm^{-2}. Intensities are then low, approaching background, and sensitivity is reduced. In such cases the thin film method is obviously of limited value, but in other cases it may be very successful; for example Pik *et al.* (1981) reported detection limits in the parts per billion or lower ranges for trace heavy metals in river and estuarine waters, using a coprecipitation technique and collecting the sample precipitate on a membrane filter.

8.3.3 The use of numerical correction ('influence') coefficients

Consider the determination of concentrations of element A in a simple binary system AB, utilizing a series of intermediate compounds as calibration standards. The *calibration curve* (or '*working curve*') will only plot as a straight line (Fig. 8.11) if there is a linear relationship between c_A (the concentration of element A) and $I_{\lambda(A)}$ (the intensity of the characteristic radiation emitted by element A). Equation 8.9 indicates that this can only be the case if A_2 is constant, which requires that the mass absorption coefficients of A and B for $I_{\lambda(A)}$ be identical. Substitution of varying amounts of B for A in the samples then has no effect on the sample mass absorption coefficients.

Generally this will not be the case, and the calibration curve will be non-linear (hyperbolic) – bowed upwards or downwards depending on the relative magnitudes of μ_A and μ_B for $I_{\lambda(A)}$ and on possible secondary enhancement effects (Fig. 8.11). The *degree of curvature* is thus a measure of the magnitude of the effect of B on the emission intensity of $I_{\lambda(A)}$, and a parameter expressing the degree of curvature can be used to correct the measured intensities of $I_{\lambda(A)}$ for the matrix effects of B (i.e. it can be used to 'straighten' the calibration curve).

If Equation 8.2 is simplified by assuming

(a) an effective wavelength λ_{eff} in place of the integral of $J(\lambda)$; and
(b) that secondary fluorescence can be either neglected or treated as negative absorption

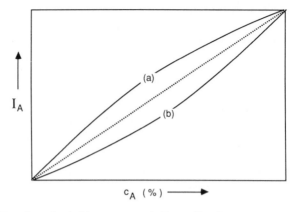

Figure 8.11 Even in a simple binary system A–B, a calibration curve will only be linear if there are no composition-dependent absorption/enhancement effects on the intensities of the analytical emissions. In most cases the curve for element A will be hyperbolic, bowed in one direction (a) or in the other (b), depending on the relative mass absorption coefficients of A and B for the emissions from A and on the relative magnitudes of absorption and enhancement effects. For practical purposes, portions of the calibration curve may usually be regarded as effectively linear over limited compositional ranges, but the assumption should always be verified experimentally.

then we derive

$$I_{\lambda(A)} = (k_A \, c_A) / \Sigma \, (\alpha_j \, c_j) \tag{8.13}$$

which for the binary system AB becomes

$$I_{\lambda(A)} = k_A \, c_A / (\alpha_A \, c_A + \alpha_B \, c_B) \tag{8.14}$$

α_A and α_B are coefficients which represent the *total* absorption effects (i.e. on both $\lambda_{\text{effective}}$ and $\lambda_{(A)}$) due to the elements A and B respectively. Since $(c_A + c_B) = 1$, then

$$I_{\lambda(A)} = (k_A \, c_A) / [\alpha_A + c_B (\alpha_B - \alpha_A)] \tag{8.15}$$

If we define a parameter $\alpha_{A(B)} = (\alpha_B - \alpha_A) / \alpha_A$, together with a new proportionality constant $K_A = \alpha_A / k_A$, then

$$I_{\lambda(A)} = k_A \, c_A / [\alpha_A + c_B \, \alpha_A \, \alpha_{A(B)}]$$
$$= (c_A / K_A) / [1 + c_B \, \alpha_{A(B)}]$$

or

CORRECTION PROCEDURES

$$c_A = I_{\lambda(A)} K_A [1 + c_B \alpha_{A(B)}] \quad (8.16)$$

The parameter $\alpha_{A(B)}$ represents the effect of element B on the emitted characteristic intensity of element A, under specified excitation conditions embodied in the proportionality constant K_A (note that a parameter measured or calculated for a particular set of experimental conditions will generally not be appropriate if the conditions are changed). Equation 8.16 is a linear expression which can be used to correct the measured intensities for A providing $\alpha_{A(B)}$ is known or can be established, and also providing c_B is known. In normal analytical circumstances the latter condition is not met, but uncorrected ('raw') estimates of the concentrations of A and B can provide initial values which can then be refined by iteration.

So-called *alpha-factors* have been defined and applied in several slightly different ways, but they all amount to the derivation of numerical coefficients to express the nature and magnitude of the influence that one element in a sample has on the emission intensities of another. For this reason they are sometimes described as *influence coefficients*.

α-factors can be determined for suitable binary systems in any of several different ways (or by combinations of those ways):

(a) *Graphical*: if we replace $I_{\lambda(A)}K_A$ (i.e. the product of the measured intensity – or the ratio of the measured intensity to the constant intensity of a reference standard – and the proportionality constant) by R_A, then Equation 8.16 can be rewritten in the '$y = mx + b$' form of a straight line as

$$c_A / R_A = \alpha_{A(B)} c_B + 1 \quad (8.17)$$

Hence, using data from a set of calibration standards, a plot of c_A / R_A against c_B should yield a straight line of unit intercept and slope of $\alpha_{A(B)}$.

(b) *Regression*: an equivalent to the graphical solution is to perform a linear regression analysis on the intensity data obtained from a series of calibration standards, based again on a suitable linear form of Equation 8.16.

(c) *Calculation*: since the factor $\alpha_{A(B)}$ has been defined as $(\alpha_B - \alpha_A) / \alpha_A$, then

$$\alpha_{A(B)} = \left[\frac{\mu_{B_{(\lambda_{\text{effective}})}} + A\, \mu_{B_{(\lambda_{(A)})}}}{\mu_{A_{(\lambda_{\text{effective}})}} + A\, \mu_{A_{(\lambda_{(A)})}}} \right] - 1 \quad (8.18)$$

As noted earlier, $\lambda_{\text{effective}}$ is normally taken as about two-thirds of the absorption edge wavelength for $\lambda_{(A)}$. A is a geometrical constant which is fixed for a given spectrometer design (see Eqn 8.2), and the various mass absorption coefficients may be obtained from tables (e.g. Heinrich 1966) or by calculation (see Appendix A.2) in order to solve Equation 8.18 for $\alpha_{A(B)}$.

In multicomponent systems the situation is rather more complex than the simple binary case, but it is common to make the simplifying assumption that binary α-factors are additive in linear fashion, i.e. that in a ternary system ABC, the net influence coefficient of the matrix (B + C) on element A can be calculated by linear addition of the weighted coefficients of B on A and C on A:

$$I_{\lambda(A)} = (K_A \, c_A) / [\alpha_A \, c_A + \alpha_B \, c_B + \alpha_C \, c_C] \qquad (8.19)$$

or

$$c_i = R_i \, [1 + \Sigma \, (\alpha_{i(j)} \, c_j)] \qquad (8.20)$$

It can be shown that the assumption of linear additivity is not strictly valid, and it is certainly not applicable if the further assumptions inherent in the simplifications of Equation 8.2 are invalid. However, the use of influence coefficients has been found to provide a satisfactory solution for a remarkably extensive range of analytical problems, particularly if it is used in conjunction with other techniques such as dilution to reduce the overall magnitude of the matrix errors, and if it is applied only to a reasonably restricted range of compositions.

The coefficients used in multicomponent systems can in some cases be determined separately from the constituent binaries, e.g. in the system ABC, the necessary α-factors may be determined by experimental study of the binary systems AB, AC and BC. However, this may not be practicable in binary systems lacking stoichiometric intermediate compositions, and it becomes very difficult in systems of four or more components. In such cases the coefficients may be calculated from tabulated mass absorption coefficients and then refined if necessary in the light of experimental observations. Alternatively, they may be determined by multiple linear regression techniques, using intensity data from a range of multi-element standards (Section 8.4.1). In the latter event, theoretical support for the coefficients is dubious, and they are better regarded as empirical factors in the sense that their actual physical significance is obscure. Despite this uncertainty, they may be used with considerable success, providing they are thoroughly tested.

The more popular influence coefficient algorithms have been summarized and evaluated by Tertian and Claisse (1982). It is worth noting that there are small but significant differences between the algorithms, depending on whether the influence coefficients are applied to the correction of apparent *compositions* or to the measured X-ray *intensities*.

The potential power of the influence coefficient approach is emphasized by the observation of Tertian and Claisse that many of the algorithms have been successfully tested on compositions in the ternary system Fe–Ni–Cr, in which relatively strong effects of both absorption and enhancement are encountered. Tertian and Claisse have also noted that ternary systems tend to be among the most difficult cases; in more complex multi-element systems the relative effects of the constituent elements tend to diminish as they each tend to lower concentrations and in effect dilute each other. Errors due to one or two inadequate corrections become proportionally smaller. Perhaps unexpectedly, systems with eight or ten elements (e.g. silicate major elements) often prove to be easier to handle than ternary or quaternary systems such as steels, bronzes or other alloys.

8.3.4 Measurement or compensation of absorption effects

The simplified form of Equation 8.9 indicates that for the many systems that satisfy its bounding assumptions (specifically including the assumption of an effective wavelength for the primary beam), absorption effects are the most serious source of matrix error, but can be compensated if the absorption term A_2 is known (or can be calculated or measured) and substituted in the equation.

If the major-element composition of a sample is known, A_2 can be calculated from published or calculated mass absorption coefficients of its individual elements. Minor or trace elements have little effect on the gross sample absorption and can usually be ignored. Calculations of this type are essential features of many analytical procedures, e.g. of the α-factor approach described in Section 8.3.3, in which calculation of the α-factors is equivalent to the calculation of absorption coefficients. However, these techniques do not always yield satisfactory solutions to the problem. In some cases the necessary absorption coefficients may not be known with sufficient accuracy (or α-factors may not have been derived for one or more critical elements). This often proves to be the case, e.g. if the analyst is faced with the problem of determining one or more trace elements in samples which vary appreciably in major element composition. Experimental determination of all of the necessary α-factors (say for 4 or 5 trace elements in a matrix containing 10 or 12 major elements) becomes a daunting task of doubtful value.

Further, the major element composition of the sample may not be known, and it is clearly inefficient to be compelled to make independent determinations of ten or more major elements in a sample in order to correct the data for two or three trace elements when it is only the latter that are required. Techniques have been developed to resolve this dilemma, by

(a) either measuring the absorption coefficients, directly or indirectly, *on the samples themselves*; or
(b) compensating for the absorption effects without actually determining numerical values for them.

DIRECT MEASUREMENT OF ABSORPTION COEFFICENTS

Suppose, for example, that Sr is to be determined in a suite of rock or soil samples. Sr abundances in such samples are commonly of the order of 100 ppm, so that the loss of sensitivity inherent in dilution or fusion techniques is unacceptable and the analyses would normally be made on pressed pellets prepared from finely ground powder samples. Measured Sr Kα intensities then require correction for the effects of sample/standard matrices of variable composition, and this in turn requires knowledge of the effective mass absorption coefficient of each sample/standard for Sr Kα.

The simplest procedure is to duplicate the experimental configuration of Figure 3.3, i.e. to measure the attenuation of a beam of Sr Kα X-rays produced by a parallel-sided sample slab of known mass thickness. This can be done by adapting a technique used in X-ray diffraction studies to measure mass absorption coefficients:

(1) A suitable quantity of the powdered sample is pressed into a circular hole in a plastic plate, using a special piston and die (Fig. 8.12). The cross-sectional area of the circular hole is easily established (e.g. 0.79 cm^2 for a hole of 1 cm diameter), and if the plastic holder is weighed before and after the pellet is pressed into it, the weight of the pellet can be obtained by difference. The mass thickness ϱl of the pellet is then given by

$$\varrho l = \frac{\text{weight of pellet}}{\text{cross-sectional area}}$$

(2) A wavelength-dispersive X-ray spectrometer is modified by attaching a slotted holder in front of the secondary collimator of the scintillation detector. The slot is arranged to accept the pellet holder so that the pellet lies in the path of the beam of X-rays entering the detector, and the slot fitting contains one or more lead apertures to limit the radiation entering the detector to that passing through the sample pellet when the latter is in place (Fig. 8.12). The same configuration can easily be adapted to an energy-dispersive spectrometer.

CORRECTION PROCEDURES

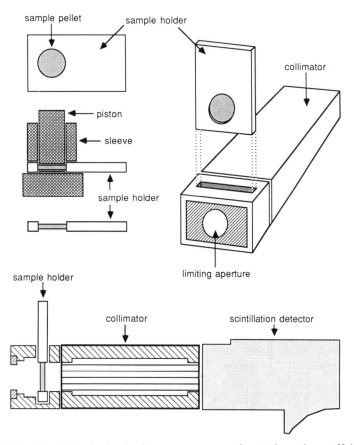

Figure 8.12 A simple device for the direct measurement of mass absorption coefficients. A pellet of the sample itself is pressed into a holder which can be inserted into a slot attached to the front of the secondary collimator and scintillation detector in a normal XRF spectrometer. Dispersed X-ray intensities emitted from an appropriate calibration sample are measured with the pellet in place and removed, and the attenuation (I / I_0) is calculated. If the pellet weight and cross-sectional area are known, the mass thickness (ρl) can be calculated and substituted in Equation 8.19 to determine the mass absorption coefficient (see Norrish & Chappell 1977).

(3) A high-Sr sample (e.g. of $SrCO_3$) is placed in the normal spectrometer sample holder, and the spectrometer is adjusted to the correct Bragg angle for measurement of Sr Kα. I_0, the unattenuated intensity of Sr Kα with no sample pellet in the holder, is then measured.

(4) The pressed sample pellet is placed in the holder slot and the attenuated intensity I is measured under the same operating conditions. From Equation 3.1,

$$I = I_0 \exp^{-\mu\, l}$$
$$= I_0 \exp^{-(\mu\,/\,\varrho)\,\varrho\, l} \tag{8.21}$$

Since I, I_0 and $\varrho\, l$ are all known, it is easy to calculate $(\mu\,/\,\varrho)$.

It is necessary to prepare a special second pellet for measurement of the mass absorption coefficient because the first (used to measure the analytical Sr Kα emission) is normally deliberately made to be more than 'infinitely thick'. It would thus completely or almost completely absorb the Sr Kα beam in the attenuation measurement, so that the mass absorption coefficient would be indeterminate, or at best determined with poor precision.

Once the experimental configuration has been established, this is a simple technique that has been shown to yield good results for the range of sample types and analyte elements for which it is appropriate. Its principal disadvantages are:

(a) It requires an additional, time-consuming set of measurements for each element on each sample/standard.
(b) It may not be easy to modify the spectrometer to incorporate the pellet holder and limiting apertures, and routine insertion and removal of the pellets may be difficult. This was not a problem with earlier generations of XRF systems, in which the spectrometer was usually mounted in an exposed and readily accessible position. In later instruments, however, the spectrometer is often enclosed within a cabinet fitted with safety switches as protection against accidental radiation hazards. It may then not be easy to modify, and insertion or removal of the pellets may require bypassing of the safety switches; this procedure cannot be recommended, particularly for relatively unskilled operators.
(c) The method is often inappropriate for heavy-element samples whose mass absorption coefficients are high enough to make it very difficult to prepare durable pellets of less than 'infinite' thickness. In such cases the attenuation can be reduced by mixing the sample powder with a known proportion of a low-absorbing diluent of known mass absorption coefficient, and then applying a correction for the dilution; however, this involves additional weighing and mixing, consuming more time and presenting further sources of potential error.
(d) Likewise the method is usually inappropriate for light elements with low-energy, long-wavelength characteristic emissions for which mass absorption coefficients are relatively high. In this case dilution with a low absorber may be precluded by particle-size effects, so that the direct measurement technique is of only limited value for elements

CORRECTION PROCEDURES

whose characteristic wavelengths are longer than about 3.0 Å (i.e. for elements lighter than Ca).

(e) Clearly it is inappropriate for samples which cannot be pressed into absorption pellets because of their physical form (solutions, castings etc.), although it may be possible to adapt it successfully in some specific cases.

COMPENSATION TECHNIQUES

The scattered radiation method X-radiation entering the spectrometer from the sample includes the following:

(a) Characteristic emission spectra from the elements present in the sample, including the line $\lambda(i)$ used for determination of the analyte element i. Since part of the characteristic spectrum is generated at depth, the generated intensity of $\lambda(i)$ will be attenuated by absorption, to an extent determined by the mass absorption coefficient of the sample for $\lambda(i)$, such that

$$I_{\lambda(i)} = k \; c_i \:/\: A_2(i) \qquad \text{(Eqn 8.9)}$$

(b) Primary tube radiation *scattered* by the sample in the direction of the spectrometer. This includes both continuous and characteristic radiation from the tube anode, and each wavelength in the tube spectrum is scattered both coherently and incoherently (Ch. 3). Scattering takes place at the surface of the sample *and also at depth within it*, and the depth-scattered X-rays are attenuated by absorption in the same way as the analyte emissions; their intensities measured in the spectrometer will also depend on the appropriate sample mass absorption coefficients.

In fact it can be shown that scattered intensities also vary as the reciprocal of their mass absorption coefficients, in a fashion which can be described by an analogue of Equation 8.9, viz.

$$I(\lambda_{\text{scattered}}) = (k_2 \; \sigma) \:/\: A_{\text{scattered}} \qquad (8.22)$$

in which $A_{\text{scattered}}$ is the sample mass absorption coefficient for the scattered wavelength and σ is a composition-dependent parameter called the *mass scattering coefficient* (which has different values for coherent and incoherent scatter of a given wavelength).

If mass scattering coefficients for a selected scattered wavelength λ_s are measured or calculated for a range of sample compositions, and plotted against mass absorption coefficients of the same samples for an analytical wavelength $\lambda(i)$ (Fig. 8.13), it is found that $\sigma_{\text{incoherent}}$ varies linearly as μ for $\lambda(i)$, provided that none of the samples has a significant absorption

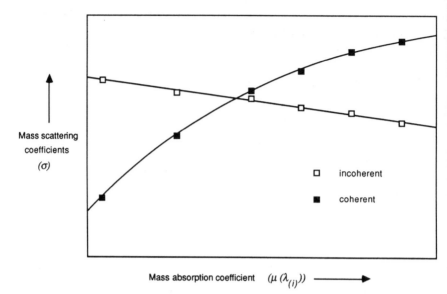

Figure 8.13 A typical plot of mass scattering coefficients against mass absorption coefficients for one analytical wavelength, for a series of samples of different compositions. Incoherent mass scattering coefficients are essentially linear functions of the corresponding mass absorption coefficients. The variation in coherent mass scattering coefficients is non-linear, but approximately linear over limited but useful compositional ranges. Subject to certain limitations (see text), calibrated measurements of tube radiation scattered from samples can therefore be used to estimate mass absorption coefficients (after Tertian & Claisse 1982).

edge between λ_s and $\lambda_{(i)}$. $\sigma_{coherent}$ varies as μ in a fashion which is non-linear overall, but quasi-linear over a limited range of μ, i.e. it is effectively linear for a group of samples which do not vary greatly in composition (and which likewise do not have absorption edges between λ_s and $\lambda(i)$).

In the case of incoherent scattering, the linear relationship may be expressed by

$$\sigma_{incoherent} = a + b\, A_{sample}(i) \qquad (8.23)$$

in which $A_{sample}(i)$ is the mass absorption coefficient of the sample for the analytical emission $\lambda(i)$, and $\sigma_{incoherent}$ is the mass scattering coefficient of the sample for an incoherently scattered wavelength λ_s. Combining Equations 8.22 and 8.23,

$$I)(\lambda_{scattered\ (incoherent)}) = [k_2\,(a + b \cdot A_{sample}(i))] / A_{scattered}$$

CORRECTION PROCEDURES

Because of the smooth variation of mass absorption coefficients between absorption edges, and hence providing there are no such edges between $\lambda_{\text{scattered}}$ and $\lambda(i)$,

$$A_{\text{scattered}} = k' A_{\text{sample}}(i)$$

and therefore

$$I_{\lambda(i)} / I_{\lambda_{\text{scattered (incoherent)}}} = [k_3\, a) / A_{\text{scattered}}] + k_3\, b \qquad (8.24)$$

Because b, the slope of the incoherent scatter curve in Figure 8.13, is very small, the term $(k_3\, b)$ is also very small and can be neglected. Combining Equations 8.9 and 8.24,

$$I_{\lambda(i)} / I_{\lambda_{\text{scattered (incoherent)}}} = k_4\, c_i \qquad (8.25)$$

in which the new proportionality constant (k_4) indicates that, *subject to the limiting assumptions*, the intensity ratio of the analyte emission line to the scattered tube line is independent of matrix absorption and linearly related to the analyte concentration. A linear analytical calibration curve, compensated for absorption, can therefore be prepared by selecting an appropriate incoherently scattered tube line (close to the analyte line $\lambda(i)$ and with no intervening absorption edges) and plotting this ratio against c_i for a series of calibration standards of known c_i.

The scattered radiation method is most commonly used for trace-element determination in groups of samples or standards that do not vary too substantially in composition (remember that it is based on simplifying assumptions that may not hold in cases of gross compositional variation). In principle it is possible to use either incoherently or coherently scattered tube characteristic radiation, or even scattered tube continuum. In practice incoherently scattered radiation is normally preferred because it is relatively easy to measure and to correct for background, and because of its linear rather than quasi-linear relationship between mass absorption and mass scattering coefficients. It is a versatile technique that has been used to solve many analytical problems, its only major drawback being that it requires additional measurements for the scattered line (peak, background and possibly line overlap) and hence approximately doubles the total analysis time for each element (though one scattered tube line may serve as a convenient reference for several analyte lines).

Use of an internal standard This method is based on modification of the sample by addition of a known amount of another element having a suitable emission line and absorption edge close to those of the analyte element. The principle, which is extensively used in other spectrographic procedures such as optical emission spectrography, is that matrix effects will be similar for the closely spaced analyte and internal standard emissions, and the *ratio of their measured intensities* will then be a linear function of the *ratio of their concentrations*:

$$\frac{I_{\lambda(i)}}{I_{\lambda,\text{internal standard}}} = k \frac{c_i}{c_{\text{internal standard}}} \qquad (8.26)$$

Equation 8.26 is derived from Equation 8.9 on the same basis as Equation 8.25, and is limited by the same assumptions (including the absence of significant absorption edges between the analyte and internal standard emissions).

It is apparent that the element added as the internal standard should not be present, at least in significant proportions, in the samples or standards. It should be added in concentrations which will yield comparable intensities to those obtained from the analyte. If a K emission is to be compensated, it is best to use an internal standard yielding a suitable K line for comparison in order to match excitation efficiencies, enhancement factors etc. (although there are in fact many examples of the successful use of K lines to compensate L emissions, and vice versa).

A limiting factor is the difficulty of accurate and homogeneous addition of the internal standard (unless the samples are actually prepared as liquid solutions, to which this method is ideally suited). If the internal standard is added in solid form to powdered samples, uniform dispersion may be difficult to achieve and microabsorption ('particle-size') effects may be pronounced. For this reason the use of internal standards is not favoured for analyte emissions of wavelengths greater than about 3.0 Å. Although often recommended, addition of the internal standard as a solution, which can be accurately dispensed with a pipette or burette and then allowed to crystallize by evaporation, does not necessarily eliminate the microabsorption problem (because large crystals may form), and it does not ensure homogeneous distribution.

In practice, it can often be surprisingly difficult to find suitable internal standards – particularly for samples of complex bulk composition which have many absorption edges, or if the method is to be used for a number of different elements in the same sample. Although one internal standard may, with luck, serve for two or even three analyte elements, the procedure can become complex if multiple internal standards are necessary. Overall the method is best suited to the determination of one or two relatively heavy elements occurring at low to moderate concentrations in a relatively uniform matrix; it has been widely and successfully used for applications such as the determination of heavy metals (Cu, Ni, Zn etc.) in lubricating oils.

CORRECTION PROCEDURES

Analyte addition ('spiking') A logical extension of the internal standard method is to assume that the best element to add as an internal standard is the analyte element itself; it will obviously be subject to identical matrix effects and there is no need to worry about intervening absorption edges etc. If I_1 and I_2 are the measured analyte fluorescence intensities before and after addition of a known concentration of the element itself, then it can be shown that

$$c_i = \frac{I_1 \times \Delta c_i}{I_2 - I_1} \tag{8.27}$$

The added concentration should be sufficient for adequate precision in the measurement of I_2, which generally means that $\Delta c_i \approx c_i$, but clearly should not be sufficient to alter significantly the mass absorption coefficient of the sample for the analyte emission. Hence this technique is best suited to the determination of low (trace) concentrations. It is subject to the same homogeneity and microabsorption problems as the internal standard method, but it is free of the restrictions on sample complexity because the addition is the analyte itself, and therefore it does not introduce any undesirable interferences, and because there is no problem with intervening absorption edges. It is surprising that this approach has not been more widely used, particularly in combination with heavy absorber dilution/fusion techniques which could extend its useful concentration range beyond the trace level.

Double dilution A variant of simple dilution, known as *double dilution*, has been successfully used to handle the problem of matrix absorption effects by compensation rather than by simple attenuation. Its principles may be briefly summarized as follows.

Suppose that a sample containing an analyte concentration c_i is diluted (e.g. with a fusion flux) to two or more dilution levels yielding *sample* (i.e. not analyte) concentrations of $y_1, y_2. \ldots$ Suppose further, in the first instance, that the diluent (e.g. the fusion flux) has exactly the same mass absorption coefficient for the measured analyte emission as the sample itself. The absorption term A_2 in Equation 8.9 will then be constant, regardless of the extent of dilution, and Equation 8.9 will then reduce to

$$I_{\lambda(i)} = k' c_i y \tag{8.28}$$

A plot of $I(\lambda(i))$ against y will then be a straight line, such as the line A in Figure 8.14. Given this linear relationship, the concentration of the analyte element i could then be determined simply by calibration with a

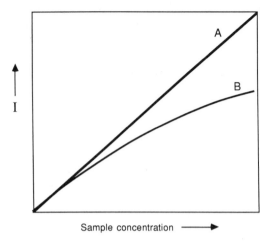

Figure 8.14 Typical plots of emission intensities against the compositions of a series of diluted samples (i.e. the variation in I is plotted as a function of the *dilution*). If the diluent has the same mass absorption coefficient as the sample, the dilution attenuation will be linear (A). More generally, the diluent will either reduce or increase the net absorption coefficient, and the curve will be bowed either downwards (B) or upwards. If the mass absorption coefficient of the diluent for the measured wavelength is known, that of the sample can be determined by making calibration measurements on two or more samples of different dilution to establish the direction and degree of curvature (after Tertian & Claisse, 1982).

standard of known c_i, diluted to two or more levels to establish the slope of the intensity vs. dilution curve.

In general, however, the mass absorption coefficient of the diluent will be different from that of the sample. If the sample is more highly absorbing than the diluent (the more common case), the intensity vs. dilution curve will approximate to the simple linear case only as $y \to 0$ (i.e. at near-infinite dilution). As the dilution level is reduced (the sample concentration is increased), the absorption term A_2 of Equation 8.9 will increase, and the ratio

$$I_{\lambda(i)} / (c_i \, y) = (k' / A_2)$$

will decrease. The intensity vs. dilution curve will then become hyperbolic (e.g. B in Fig. 8.14), and its curvature will be a measure of the *difference* between the mass absorption coefficients of the sample and the diluent. The curvature may be expressed as a parameter φ, defined as

$$\varphi = \{A_2(\text{sample}) / A_2(\text{diluent})\} - 1 \qquad (8.29)$$

EXAMPLES OF MATRIX CORRECTION PROCEDURES

which Tertian and Claisse (1982) have called the *differential matrix factor*. It can be incorporated in an equation for the intensity vs. dilution curve, written in the form

$$I(\lambda(i))_y = (k' \, c_i \, y) / (1 + \varphi \, y) \quad (8.30)$$

Note that if A_2(sample) = A_2(diluent) then $\varphi = 0$ and Equation 8.30 reduces to the simple linear case of Equation 8.28.

If A_2(diluent) is known, then A_2(sample) could be determined by establishing φ from Equation 8.30, using $I(\lambda(i))$ measurements made at two or more dilutions y (c_i, the concentration of i in the sample, is constant). However, it can be shown (e.g. Tertian & Claisse 1982, pp. 243–56) that it is in fact not necessary to know or even to estimate A_2(diluent) if a parallel series of two or more dilution measurements is made on the sample and on a calibration standard containing known c_i. The common effect of A_2(diluent) is eliminated by ratioing the sample and standard data, and c_i for the sample can then be obtained directly without need of a calibration curve. In effect, the absorption errors in both sample and standard are simultaneously compensated by measuring their differential matrix factors for a common diluent.

Double dilution is a versatile technique that can be applied to almost any element in any sample that can be effectively diluted by solution, fusion or even powder mixing (subject, as usual, to possible homogeneity and/or microabsorption problems). High concentrations can be used to enhance sensitivity and counting precision; by contrast simple dilution demands low diluted concentrations to be effective.

On the other hand, extra sample preparation and measurement are involved, and weighing, mixing and intensity measurement must all be performed with high standards of accuracy to obtain reliable results. Again it must not be forgotten that the method depends on the simplifying assumptions inherent in Equation 8.9 (e.g. the concept of an effective excitation wavelength, or the assumption that enhancement effects are negligible or can be treated as negative absorptions). The potential errors involved in these assumptions are usually small, but not necessarily negligible.

8.4 Examples of matrix correction procedures

In the following sections, two commonly used analytical procedures are described in a little more detail, to illustrate some of the principles discussed above. It is emphasized that the chosen examples are intended only as illustrations, and there is no intention to imply that these methods are superior to or more versatile than any of the others mentioned in the previous discussion or described elsewhere in the literature of X-ray spectrometry.

8.4.1 The use of empirical interelement correction factors

It has already been noted that a calibration curve will only plot as a straight line if there is a linear relationship between c_i (the concentration of element i) and $I(\lambda(i))$ (the intensity of the analytical characteristic radiation emitted by element i), i.e. if there are no significant interelement absorption or enhancement effects. Of course there will always be some scatter of calibration points about the line because of statistical counting errors, but it is assumed that these can be adequately compensated by simple regression techniques and adoption of proper counting strategy, and they will not be further considered in this part of the discussion.

In general, matrix effects will lead to significant departures from linearity. In systems of only two or perhaps three components, the straight line will become a curve, bowed upwards or downwards depending on relative mass absorption coefficients and on whether absorption or enhancement effects predominate for the elements concerned (Fig. 8.11).

In more complex multicomponent systems, the form of the experimental calibration curve becomes more irregular. While a straight line can still be fitted to the calibration points by simple linear regression, some points will lie above the curve and some below it, since varying proportions of the other elements present in the sample can result in either net absorption or net enhancement of the true analyte intensity (Fig. 8.15). Specifically, it is possible for two or more calibration standards containing the same concentration of the analyte element i to plot on either side of the simple best-fit calibration line, depending on the relative concentrations of other elements which variously attenuate or enhance the emitted intensity of $\lambda(i)$.

The basis of several empirical approaches to the problem of matrix errors is to assume that the effect of element j (either attenuation or enhancement) on the emission of $\lambda(i)$ can be expressed as a negative or positive influence coefficient or α-factor (Section 8.3.3). The term α-factor is also used in some other, slightly different contexts, requiring some caution in interpretation (see also Ch. 9). It follows that in this simple approach, secondary fluorescence (enhancement) effects are treated simply as negative absorptions, which is a convenient but not always valid approximation.

The equation for the calibration line

$$c_i = D + E\, I(\lambda(i))$$

EXAMPLES OF MATRIX CORRECTION PROCEDURES

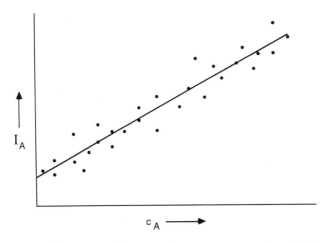

Figure 8.15 In multi-element samples, some elements may attenuate and others enhance the measured emission intensity from an analytical element A. If concentrations of the interfering elements vary more or less independently, it is possible for some samples containing a given concentration of A to yield a net attenuation while others show net enhancement. In other words, the degree of attenuation or enhancement of the emission intensity of A depends not only on the total concentration of the other elements but also on their *relative* concentrations. A working curve prepared from a series of calibration standards must then be fitted by multiple rather than simple linear regression. In general, such a working curve will not be truly linear, but in practice it can often be assumed to be so over a reasonably limited range of compositions.

can then be replaced by

$$c_i = \{D + E\ I(\lambda(i))\} \times \{l + \Sigma\ (c_j\ \alpha_j)\} \qquad (8.31)$$

If the α-factors for all of the elements present in a sample are known, and if the concentrations of those elements can be established with reasonable accuracy (e.g. from raw intensity data), then this equation can be used to correct the apparent concentration of each element for matrix errors. This will yield a better estimate of concentrations, which can then be used to recalculate the correction factors, and the whole procedure can be reiterated to a satisfactory level of convergence.

If absorption effects predominate, the α-factors can be calculated from known (or often, in the case of light elements, extrapolated) mass absorption coefficients. However, because of uncertainties in the mass absorption coefficient data and because enhancement effects can seldom be overlooked altogether, the resulting α-factors are often significantly inaccurate and can be improved considerably by empirical modification in the light of practical experience.

Alternatively, the α-factors can be determined experimentally from a series of calibration standards, of known compositions which should preferably span the complete range of compositions expected in the analytical samples. One of the great advantages of the XRF technique is that it is often possible to synthesize many or all of a wide range of calibration standards, particularly when dilution procedures such as flux fusion are to be employed. Silicate major-element standards, for example, can be prepared by fusion of an appropriate series of oxide mixtures.

The fluorescent intensity emitted under analytical conditions from each element in each calibration standard is measured, ensuring that counting statistical errors are contained within suitable limits. The slope and intercept of the best-fit linear calibration curve and the best-fit α-factors are determined by multiple linear regression, and they can then be used for subsequent correction of the raw data from analytical samples.

Multiple regression analysis requires that the number of calibration standards employed significantly exceeds the number of elements to be determined, without compositional duplication. Clearly the success of this method will depend in a large part on the intelligence with which the calibration standards are selected and the skill with which they are prepared and analysed.

Given adequate facilities for the calculations involved, this approach is capable of satisfactory solution of a remarkably wide range of matrix error problems; subject, as always, to the simplifying assumptions on which it is based (principally that the matrix effects of different elements on the analytical wavelength are additive in simple linear fashion, which cannot be strictly true since the interfering elements will also have some effect on the X-ray emissions from each other, and that fluorescent enhancement can be regarded as a negative absorption, which also cannot be strictly true).

Of course, the use of matrix coefficients to correct for absorption and enhancement cannot cope with other sources of potential error, such as particle-size or counting statistical effects, and like any other simplified approach it works best when the magnitude of the corrections required is not too large. It may thus still be helpful to reduce the magnitude of potential errors, as much as possible, by a suitable sample preparation technique, such as dilution or fusion.

Among other successful applications, this has proved to be a suitable method for routine major-element silicate analysis. Samples and standards can be prepared by lithium tetraborate fusion, according to the method of Norrish and Hutton (Section 8.4.2); this reduces the magnitude of matrix corrections, overcomes microabsorption (particle-size) effects on the determination of light elements, and also makes it possible to prepare a wide range of calibration standards. Addition of lanthanum oxide to the fusion mix provides a heavy absorber which further reduces the relative variation in absorption effects from one sample (or standard) to another.

In silicate analysis, up to 12 major and minor elements are routinely determined. This requires solution for 14 coefficients (slope, intercept and 12 α-factors for each element), and hence demands more than 14 calibration standards; because of experimental and counting statistical errors, better results are obtained if considerably more standards are used for the regression. Stephenson (1971) recommends at least $2N$ standards, where $N = ((n + n^2) / 2)$ is the number of coefficients required for each element in a system of n elements, but in complex systems N rapidly becomes very large and satisfactory results can usually be obtained from a somewhat smaller number.

Silicate standards can be prepared from artificial oxide mixtures fused in borate glass in the usual way (see below), and synthetic standards can also be prepared for many other sample types. They should be designed to span the complete compositional range expected in the analytical samples.

The multiple regression calculations are tedious and best performed by computer. Some commercial XRF equipment manufacturers provide the necessary facilities in on-line mini- or microcomputers which are also used for instrument control during routine analysis. Such on-line systems can compute and store the appropriate coefficients and subsequently access them for on-line processing. Corrected compositions can then be printed directly at the end of each analysis, eliminating the errors and the tedium involved in transferring the raw data to an external calculator or computer for processing. This is a great convenience, but it is also necessary to repeat earlier warnings that on-line processing can mask severe errors unless the analyst is thoroughly conversant with the procedures involved and maintains conscientious supervision.

It is also necessary to stress once again that this approach is based on certain simplifying assumptions, and that although it has proved to be successful in many analytical situations, it cannot safely be assumed always to be so. Like any other analytical procedure, it must be thoroughly tested (e.g. by using it to analyse a comprehensive range of 'samples' of known composition, preferably different from the standards used for calibration) before it is adopted for routine applications. It is also desirable to limit the compositional range over which empirically determined α-factors are assumed to be constants.

One further point also needs emphasis at this stage. Novice analysts in particular are often inclined to place excessive faith in analysis totals and/or agreement between replicates as criteria of reliable analytical procedures. Of course, poor totals or poor replicates are convincing evidence of unsatisfactory results; the reverse, unfortunately, is not necessarily true. Good totals can often turn out to be due to one or more normalizing steps in the calculation of corrections (particularly those performed by computers, on- or off-line, whose fundamental algorithms are not clearly understood). They may have little or no significance for some samples, e.g. those in which volatile components such as water are lost in sample preparation. Good replicates may just be accurately reproducing systematic errors in sample preparation, analysis or calculation. Due consideration must be given to these possibilities whenever an analytical technique is being assessed.

8.4.2 The Norrish and Hutton method for silicate analysis

This method is at least partly a variant on the empirical approach outlined above, in which the correction coefficients are determined and applied in a somewhat different fashion. It is again based on some critical assumptions which have been discussed earlier in reviewing the 'simplified theory' approach – specifically those of an equivalent wavelength for the primary beam (which implies the absence of significantly interfering absorption edges), of the predominance of A_2 over A_1 (Eqn 8.9) in the correction procedure (which implies appropriate spectrometer geometry), and of a lack of other systematic errors such as those due to microabsorption effects.

The latter assumption is ensured by fusion of the samples and standards, prior to analysis, in a flux based on lithium tetraborate. Flux fusion procedures are very versatile and have been extensively used for a wide variety of analytical applications in which various different flux compositions have been proposed for specific applications. Although considerable variation can be tolerated, and fluxes of special composition may be more appropriate for some samples (e.g. highly siliceous or highly refractory samples), Norrish and Hutton (1969) have recommended a suitable general-purpose 'recipe' for silicate analysis, as follows:

$$\begin{array}{l} 1.50 \text{ g flux (see below)} \\ 0.02 \text{ g LiNO}_3 \\ \underline{0.28 \text{ g sample or standard}} \\ \underline{1.80 \text{ g}} \end{array}$$

The recommended flux is prepared in advance by prefusion, grinding and thorough mixing of appropriate proportions of the following mixture:

38.0 g lithium tetraborate (anhydrous)
29.6 g lithium carbonate
13.2 g lanthanum oxide

EXAMPLES OF MATRIX CORRECTION PROCEDURES

Lanthanum oxide is added as a heavy absorber to reduce the relative magnitude of matrix absorption variations between samples and standards. Details of the flux preparation procedure are given by Norrish and Chappell (1977); alternatively, prepared flux may be obtained commercially. A technique for fusing the sample/flux mixture and casting the melt into glass discs ('buttons') is described by Norrish and Hutton (1969) and summarized by Norrish and Chappell (1977). It consists essentially of fusing the mixture in a platinum or gold/platinum crucible at about 1000 °C, casting the melt in a preheated mould to form a disc of appropriate dimensions, and controlled cooling ('annealing') of the disc to prevent shattering due to thermal stresses. The fusion may be performed over a Meker or similar burner, or it may be carried out in an electric furnace (preferably with provision for accurate control of the heating rate). In either case, special effort must be made to ensure homogeneity of the melt before casting.

Most geochemical laboratories have now adopted some variant of the basic procedure described by Norrish and Chappell. If large volumes of samples are to be processed, an automatic fusion apparatus may be appropriate; these are available from commercial manufacturers.

The 1.80 g mixture originally proposed by Norrish and Hutton may be insufficient to provide discs of 'infinite thickness' for some X-ray spectrometers utilizing wide sample holders. In this case the proportions can simply be scaled up as required.

Lithium metaborate ($LiBO_2$) can be used instead of lithium tetraborate ($Li_2B_4O_7$); the metaborate melts at a slightly lower temperature (850 instead of 920 °C) and appears to be more effective for rocks which are highly 'acid' in the geological sense (i.e. with high contents of SiO_2), and it yields a more fluid melt which is easier to cast. On the other hand, tetraborate is somewhat more effective for rocks with high CaO, K_2O or Al_2O_3 contents. Other fluxes that have been successfully used include sodium tetraborate (which melts at 740 °C, but of course is undesirable if sodium is to be determined in the samples and preferably should be avoided if other light elements are to be analysed because of its relatively high absorption) and sodium hexametaphosphate ($(NaPO_3)_6$). Sodium carbonate fuses readily to a clear melt, but often yields inhomogeneous cast discs. However, mixtures of sodium carbonate with sodium metaborate have been found to be effective for some refractory materials (e.g. SnO_2) that do not fuse readily in lithium tetraborate.

For specific analytical problems there may therefore be considerable scope for experiment. The Norrish and Hutton 'recipe' was proposed as a good all-purpose approach for most silicate analysis, and it is now used in many different laboratories with considerable success. It should be noted that if it is modified for any reason, the correction coefficients suggested by Norrish and Hutton (see below) must be recalculated or redetermined, since they are dependent on flux composition.

The fusion procedure dilutes the samples (and the standards) to a point where residual matrix errors may be tolerable or even insignificant, particularly if there is little compositional variation within and between samples and calibration standards. In such cases a simple linear calibration curve can be used for each element. In general, however, for more versatile routine applications matrix absorption/enhancement errors are not sufficiently compensated by the dilution, and supplementary correction of the residual errors is required. This is performed by an iterative technique in which an approximate composition is first determined by linear calibration and used to calculate a correction factor for each element. These factors are then applied to the raw data to yield a corrected composition, which is used to calculate new, refined correction factors. The calculation may be reiterated as often as desired, but usually converges satisfactorily after two or three cycles.

The philosophical basis of the method is that the samples or standards actually constitute only minor proportions of the fused discs which are analysed. In effect, all samples can be regarded as borate glass, modified by relatively minor substitution of the sample elements. The effective absorption or enhancement coefficient for each element in the glass disc will thus be the appropriate coefficient of borate glass for that element, modified slightly (but not insignificantly) by substitution of the sample elements.

Hence we can write

$$A_{\lambda(i)} = X_i + c_1 Y_1 + c_2 Y_2 + c_3 Y_3 + \ldots + \text{loss } Y_L \qquad (8.32)$$

in which $A_{\lambda(i)}$ = mass absorption coefficient of the fused sample disc for the emission wavelength of the analyte element i,

X_i = mass absorption coefficient of pure borate glass for the wavelength $\lambda(i)$ (including the lanthanum oxide, if added),

c_j = weight fraction of element j,

Y_j = the *difference* between the mass absorption coefficient of pure borate glass (X) and the mass absorption coefficient of element j for wavelength $\lambda(i)$,

loss = loss on fusion (e.g. due to the loss of volatiles such as water or CO_2), which effectively decreases the total disc weight and is thus a 'negative' element, i.e. it increases the effective concentrations of all elements and hence enhances their emitted intensities; it must therefore have a negative Y_L.

EXAMPLES OF MATRIX CORRECTION PROCEDURES

The mass absorption coefficients (X_i, Y_j) in this expression can be calculated from tables of mass absorption coefficients, or they can be determined experimentally by an empirical technique similar to that described above for the α-factor approach. For simplicity of routine calculation it may be convenient to normalize the mass absorption coefficients for each analytical wavelength (i.e. for each analyte element) so that $A_{\lambda(i)}$ is approximately equal to unity for a so-called 'average' rock, which Norrish and Hutton defined as a 50 : 50 mixture of the international rock standards G1 and W1. Alternatively, the coefficients may be normalized to unity for pure borate glass. Normalizing is not essential and is not of great practical value if corrections are computer-based; however, Norrish and Hutton used it to calculate their tables of suggested coefficients, which are still widely used in other laboratories.

Using the 'simplified theory' approach, and ever-mindful of its specific assumptions, we can write

$$c_i = k_i\, I_{\lambda(i)}\, A_{\lambda(i)} \qquad \text{(Eqn 8.9)}$$

in which C_i = weight concentration of element i,

k_i = calibration constant for element X,

$I_{\lambda(i)}$ = emitted intensity of characteristic wavelength $\lambda(i)$ of element i,

$A_{\lambda(i)}$ = mass absorption coefficient of analysed sample (i.e. of the borate disc) for $\lambda(i)$.

For a measured intensity I,

$$I = I_{\text{peak}} - I_{\text{background}}$$

Therefore, omitting the 'i' and 'λ' subscripts for convenience,

$$c = k\,(I_{\text{peak}} - I_{\text{background}})\,A$$
$$= (k\,I_{\text{peak}}\,A) - (k\,I_{\text{background}}\,A) \qquad (8.33)$$

For diluted samples, background intensity will be relatively insensitive to compositional variations, and the term $(k\,I_{\text{background}}\,A)$ will be effectively constant (particularly in the light of counting statistical errors in the measurement of relatively low background intensities). Hence

$$c = (k\,I_{\text{peak}}\,A) - B' \qquad (8.34)$$

where B' is a constant, physically equivalent to the apparent weight percentage of the analysed element corresponding to the background intensity.

Now substituting the general expression for $A_{\lambda(i)}$,

$$c = k\, I_{\text{peak}}\, (X + \Sigma\, c_j\, Y_j) - B' \qquad (8.35)$$

in which $k\, I_{\text{peak}}$ = measured (linear approximation) concentration,

X = mass absorption coefficient of pure borate glass,

B' = background correction, which can be determined experimentally from a fusion sample free of the element sought,

c_j = weight concentration of element j, determined experimentally by linear approximation,

Y_j = the *difference* between X and the mass absorption coefficient of element j for the analytical wavelength, calculated from tables of mass absorption coefficients or determined experimentally and including a term for fusion loss.

Norrish and Hutton (1969) presented a table of experimentally determined and normalized absorption coefficients appropriate to routine silicate major-element analysis, and a revised set of coefficients (slightly modified empirically in the light of experience) was given by Norrish and Chappell (1977; see Appendix A, which also includes a table of equivalent coefficients calculated by Dr M. J. Hough and normalized to unity for pure borate glass instead of the G1 : W1 standard of Norrish and Hutton).

It is emphasized that these coefficients are not mass aborption coefficients in the conventional sense; instead they represent the effect on the mass absorption coefficients of borate glass produced by substitution of the elements concerned. As such they refer specifically to samples/standards prepared according to the Norrish and Hutton flux formula; they must be recalculated or redetermined for other fusion mixes or procedures.

Their practical application is illustrated by the following example, from Norrish and Chappell (1977). A linear approximation (i.e. uncorrected) analysis of a sample of the mineral sphene gave the following results:

Fe_2O_3	1.32 wt per cent
TiO_2	34.98
CaO	29.01
SiO_2	31.55
Al_2O_3	0.52
loss on fusion	—
	97.38

EXAMPLES OF MATRIX CORRECTION PROCEDURES

Background concentration equivalents (B') measured experimentally on a series of fusion discs, each containing none of the element concerned, were determined as follows:

Fe	0.18
Ti	0.16
Ca	0.06
Si	0.24
Al	0.12

X and Y values, from the table of Norrish and Chappell (1977), are as follows:

Element	X	Fe	Ti	Ca	Si	Al	loss
				Y			
Fe	1.046	−0.027	0.146	0.134	−0.065	−0.074	−0.163
Ti	0.851	0.081	0.179	0.647	0.110	0.078	−0.132
Ca	0.865	0.090	0.065	0.110	0.128	0.105	−0.134
Si	1.014	0.082	−0.034	−0.042	−0.061	0.122	−0.158
Al	1.056	0.112	−0.032	−0.037	−0.088	−0.072	−0.164

Consider the case of Fe_2O_3:

$c = k\, I_{peak}\, (X + \Sigma c_j\, Y_j) - B'$

$ = k\, I_{peak}\, (X + c_{Fe}\, Y_{Fe} + c_{Ti} + Y_{Ti} + \ldots + \text{loss} \cdot Y_{loss}) - B'$

$ = 1.32 \times [1.046 + (0.013 \times -0.027) + (0.3498 \times 0.146) +$
$(0.2901 \times 0.134) + (0.3155 \times -0.065) +$
$(0.0052 \times -0.074) + (0 \times -0.163)] - 0\text{:}18$

$ = (1.32 \times 1.114) - 0.18$

$ = 1.29$

When similar corrections are applied to all elements (note that, as is conventional in silicate analysis, oxides are treated as 'elements' and the Norrish and Chappell coefficients were calculated for oxides rather than pure elements):

Oxide	Uncorrected composition	A	B'	Corrected composition
Fe_2O_3	1.32	1.114	0.18	1.29
TiO_2	34.98	1.138	0.16	39.63
CaO	29.01	0.962	0.06	27.84
SiO_2	31.55	0.972	0.24	30.44
Al_2O_3	0.52	1.007	0.12	0.40
Total	97.38			99.60

The improvement in analysis total is not necessarily proof of the success of the method – but calculation of the ideal composition of sphene $((Ca,Fe)TiSiO_5)$ and comparison with the raw and corrected compositions is more convincing.

If necessary, the corrected compositions could now be utilized to calculate refined values of A, which in turn could be used for further correction and so on. In this case, however, convergence is rapid and a further cycle does not yield significantly different corrections (e.g. Fe_2O_3 becomes 1.30, not significantly different from 1.29). The reader is urged to make the calculation as a useful exercise.

This is a simple technique to apply, once its essential concepts have been grasped (and providing its limitations are understood). In one form or another, it is currently the most widely used of all XRF silicate analysis procedures. However, it is emphasized that it cannot be used for sample types that do not satisfy its implicit assumptions. Nor is it recommended for the analysis of low concentrations ('trace elements') because of the loss in sensitivity and the possibility of contamination arising from dilution with the fusion mix.

A point often overlooked by beginners is that derivation of the uncorrected composition by reference to one or more calibration standards implies determination of the calibration constant k from one or more relationships of the form

$$\frac{c_{sample}}{c_{standard}} = \frac{k\, I_{sample}\, (X + \Sigma\, c_j\, Y_j)_{sample} - B'}{k\, I_{standard}\, (X + \Sigma\, c_j\, Y_j)_{standard} - B'}$$

In other words, the measured *standard* intensities must also be corrected for matrix effects (using the same procedure and coefficients) before they are used to calculate the apparent compositions of the samples. If calibration is made with a group of standards, as is usually the case, the measured standard intensities must be corrected before the calibration constant is calculated.

EXAMPLES OF MATRIX CORRECTION PROCEDURES

In cases such as trace-element analysis, for which the dilution inherent in flux fusion is unacceptable, it is often satisfactory to use Equation 8.11, calculating the relevant mass absorption coefficients from the known (standard) or approximate (sample) compositions. Of course this requires determination or estimation of the major-element composition of each sample. In silicate analysis, trace elements are usually determined in pressed powder pellets, and the major-element compositions in fused discs. Mass absorption coefficients required for the trace-element correction calculations, using Equation 8.11, are obtained from standard sources, such as the tables of Heinrich (1966), or they may be computed from polynomial expressions fitted to experimentally measured and/or extrapolated data (see Table A.2).

9 Composition-dependent errors in electron microprobe analysis

As in XRF analysis, the major composition-dependent sources of error in electron microprobe analysis may be conveniently divided for discussion into two groups:

(a) *Interference* effects, including interference from background radiation and characteristic line spectra.
(b) *Interelement attenuation/enhancement* ('matrix') effects.

Although there are many similarities between the problems encountered in the two analytical methods, there are also some critical differences, as the following discussion will demonstrate.

9.1 Interference

The principal source of interference in microprobe analysis, as in XRF, is continuum radiation on which the analytical line is invariably superimposed. Estimation of the true peak intensity requires that the background be accurately estimated and subtracted from the measured (peak plus background) intensity.

The principles of background estimation are generally similar to those described for XRF (Ch. 8), but some key differences should also be noted. The most important of these is that because microprobe spectra are electron-generated, continuum and hence background intensities are proportionally much higher than they are in XRF (in which the continuum background is limited to tube continuum scattered from the sample and the dispersing crystal). Peak-to-background (P/B or P : B) ratios are therefore lower, with several important consequences:

(a) The sensitivity of microprobe analysis, expressed as the lowest weight concentration of an element yielding a peak of sufficient intensity to be significantly distinguished from the background, is generally poorer than the sensitivity of the same element/sample in XRF analysis. Whereas optimum XRF detection limits are commonly of the order of 1–10 parts per million or better for all except the lightest elements, microprobe detection limits are seldom better than about 100 ppm and are often much poorer. However, it should be remembered that the actual analysis volume is very much smaller in microprobe analysis, and its absolute mass detection limits (of the order of 10^{-15} g) are usually much better than in XRF, which normally requires much larger sample analysis volumes.

(b) Because of this limitation on sensitivity as a result of lower P : B ratios, microprobe analysis requires even more emphasis on accurate estimation of background intensities. In XRF, P : B ratios are often high enough for background estimation to be omitted altogether, at least in major-element analysis. This is seldom the case in microprobe analysis.

(c) It follows that in microprobe analysis it is essential for the analyst to keep background intensities as low as possible. This depends partly on good spectrometer design and maintenance to keep extraneous background radiation away from the detector (e.g. the use of magnetic or similar electron traps at the spectrometer ports), and partly on the operator's skills and strategies. For example, unnecessarily high accelerating potentials on the electron gun are undesirable, since they produce unnecessarily high continuum intensities. In particular, in wavelength-dispersive systems proper adjustment of the pulse height discriminator is mandatory – but is a factor often overlooked by the novice.

(d) As in XRF, background is often estimated by averaging intensity measurements made at one or more off-peak positions (Figs 8.1–8.3), correcting if necessary for sloping and/or non-linear background (Fig. 8.4). However, it is usually rather more difficult, or impossible, to synthesize profile calibration standards similar in composition to the analytical samples but containing none of the element sought, since probe standards must be homogeneous at the sub-micron level and are therefore not easy to prepare from pure starting materials.

Ideally a reference standard used to establish a background profile should be of the same (mean) atomic number as the analytical sample, since continuum intensity is Z-dependent, and it should be of the same or very similar major-element chemical composition to avoid errors due to differences in absorption characteristics. As far as the Z-dependence is concerned, good use can often be made of a reference library of 'pure' standards with a reasonable span of mean atomic numbers. These may be elements or simple compounds, such as oxides, sulphides or halides, that are available in 'spectrographically pure' quality. Most of the common oxides, sulphides and silicates have mean atomic numbers in the range from about 10 to about 25, and it is often not difficult to find one or more appropriate pure elements or simple compounds that can at least be used to establish the background profile in the vicinity of an analytical peak and select suitable off-peak measurement positions and procedures. In some cases these 'atomic number' standards can even be used for direct measurement of the background at the peak position, but it is always necessary to ensure that they really do contain none of the element sought and that there is no possibility of errors arising from composition-dependent differences in absorption. These criteria are usually very difficult to meet.

(e) Background and spectral interferences are both particularly severe if an energy-dispersive spectrometer is used, because of the relatively poor overall resolution of solid-state systems. In such cases background can never be ignored, even for major elements, and special, relatively sophisticated background correction procedures are required. Although these are normally performed on-line by the computer which is an essential component of most EDS systems, the analyst must understand the basis of whatever procedure is used, and must verify that it is both appropriate and effective.

Of all the factors that contribute to successful microprobe analysis, particularly in the lower concentration ranges, proper control and accurate measurement of background are possibly the most important – yet often the most neglected. Extra time spent in planning and investigating analytical strategies to cope with the background problem is nearly always amply repaid.

9.2 Interelement ('matrix') effects

As in XRF, quantitative microprobe analysis is based on the assumption of proportionality between the concentration of an element in a sample and the intensity of emission of the lines in its characteristic spectrum. If the proportionality is linear, the relationship can be expressed in the now familiar form of

$$c_i = k_i I_{\lambda(i)} \tag{9.1}$$

Experiment has shown, however, that this simple linear relationship only holds if the analytical samples and their calibration standards are all very similar in composition, e.g. if K-feldspar samples are analysed by reference to K-feldspar standards. In most other cases the composition–intensity relationship is demonstrably non-linear. As in XRF, this situation arises because the measured characteristic X-radiation is generated not only at the surface of the sample but also to some finite depth within it (Fig. 9.1). The nature and extent of interaction between the sample and the electron beam, and between the sample and the generated X-rays, are both composition-dependent. If there are appreciable differences in composition between samples and standards, the composition–intensity relationship becomes more complex, and

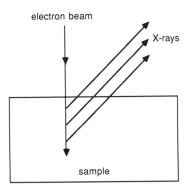

Figure 9.1 Depth generation of X-rays in electron microprobe analysis. X-rays generated beneath the sample surface interact with atoms along a finite path length within the sample before emerging at the surface.

appropriate correction procedures must be developed to compensate for these additional complexities.

The characteristic X-radiation may be considered to be generated in a series of thin layers dl, each at a depth l beneath the surface (Fig. 9.2). The total generated intensity of a particular X-ray line will be the sum of the intensities generated in all of these layers down to the 'infinite' depth L, which in practice is the depth at which no electrons retain sufficient residual energy to excite the spectral series to which that line belongs.

The total measured intensity will be that portion of the total generated intensity which is emitted towards the spectrometer, less the radiation absorbed by the sample itself (neglecting the usually minor additional absorption along the spectrometer path). Depending on the extent of the absorption, the effective 'infinite' thickness for the measured radiation may or may not be less than L; if the self-absorption is pronounced, radiation generated in the deeper layers may be totally absorbed before it reaches the sample surface.

The volume of sample contributing to the net measured intensity is thus determined either by the rate of electron energy loss with increased depth or by the extent of self-absorption. Both of these functions are composition-dependent.

This situation is further exacerbated by the possibility of *secondary fluorescence* – the intensity of the electron-generated X-radiation may be enhanced by fluorescence produced either by the continuum or by characteristic radiation from other elements in the sample. This effect is also clearly composition-dependent.

Once again, it is possible in theory to derive a set of equations to describe the effects of variable composition on each of these processes, and to combine the equations into a mathematical description of the

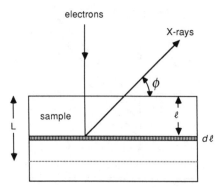

Figure 9.2 The geometry of electron microprobe X-ray generation. The electron beam penetrates the sample to an 'infinite' depth L, which is the depth at which the electrons no longer have sufficient energy to excite the analytical X-rays. The latter are produced in a series of thin layers, each of thickness dl at a depth l between the surface and L, and they emerge at the surface on their way to the spectrometer at a 'take-off' angle ϕ. Each X-ray thus traverses a path length $l \sin \phi$ within the sample. In this example the electron beam is vertically incident upon the sample surface, but in some designs electron incidence and X-ray take-off are both inclined.

generation of the measured emergent characteristic radiation. This mathematical model could be used, in principle, to apply corrections to the observed intensities to compensate for all of the composition-dependent effects. However, although considerable effort has already been expended on this fundamental approach, it has not yet yielded a satisfactory universal solution. The descriptive equations are complex and difficult to work with, and some of the physical phenomena involved are not completely understood. Many of the essential correction parameters (e.g. mass absorption coefficients for low-energy, long-wavelength radiations) are not yet accurately established.

For the time being, analysts have therefore necessarily adopted simplified approaches which have been shown to yield satisfactory results in cases where their limitations are appreciated and the assumptions on which they have been based are not overlooked. Most of the more pragmatic solutions to the problem fall into one of three overlapping groups:

(a) The errors may simply be recognized and tolerated. They are often not extreme, although it can never be safely assumed that this is the case. In general, matrix errors are more of a problem in microprobe analysis than in XRF, partly because of the physical factors involved and partly because compensating sample modification techniques such as dilution are usually inappropriate or impossible. Nevertheless, specific applications do not always demand high accuracy,

THE 'SIMPLIFIED THEORY' APPROACH

particularly if relative rather than absolute compositional data are required.

(b) The theory can be simplified to yield less complicated correction procedures appropriate to applications in which the simplifications are justified.

(c) Experimental observations can be used to derive empirical correction ('influence') coefficients that can be used within specific systems with theoretically limited but practically useful compositional ranges.

At present the most widely used correction procedures represent various combinations of (b) and (c). Simplified theoretical approaches have been used in conjunction with experimentally determined functions and parameters to develop procedures that have proved to be remarkably successful in handling a wide range of analytical problems. However, none of them is universally applicable, and it is still necessary for the analyst to understand the physical basis for their derivation and to ensure that they are not used in cases where their bounding assumptions or empirical components are not justified.

9.3 The 'simplified theory' approach

X-ray photons are produced in the electron microprobe as a result of interactions between energetic electrons and the atoms of which the sample is composed. Some of the energy lost by the electrons in such interactions (or 'collisions') is emitted as X-ray photons belonging to the continuum or to characteristic spectra, and quantitative analysis is based on the measurement of intensities of selected emissions in the characteristic spectra. A single electron typically loses its energy progressively as a result of a series of 'collisions' with the target atoms, with the generation of characteristic X-rays continuing for as long as

(a) the electron remains within the sample, and
(b) its energy remains greater than the *critical excitation energy* of the characteristic X-rays used in the analysis.

X-rays generated within the sample can themselves subsequently interact with sample atoms, so that their intensities may be reduced by absorption. If the conditions are appropriate, absorption may lead to the emission of fluorescent X-radiation characteristic of the absorbing atom, whose emitted intensity will then be enhanced. In a sample containing Cu and Fe atoms, for example, Cu K X-rays will be relatively heavily absorbed by Fe atoms. Cu Kα photons have energies of 8.040 keV, somewhat higher than the Fe K critical excitation potential of 7.110 keV; the mass absorption coefficient of Fe for Cu Kα radiation is 313

(Table A.6), compared to 52 for Cu Kα absorbed by Cu itself. In such a sample, the generated intensity of Cu Kα will be attenuated by Fe absorption, and the generated intensity of Fe Kα will be enhanced by Cu K-induced fluorescence (Cu Kβ, with an energy of 8.904 keV, will also contribute to the fluorescent enhancement of Fe Kα and Fe Kβ).

The actual extent of both attenuation and enhancement processes will clearly depend on the relative abundances of Cu and Fe atoms in the sample, and will therefore be composition-dependent.

In order to correct for these disturbing effects, it is necessary to consider the manner and rate at which the incident electrons lose their energy, the depth distribution of X-ray generation within the sample, the manner in which electron/atom interactions produce X-rays, and the ways in which X-ray/atom interactions modify the intensities of the electron-excited X-rays. In each case, the possible effect of compositional variations must be investigated.

It is convenient to discuss the complex phenomena involved under three broad headings, although it will become plain that they are more closely interrelated than this arbitrary division implies. We will consider

(a) electron paths and the question of electron energy transfer to X-ray generation;
(b) absorption of X-rays within the sample; and
(c) fluorescent enhancement within the sample.

All three are composition-dependent. The first is most strongly affected by the atomic numbers of the target atoms, since X-ray generation in the microprobe is primarily the result of electron–electron interactions, and the atomic number Z is equal to the number of electrons in a target atom. A correction factor to compensate for composition-dependent variations in the generation of characteristic X-rays is often called an *atomic number*, or Z, factor. However, it involves more than just atomic number effects, so that the term *generation factor* may be more appropriate. Similarly, absorption effects are handled in terms of an absorption, or A, factor and fluorescence effects as a fluorescence, or F, factor. This approach to the complete analysis of composition-dependent effects is then often described as the 'ZAF' procedure. It should be noted, however, that the separation of Z, A and F components is artificial and adopted for convenience in discussion and calculation. Absorption, for example, is a function of mass absorption coefficients, which in turn are functions of atomic number.

9.3.1 *The atomic number (Z) factor*

Energetic electrons can produce ionizations, which in turn may lead to the generation of X-rays, as long as they retain sufficient energy and remain within the sample. Hence it is necessary to consider the paths

THE 'SIMPLIFIED THEORY' APPROACH

followed by electrons within the sample, and the manner and rate at which the electrons lose energy along those paths.

When electrons in the incident microprobe beam encounter the sample ('target') atoms, they undergo complex interactions which may result in changes in their direction and energy. The principal effects involved are *elastic scattering*, in which an electron changes its direction of movement with relatively little energy loss, and *inelastic scattering*, which involves a loss of energy but usually negligible or relatively small change in direction. The first of these determines the paths of electrons within the target; the second, their rate of energy loss.

The energy lost in inelastic scattering may be consumed in the excitation of free or bound electrons or in the emission of *continuum X-ray photons*. The excitation of bound electrons may result in the emission of

(a) low-energy *secondary electrons*, which are those that absorb sufficient energy from the incident electron to 'escape' from the atom to which they were originally bound;
(b) characteristic *X-rays*, produced by orbital electron transitions following the ejection of an inner-shell bound electron (Ch. 2); and
(c) *Auger electrons*, produced by further energy transitions among the bound electrons, induced by some of the X-ray photons produced by (b).

The amount of incident electron energy lost in a single 'collision' may thus range from a few electron volts (required to free a valence-shell electron as a secondary electron) up to the total energy of the incident electron. The low-energy losses are favoured statistically, and large losses due to a single interaction are relatively rare. Hence electrons typically lose their energy progressively in a sequence of interactions which can be regarded as a more or less continuous process (the 'continuous slowing down' approximation of nuclear physics), and it can be assumed that there will be some functional relationship between the decreasing kinetic energy E of the electron and the path length that it has traversed.

We might thus hope to treat the problem of depth generation of X-rays in the microprobe by an analogue of the simplified treatment that we used for excitation by a primary X-ray beam in XRF (Fig. 9.3).

This simple model would require no more than the derivation of a linear function describing the changing rate of generation of X-rays, which could then be integrated over the depth/energy range corresponding to E_0 (the energy of the incident electron) to E_c (the depth at which the electron energy has been reduced to the critical excitation potential of the measured characteristic X-radiation), and then to treat the sample attenuation/enhancement effects on the X-rays generated from the surface down to the limiting depth.

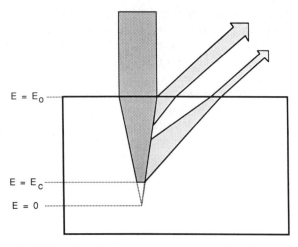

Figure 9.3 Simplified schematic representation of electron energy loss and sample attenuation of generated X-ray intensities in electron microprobe analysis. 'Infinite' depth (corresponding to L in Fig. 9.2) is the depth at which electron energy is reduced to E_c, the critical excitation potential of the analytical radiation. However, electrons may still have sufficient energy below this depth to excite continuous and lower-energy characteristic radiation.

Unfortunately, in the case of electron excitation there is an additional complication arising from the directional scattering of the electrons – relatively few electrons will travel along a straight path from the surface to the limiting depth. Equivalent scattering of primary X-rays in XRF is a comparatively trivial effect which can be ignored, but elastic scattering of the incident electrons (Fig. 9.4) has several important consequences:

(a) Some of the electrons will be scattered through large angles in a relatively small number of collisions, and

(i) some of these will be backscattered at or very close to the sample surface, so that they can leave the sample without undergoing significant energy loss and without producing any X-ray-generating ionizations;

(ii) some will penetrate the sample to shallow depths, losing some energy and producing some ionizations, before being scattered back out at the surface with significant residual energy (i.e. having failed to produce all of the ionizations of which they were capable); and

(iii) some will penetrate deeper into the sample before being backscattered, so that they lose all or almost all of their effective energy in elastic and inelastic collisions before escaping at the

THE 'SIMPLIFIED THEORY' APPROACH

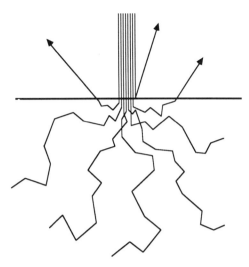

Figure 9.4 Schematic representation of the paths of scattered electrons within an electron microprobe sample. Scattering results in the excitation of analytical X-rays within a truncated, approximately spherical volume of the sample, and also in the loss of portion of the potentially ionizing energy of the electron beam through backscattering of some of the electrons before their energy has been fully utilized in X-ray generation. Since scattering results from interaction with the sample atoms and is composition-dependent, the size and shape of the excitation volume and the extent of backscatter energy loss are also both composition-dependent sources of potential analytical error.

surface as low-energy electrons that are indistinguishable from secondary electrons.

(b) Other electrons will be scattered at (mostly) lower angles; however, after a relatively small number of collisions their directions of movement will become effectively random, i.e. they will have 'lateral' as well as 'vertical' components. They will continue this random movement until they lose all of their kinetic energy, or 'come to rest'. From the viewpoint of X-ray generation, the boundary condition is not that of total energy loss, but of energy loss to E_c. Electrons of lower energy than E_c can contribute only to the generation of lower-energy spectral series or to the longer-wavelength continuum.

(c) The depth distribution of energy loss, and hence of X-ray generation, will not be a simple function because of the directional scattering of the electrons. Some of the potential excitation energy of the electron beam will be lost in backscattered electrons, but the loss associated with a single electron will depend on the depth at which high-angle scattering occurs; there will be a finite depth below which backscattering losses will be negligible (Fig. 9.5).

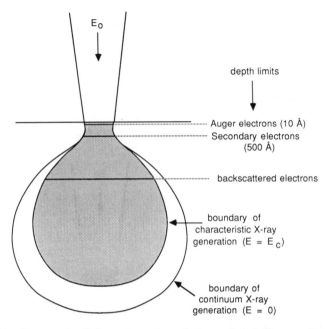

Figure 9.5 A more detailed representation of the analytical X-ray excitation volume (shaded) and some related surfaces. Electrons with energies below E_c may still have sufficient energy to excite continuum X-rays, so that the analytical excitation volume is enclosed in an envelope of additional continuum excitation. Detectable Auger electrons have low energies and are derived only from a thin surface layer, approximately 10 Å deep; similarly, secondary electrons do not escape at the surface if they are generated at depths greater than about 500 Å. Higher-energy backscattered electrons can reach the surface from somewhat greater depths, depending on their energies and on the sample composition. Thus the spatial resolution of the various signals generated by a focused electron beam is best for Auger and secondary electrons, significantly poorer for backscattered electrons, and (mostly) poorer again for characteristic X-rays. Secondary electron scanning images are typically sharper than backscattered electron images, and both are usually superior in their definition to X-ray images.

Since the nature and extent of electron scattering is strongly dependent on the atomic numbers of the scattering atoms, it follows that the scattering behaviour will vary between samples and/or standards of different chemical composition, and so will the depth distribution of energy transfer and hence of X-ray generation (Fig. 9.6).

The number of characteristic X-ray photons that will be produced by an electron along its actual path in the target will depend on:

(a) the number of potentially emitting atoms that it meets along this path, which in turn depends on the density of the target (ϱ), the mass

THE 'SIMPLIFIED THEORY' APPROACH

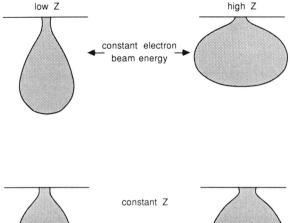

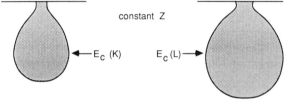

Figure 9.6 Variation in the shape and size of the X-ray excitation volume as functions of the mean atomic number (Z) of the sample and of the critical excitation potential ($E_c(L) < E_c(K)$).

concentration of those atoms in the target (c), the atomic weight of the emitting atoms (A) and Avogadro's number (N);

(b) the probability that ionization will actually take place when a potentially emitting atom is encountered, which is called the *ionization cross-section* (Q) and is a function of the electron energy E (so that it varies along the electron path as the electron loses energy); and

(c) the probability that an ionization will result in the emission of an X-ray photon rather than a photoelectron, which is expressed as the fluorescent yield (ω) and is a function of the atomic number Z of the ionized atoms (Ch. 2) and of the spectral series (K, L, . . .) involved.

For a given spectral series emitted by a specific element, the fluorescent yield and the relative probabilities of emission of the various lines belonging to that series are constant and not composition-dependent. The number of X-ray photons produced, per electron, along an incremental path length dx is thus given by

$$dn = \omega \left(\frac{N \varrho c}{A} \right) Q(E) \, dx$$

$$= \left(\frac{\omega N c}{A}\right) \left(\frac{\varrho\, Q(E)}{\mathrm{d}E/\mathrm{d}x}\right) \mathrm{d}E \qquad (9.2)$$

The total number of X-ray photons produced by an electron is then obtained by integrating this expression over the energy range from the incident electron energy E_0 down to the critical excitation potential E_c (for the appropriate spectral series). If we put

$$S = -\left(\frac{\mathrm{d}E}{\mathrm{d}(\varrho x)}\right) \qquad (9.3)$$

then

$$n = \left(\frac{\omega N c}{A}\right) \int_{E_c}^{E_0} \frac{Q(E)}{S}\, \mathrm{d}E \qquad (9.4)$$

S is a parameter called the *stopping power*, which expresses the energy loss in terms of the mass penetration of the electron into the target, measured along the actual electron path.

Evaluation of the integral of Equation 9.4, in order to calculate the total number of ionizations produced by a single electron, requires knowledge of the relationship between Q and E, and evaluation of the stopping power S for a given target composition. It will also be necessary to determine the possibility that some or all of the electron energy will be lost by *backscattering*; this question will be considered shortly.

Q may be calculated from a complex formula derived originally by Bethe, and often summarized in a simplified form as

$$Q = \frac{a}{E_c^2} \frac{\ln(b\, U)}{U} \qquad (9.5)$$

in which U is the *overvoltage ratio* ($= E / E_c$), E_c is the critical excitation potential for the excited radiation, and a and b are constants which depend on the excitation shell (K, L, ...) and on the number of electrons in that shell. For a given spectral line, Q thus depends only on E_c (which is constant for that line) and on E, which varies continuously between the limits of E_0 and E_c as the electron progressively loses its energy.

Experimental data suggest that the constant b can be taken as unity for $U<20$ (which is usually the case in microprobe analysis), and that a has

THE 'SIMPLIFIED THEORY' APPROACH

the values of 7.92×10^{-14} and 41×10^{-14} for K and L shell excitations respectively (for E_c in eV and Q in cm^2). If $U>2$, Q will thus vary only slowly as a function of U (Fig. 9.7), and little error arises from the assumption that it is constant, which then allows its numerical integration to be avoided. This and other simplifying assumptions are supported by the fact that microprobe analysis is a comparative rather than an absolute technique in which only intensity *ratios* are considered. The errors involved in neglecting the variation in parameters such as Q, which are not strongly composition-dependent, are reduced in magnitude by the ratioing process.

S may be calculated from the Thomson–Whiddington Law, which relates the energy E of an electron of initial energy E_0 to the mass thickness that it has traversed:

$$E_0^2 - E^2 = c \varrho x \qquad (9.6)$$

in which c has the value of approximately 3×10^{11} eV^2 cm^2 g^{-1} if E_0 and E are in eV. In its differential form, this expression reduces to

$$-\frac{dE}{d(\varrho x)} = S = \frac{c}{2E} \qquad (9.7)$$

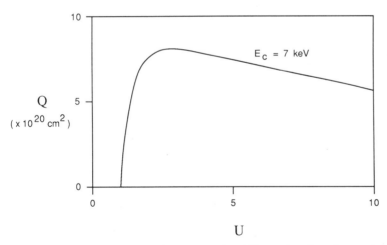

Figure 9.7 Variation of the ionization cross-section (Q) as a function of overvoltage (U) ($U = E / E_c$). Above a minimum value of U (typically about 2.5), Q varies only slowly with U and can be assumed to be constant without introducing serious error. (After Reed, 1965; values of Q have been computed from Eqn 9.5, using the values for a and b suggested by Reed – see text.)

However, the simplified form of Equation 9.7 implies that 'stopping power' is independent of the atomic numbers or atomic weights of the target atoms. A more accurate relationship is that derived by Bethe:

$$S = [g \times (B / A)] / E \tag{9.8}$$

in which $g = 7.85 \times 10^4$, A is the atomic weight of the target element, and B (the *stopping number*) is given by

$$B = Z \ln [k (E / J)] \tag{9.9}$$

in which k is a constant and J is the *mean excitation energy* of the target atoms, usually calculated in practice from an empirically derived expression such as

$$J = 9.76 \times Z + 58.8 \times Z^{-0.19} \tag{9.10}$$

For convenience in calculation, an even simpler approximation

$$J = 11.5 \times Z \tag{9.11}$$

has often been used, but appears to be significantly less accurate. Alternatively it may be easier to use tabulated values of J.

Since S is a function of E, Equation 9.4 requires integration over the electron energy range from E_0 to E_c. This is inconvenient in routine calculation, and is often avoided by computing S for a constant mean energy intermediate between the integration limits. In fact, the ratio of stopping powers for samples and standards varies only slowly with E, so that the simplification is frequently justified and the actual mean energy chosen is not highly critical. Most correction procedures substitute the arithmetic average

$$E_{mean} = (E_0 + E_c) / 2$$

although Reed (e.g. 1975) has suggested that a weighted mean

$$E_{mean} = (2E_0 + E_c) / 3$$

would take better account of the proportionally greater ionization contribution of the more energetic electrons.

For simplified calculation, the expression

$$S = k \frac{Z}{A} \ln \left(\frac{1.166 \times E_{mean}}{J} \right) \tag{9.12}$$

THE 'SIMPLIFIED THEORY' APPROACH

is often used. The numerical value of the constant k is irrelevant, since it will cancel when the sample/standard ratios are taken.

For compound targets (the general case in microprobe analysis), the mean stopping power S_{mean} is calculated as the concentration-weighted sum of the stopping powers of the individual atomic species:

$$S_{\text{mean}} = \Sigma \, c_j \, S_j \qquad (9.13)$$

S is first evaluated for each element in the sample, and the concentration-weighted value of S_{mean} is then determined.

It is now necessary to consider the loss of potential ionizations (and hence of X-ray generation) resulting from backscattered electrons escaping from the target with residual energies greater than E_c. The reduction in generated X-ray intensity will be a function of a factor η, which is defined as the fraction of incident electrons that are backscattered from the sample.

It is relatively easy to measure η experimentally, since it can be determined from the difference between the electron beam current incident on the sample and the current flowing from the sample to ground ('earth'). The latter is a measure of 'absorbed' electrons, and

$$\eta = \frac{\text{incident electrons} - \text{absorbed electrons}}{\text{incident electrons}}$$

However, some special precautions must be taken if accurate results are to be obtained from such measurements, and it is not then convenient to make them routinely for all samples and standards. Furthermore, establishment of η alone will not suffice for calculation of the X-ray intensity losses due to backscattering because the latter arise not just from the loss of electrons but rather from the loss of electron *energies*, and consideration must therefore also be given to the distribution of energies among the escaping backscattered electrons.

This problem has been the subject of intensive theoretical and experimental studies, including the use of 'Monte Carlo' techniques for computer simulation of a large number of individual electron trajectories and energy loss functions. While such techniques are not well suited to routine on-line data correction, since they are time-consuming and require powerful computation facilities, they have yielded useful information which has been combined with experimental observations to develop several simplified correction procedures.

Essentially, these involve calculation for each target (sample or standard) of a parameter R, which is defined as the effective fraction actually achieved from the total possible number of ionizations, i.e. as the total possible number of ionizations less those lost as the consequence of backscattering. The backscatter ionization loss is thus equal to $(1 - R)$.

As would be expected, R is strongly dependent on Z_{mean}, the mean atomic number of the target, and also on the electron energy ratio $E_c : E_0$, which is often represented by the symbol W (the inverse energy ratio $E_0 : E_c$ is also used elsewhere in matrix correction procedures, and is usually represented by U). Various polynomial expressions, fitted to the theoretical and experimental data, have been used to calculate R as a function of overvoltage and sample composition. Some of these expressions are rather complex, so that in practice it may be more convenient to obtain R values from tables computed from them (e.g. the tables of Duncumb & Reed 1968), using interpolation when necessary; R varies fairly smoothly with both Z and W.

Alternatively, Springer (1974) has suggested that

$$R = 1 / [1 + 0.008 \times (1 - W) \times Z] \qquad (9.14)$$

is a simple expression that is well suited to hand calculation and gives reasonably accurate results as long as the atomic number effects are not unduly severe.

For compound samples, R_{mean} is again usually calculated as the concentration-weighted mean

$$R_{mean} = \Sigma\, c_j\, R_j \qquad (9.15)$$

This is in fact something of an approximation, but it usually introduces no significant error unless low overvoltages are used (particularly with heavy elements).

The backscattering and stopping power functions can now be combined into the X-ray generation equation, as follows:

$$n = \left(\frac{\omega N c}{A}\right) R \int_{E_c}^{E_0} \frac{Q(E)}{S}\, dE \qquad (9.16)$$

For a given element in a compound sample, ω, N and A can be combined into a single constant k, and the S term can be taken outside the integral by substituting the mean or weighted mean energy (E_{mean}). Thus

$$n_i = k\, c_i\, \frac{R_i}{S_i} \int_{E_c}^{E_0} \frac{Q(E)}{S}\, dE \qquad (9.17)$$

THE 'SIMPLIFIED THEORY' APPROACH

and, for the same element,

$$n_{std} = k \; c_{std} \frac{R_{std}}{S_{std}} \int_{E_c}^{E_0} \frac{Q(E)}{S} \, dE \qquad (9.18)$$

When the sample : standard ratios are taken, the constants and the integrals (which are no longer Z-dependent) will cancel, and the factor to compensate for Z-dependent variations becomes

$$Z = \frac{R_{std}}{R_i} \times \frac{S_i}{S_{std}} \qquad (9.19)$$

Apparent compositions (i.e. those calculated from the 'raw', or measured, intensity ratios using the linear approximation of Eqn 9.1) are multiplied by this factor to yield compositions corrected for the 'Z' or 'atomic number' effects. Note that the backscattering and stopping power components of this correction factor work in opposite directions and often tend to cancel each other, reducing the magnitude of the overall correction. However, exact cancellation is only fortuitous, and certainly cannot be assumed.

As an example, consider the analysis of a copper–aluminium alloy, relative to pure metal standards, using an accelerating voltage (E_0) of 20 kV. X-ray intensity ratios (corrected for dead time and background) indicate an apparent composition of $Cu_{0.775}Al_{0.076}$, and wavelength (energy) scans indicate no other elements present in the alloy in significant concentrations.

When normalized to 100 wt per cent, the apparent composition becomes $Cu_{0.911}Al_{0.089}$. However, normalization seldom suffices to correct a poor analysis, and it is clearly necessary to calculate the actual ffect of matrix errors.

!equired data

Atomic numbers: Cu = 29   Al = 13

Atomic weights: Cu = 63.5   Al = 27.0

Excitation potentials (E_c) (K):   Cu = 8.98 keV   Al = 1.56 keV

Step 1: Calculate the mean electron energies (= ($E_0 + E_c$) / 2, or use Reed's weighting):

Cu: $E_{mean} = (20 + 8.98) / 2 = 14.49$ keV

COMPOSITION-DEPENDENT ERRORS IN EPA

Al: $E_{mean} = (20 + 1.56) / 2 = 10.78$ keV

Step 2:   Calculate the stopping power coefficients (S):

(a) Calculate the J values (e.g. using Eqn 9.10):

$$J(Cu) = (9.76 \times 29) + (58.8 \times 29^{-0.19}) = 314 \text{ eV}$$

$$J(Al) = (9.76 \times 13) + (58.8 \times 13^{-0.19}) = 163 \text{ eV}$$

(b) Calculate the S values, using Equation 9.12 (ignoring the constant term, which will cancel when the sample : standard intensity ratios are taken):

(i) For Cu K generation:

$S_{\text{pure Cu}} = (29 / 63.5) \times \ln(1.166 \times 14.49) / 0.314)$

$\phantom{S_{\text{pure Cu}}} = 1.820$

$S_{\text{pure Al}} = (13 / 27.0) \times \ln(1.166 \times 14.49 / 0.163)$

$\phantom{S_{\text{pure Al}}} = 2.235$

$S_{\text{alloy}} = (0.911 \times 1.82) + (0.089 \times 2.235)$

$\phantom{S_{\text{alloy}} xx} = 1.857$

(using 'first approximation' composition from raw X-ray intensity ratios)

(ii) For Al K generation:

$S_{\text{pure Cu}} = (29 / 63.5) \times \ln(1.166 \times 10.78 / 0.314)$

$\phantom{S_{\text{pure Cu}}} = 1.685$

$S_{\text{pure Al}} = (13 / 27.0) \times \ln(1.166 \times 10.78 / 0.163)$

$\phantom{S_{\text{pure Al}}} = 2.092$

$S_{\text{alloy}} = (0.911 \times 1.685) + (0.089 \times 2.092)$

$\phantom{S_{\text{alloy}} xx} = 1.721$

Step 3:   Calculate the backscatter coefficients using Equation 9.14 (or obtain the coefficients from tables such as those of Duncumb & Reed – see text):

(a) Calculate the W factors ($W = E_c / E_0 = 1 / U$)

For Cu K, $W = 8.98 / 20 = 0.449$

For Al K, $W = 1.56 / 20 = 0.078$

(b) Calculate the R values, from Equation 9.14:

THE 'SIMPLIFIED THEORY' APPROACH

(i) For Cu K generation:

$$R_{\text{pure Cu}} = 1 / [1 + 0.008 \times (1 - 0.449) \times 29]$$
$$= 0.887$$
$$R_{\text{pure Al}} = 1 / [1 + 0.008 \times (1 - 0.449) \times 13]$$
$$= 0.946$$
$$R_{\text{alloy}} = (0.911 \times 0.887) + (0.089 \times 0.946)$$
$$= 0.892$$

(ii) For Al K generation:

$$R_{\text{pure Cu}} = 1 / [1 + 0.008 \times (1 - 0.078) \times 29]$$
$$= 0.824$$
$$R_{\text{pure Al}} = 1 / [1 + 0.008 \times (1 - 0.078) \times 13]$$
$$= 0.913$$
$$R_{\text{alloy}} = (0.911 \times 0.824) + (0.089 \times 0.913)$$
$$= 0.83$$

Step 4: Calculate the Z-correction factors for each of the measured radiations. In each case, calculate the Z-factor as a multiplier, viz. as

$$Z\text{-factor} = (R_{\text{standard}} / R_{\text{sample}}) \times (S_{\text{sample}} / S_{\text{standard}})$$

(a) Cu: Z-factor $= (0.887 / 0.892) \times (1.857 / 1.820)$
$$= 1.015$$
(b) Al: Z-factor $= (0.913 / 0.832) \times (1.721 / 2.092)$
$$= 0.903$$

Step 5: Apply the correction factors to the raw (original uncorrected) concentrations to obtain first-cycle corrected compositions:

(a) Cu: $0.775 \times 1.015 = 0.787$ or 78.7 wt per cent

(b) Al: $0.076 \times 0.903 = 0.069$ or 6.9 wt per cent

The calculation can now be repeated, using the corrected composition to calculate refined correction parameters, which again are applied to the raw concentrations; the procedure can be reiterated as often as necessary to obtain satisfactory convergence. It is not necessary to repeat the entire calculation, but only to calculate new S and R values for the sample, using the (normalized) first-correction composition for weighting; all other parameters, including the S and R values for the standards, remain unchanged:

$$Cu = 0.787 / (0.787 + 0.069) = 0.919$$
$$Al = 0.069 / (0.775 + 0.069) = 0.081$$

S: Cu: $S_{alloy} = (0.919 \times 1.820) + (0.081 \times 2.235) = 1.854$

Al: $S_{alloy} = (0.919 \times 1.685) + (0.081 \times 2.09) = 1.718$

R: Cu: $R_{alloy} = (0.919 \times 0.887) + (0.081 \times 0.946) = 0.892$

Al: $R_{alloy} = (0.919 \times 0.824) + (0.081 \times 0.913) = 0.831$

Z-factors:

$$Cu = (0.887 / 0.893) \times (1.856 / 1.820) = 1.013$$
$$Al = (0.913 / 0.832) \times (1.720 / 2.092) = 0.902$$

Revised composition:

$$Cu = 0.775 \times 1.013 = 0.785 \text{ or } 78.5 \text{ per cent}$$
$$Al = 0.076 \times 0.902 = 0.069 \text{ or } 6.9 \text{ per cent}$$

For most practical purposes, this would be regarded as a satisfactory level of convergence; the maximum difference in the new values is only 0.2 per cent.

In practice the reiteration is not usually made until absorption and fluorescence correction factors have also been calculated and applied (Sections 9.3.2 & 9.3.3). In the above example, the analysis total is still low (85.4 per cent), perhaps suggesting that the accuracy of the analysis has not been greatly improved (although it will be observed that the atomic number correction for Al is quite substantial, being approximately 10 per cent); this is because absorption and fluorescence corrections have not yet been applied, and the absorption correction for Al in particular is likely to be of considerable magnitude.

Such calculations are perhaps somewhat tedious, but not difficult when their principles are clearly understood. Since almost all microprobe data reduction is now performed by on-line computers, it is strongly recommended that novices carry out at least a few simple calculations of this type to enhance their understanding of the factors involved. Those with access to a microcomputer and a good 'spreadsheet' program will gain considerable insight and practical benefit from setting up a sheet to make trial corrections (including absorption and fluorescence corrections). Once established, the sheet can be used in the laboratory both to explore the relative magnitudes of the principal matrix errors and to investigate the effects of varying some of the operating parameters (choice of standards, E_0 etc.).

It is stressed that different options are available for the calculation of the empirically derived parameters used in the ZAF correction procedure, and that those used above are intended only as examples. In particular, the calculation of S values is very dependent on empirical relationships and is possibly the weakest link at present in the atomic number correction. It is to be expected that further refinements will be proposed from time to time.

9.3.2 The absorption (A) factor

Most of the X-rays generated in the sample are produced below the surface and must pass through a finite thickness of the sample itself to reach the spectrometer (Fig. 9.8). In so doing they will be attenuated by absorption, to an extent determined by the path length within the sample (and hence by the depth of generation) and by the mass absorption coefficients of the sample for the measured emissions. The depth distribution of X-ray generation depends on the electron paths and energy loss functions, and is therefore composition-dependent; so are the mass absorption coefficients. Differences in composition between samples and standards will therefore result in potential errors which require evaluation and, in most cases, correction. In general the worst cases will be those involving light elements (low-energy X-ray emissions and relatively high mass absorption coefficients) and/or high electron accelerating potentials (increased depths of X-ray generation).

The intensities of X-rays generated in a thin layer $d(\varrho z)$ at a depth (ϱz) within the sample will depend on the number of sufficiently energetic electrons passing through that layer and on the mean path of those

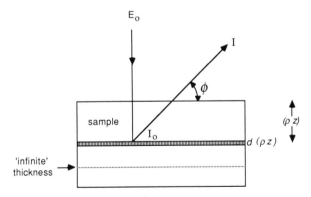

Figure 9.8 Simplified geometry of electron microprobe X-ray excitation, with path lengths expressed as mass thicknesses (ϱz), cf. Figure 9.2.

electrons *within* the layer. Both will vary with depth, the latter as a consequence of directional changes produced by scattering and the former as a result of progressive losses in energy along the electron path. Generated X-ray intensity will thus be related to a depth function which is conventionally written as $\phi(\varrho z)$. The intensity measured in the spectrometer will be the sum of all emissions in the appropriate direction, less absorption losses, generated to an 'infinite depth'. As in XRF, 'infinite depth' is the depth at which *either* the electrons no longer have sufficient energy to produce ionizations *or* the X-ray path length within the sample becomes sufficiently long for complete attenuation of the generated X-rays.

If there were no absorption, we could write

$$I = k \int_0^\infty \phi(\varrho z) \, d(\varrho z) \quad (9.20)$$

in which the constant k represents the fraction of all generated X-rays collected by the spectrometer. k is a purely geometric function which is not composition-dependent, so that it can be omitted from further consideration.

The path length (within the sample) of X-rays generated at a depth (ϱz) will be equal to $(z) \operatorname{cosec} \psi$, where ψ is the 'take-off angle' (Fig. 9.8). The intensity of X-rays generated at this depth and emerging at the surface will thus be given by

$$I_{\text{emergent}} = I_{\text{generated}} \times \exp\{-[(\mu/\varrho) \operatorname{cosec} \psi(\varrho z)]\} \quad (9.21)$$

where (μ/ϱ) is the mass absorption coefficient of the sample for the measured radiation. If we put $\chi = (\mu/\varrho) \operatorname{cosec} \psi$, the total emergent intensity will then be

$$I = \int_0^\infty \phi(\varrho z) \, e^{-\chi (\varrho z)} \, d(\varrho z) \quad (9.22)$$

The absorption correction factor (for either a sample or a standard), i.e. the factor by which the generated intensity is attenuated by absorption, may be written as $f(\chi)$, and will be given by

$$f(\chi) = \frac{\text{emergent intensity}}{\text{generated intensity}}$$

THE 'SIMPLIFIED THEORY' APPROACH

$$= \frac{\int_0^\infty \phi(\varrho z) \exp^{(-\chi \varrho z)} \, d(\varrho z)}{\int_0^\infty \phi(\varrho z) \, d(\varrho z)} \quad (9.23)$$

In some treatments, the integral expressions of Equation 9.23 are written as their Laplace* transforms, so that

$$f(\chi) = F(\chi) / F(0) \quad (9.24)$$

The 'F(0)' of Equation 9.24 represents the absorption function for zero absorption; cf. Equation 9.22.

Regardless of the notation used, evaluation of the absorption effect requires knowledge of the form of the function $\phi(\varrho z)$ in terms of such factors as sample composition, incident electron energy etc., and evaluation of χ as a function of composition. Providing mass absorption coefficients are known or can be calculated with sufficient accuracy (which is not always the case), evaluation of χ is not usually a problem. However, the depth function is more difficult.

The X-ray source volume is crudely pear-shaped, with its exact form depending on composition and incident electron energy (Figs 9.5 & 9.6). With increasing E_0, the source volume expands along the beam axis in proportion to the increasing depth penetration of the electrons, which varies approximately as $E^{1.5}$. At shallow depths within the source volume, $\phi(\varrho z)$ increases with depth (Fig. 9.9) because at these depths scattering progressively increases the mean electron path length in each successive incremental layer. At some depth, however, the mean path lengths will reach a maximum corresponding to true diffusion; thereafter $\phi(\varrho z)$ will decrease exponentially because of increased absorption of the electron energies with increasing depth. The situation is further complicated by backscattering, in that a higher proportion of electron energy is lost in the shallow layers to escaping electrons.

In addition to the depth intensity function $\phi(\varrho z)$ there will be a corresponding lateral intensity function $\psi(\varrho z)$, but its form will be most strongly dependent on the effective beam diameter, which is independent of the target composition (but critical to the resolution of X-ray scanning images).

If an adequate analytical derivation of $\phi(\varrho z)$ were available, $f(\chi)$ could be calculated by evaluating the integrals of Equation 9.24. However, this

---

* $F(x)$ is the Laplace transform of the function $g(t)$ if $F(x) = g(t) \exp^{-x \, t} dt$.

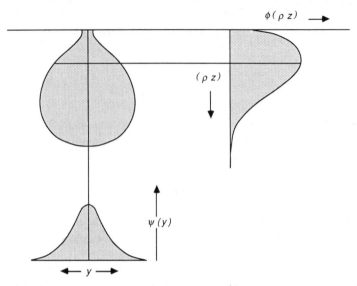

Figure 9.9 The forms of the depth and lateral generated intensity functions $\phi(\varrho z)$ and $\phi(y)$. The vertical dimension in this figure is that of depth beneath the sample surface, expressed as mass thickness (ϱz), and the horizontal dimension is lateral distance from the electron beam axis, expressed in arbitrary units, y. These functions indicate that most X-rays are generated at relatively shallow depths within the excitation volume and relatively close to the beam axis; this is the region in which electron energies have not been greatly attenuated by ionization or electron scattering.

is not yet the case; instead most correction procedures are again based more or less empirically on experimental data. Some ingenious techniques have been developed to measure the form of $\phi(\varrho z)$, e.g. by measuring the absorption-corrected intensity emitted by a thin layer of a 'tracer' element overplated by various thicknesses of another element and comparing the intensities generated at each depth with those yielded by the exposed tracer layer (Fig. 9.10). The data so obtained can then be used to fit an expression for $f(\chi)$ in terms of empirically adjusted constants.

The most widely used of these expressions is that originally due to Philibert (1963), who derived

$$f(\chi) = \frac{(1 + h)}{(1 + \chi/\sigma) \times (1 + h(1 + \chi/\sigma))} \quad (9.25)$$

in which σ, the *Lenard coefficient*, is a measure of electron absorption and depends on effective electron energy (i.e. on E_0 and E_c), and h is a parameter dependent on atomic number. Expressions for σ and h are also

THE 'SIMPLIFIED THEORY' APPROACH

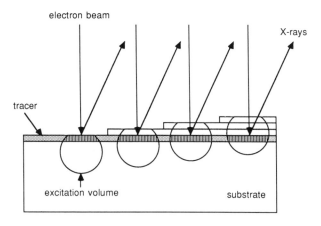

Figure 9.10 The 'tracer' technique for establishing the form of the depth generation function $\phi(\varrho z)$. The X-ray intensity emitted from a thin layer of a tracer element is measured at various depths beneath overplated layers of another element, establishing the form of $\phi(\varrho z)$ for the latter.

empirically fitted to experimental data, and various forms of each have been proposed. Those most widely used to date include

$$\sigma = \frac{4.5 \times 10^5}{E_0^{1.65} - E_c^{1.65}} \qquad (9.26)$$

and

$$h = 1.2 \times A / Z^2 \qquad (9.27)$$

The Philibert expression of Equation 9.25 is a version simplified for hand calculation by assuming that $\phi(0) = 0$ (i.e. that zero intensity is generated at zero depth), which yields a form close to but not identical with the experimental observations. The difference, and hence the validity of the simplifying assumption, is most critical for low-energy (long-wavelength) X-rays which tend to be heavily absorbed by the sample, so that only X-rays generated near the surface emerge for measurement. The assumption that $\phi(0) = 0$ is then undesirable; under these circumstances the analyst has the choice of reverting to a more complex expression which requires evaluation of $\phi(0)$, or of using an alternative expression such as that proposed by Bishop (1974), which is based on the concept of a *mean mass depth* of X-ray generation $((\varrho z)_{mean})$ and assumes that $\phi(\varrho z)$ may be taken as constant over the depth range from zero to twice the mean mass depth, and to be zero at depths greater than $2(\varrho z)_{mean}$ (Fig. 9.11).

Bishop's 'square model' amounts to a rectangular approximation to the

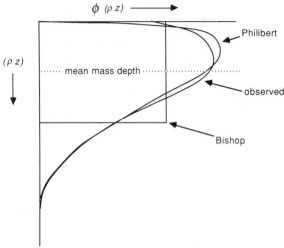

Figure 9.11 The Bishop rectangular approximation of $\phi(\varrho z)$, compared to experimental data (Fig. 9.10) and Philibert's analytical approximation (Eqn 9.25). The area of the Bishop rectangle is assumed to approximate to that under the $\phi(\varrho z)$ curve, and the continuous variation in $\phi(\varrho z)$ is replaced by the concept of a 'mean mass depth' at half the depth of the Bishop rectangle. This greatly simplifies the calculations required to correct for the effects of variable compositions on the depth generation of analytical X-rays.

area under the $\phi(\varrho z)$ curve, and has been reported to give better results than the Philibert expression in some difficult cases. The mean mass depth of generation is calculated as

$$(\varrho z)_{mean} = \frac{1 + 2h}{\sigma(1 + h)} \qquad (9.28)$$

and $f(\chi)$ is then obtained from

$$f(\chi) = \frac{1 - \exp[-2\chi(\varrho z)_{mean}]}{2\chi(\varrho z)_{mean}} \qquad (9.29)$$

Since Equation 9.28 is still based on the Philibert model, it does not always prove satisfactory in cases of high absorption; however, experimental $\phi(\varrho z)$ data are still inadequate for better refinement. If calculated values of $f(\chi)$ are less than about 0.7, the analyst should be aware of the probability of significant residual error in the absorption correction. A good on-line *ZAF* correction program should recognize this situation and issue a warning message.

For compound samples or standards, a mean value of h must be calculated from the h values of the contributing elements. The simple weighted mean

THE 'SIMPLIFIED THEORY' APPROACH

$$h_{mean} = \Sigma \, c_j \, h_j \qquad (9.30)$$

is most widely used, although some have argued a case for weighting on atom rather than weight fractions. When sample : standard ratios are taken, the differences between the two approaches are usually small.

To make the absorption correction, the relevant mass absorption coefficients must be known. Again these may be calculated from any of several expressions fitted to experimental data, such as those of Kelly (1966), Frazer (1967) or those given in Table A.2, or they may be conveniently obtained from tables such as those of Heinrich (in McKinley et al. 1966). For the Cu–Al alloy example used to illustrate the Z-correction procedure, coefficients from Heinrich are:

| | Emitters | |
Absorbers	Cu (Kα)	Al (Kα)
Cu	54 (52)	5377 (5251)
Al	50 (49)	386 (386)

(For comparison purposes, the figures in parentheses are from Table A.6, and were calculated using the polynomial coefficients of Table A.2.)

The X-ray take-off angle ψ must also be known. In the following example it is assumed to be 45°.

First the uncorrected compositions are normalized to a total of 100 weight per cent. For this example we shall take the raw data previously corrected for atomic number effects as the 'uncorrected' composition (the Z and A corrections are usually applied together in the same calculation, along with the F correction where appropriate). Thus we begin with the composition $Cu_{0.787}Al_{0.069}$, normalized to $Cu_{0.919}Al_{0.081}$.

Step 1: Calculate the required h values (Eqn 9.27):

Cu_{metal}: $h = 1.2 \times 63.5 / 292 = 0.091$

Al_{metal}: $h = 1.2 \times 27.0 / 132 = 0.192$

Alloy: $h = (0.919 \times 0.091) + (0.081 \times 0.192) = 0.099$

Step 2: Calculate the χ values ($\chi = (\mu/\varrho) \, \mathrm{cosec} \, \psi$):

(a) for Cu Kα:

Cu_{metal}: $\chi = 54 \times \mathrm{cosec}(45) = 76.4$

Al_{metal}: $\chi = 50 \times \mathrm{cosec}(45) = 70.7$

COMPOSITION-DEPENDENT ERRORS IN EPA

Alloy: $\chi = (0.919 \times 76.4) + (0.081 \times 70.7) = 75.4$

(b) for Al Kα:

Cu_{metal}: $\chi = 5377 \times \text{cosec}(45) = 7604$

Al_{metal}: $\chi = 386 \times \text{cosec}(45) = 546$

Alloy: $\chi = (0.919 \times 7604) + (0.081 \times 546) = 6979$

Step 3: Calculate the σ values (e.g. from Eqn 9.26):

(a) for Cu K: $\sigma = 4.5 \times 10^5 / (20^{1.65} - 8.98^{1.65}) = 4378$
(b) for Al K: $\sigma = 4.5 \times 10^5 / (20^{1.65} - 1.56^{1.65}) = 3258$

Step 4: Calculate the $f(\chi)$ values (e.g. from Eqn 9.25):

(a) for Cu Kα:

$$f(\chi) \text{ (Cu)} = \frac{1 + 0.091}{[1 + (75.9 / 4378)] \times \{1 + [0.091 [1 + (75.9 / 4378)]]\}}$$

$= 0.982$

$f(\chi)$ (Al) – not required (Cu Kα not measured on Al standard)

$$f(\chi) \text{ (alloy)} = \frac{1 + 0.099}{[1 + (75.4 / 4378)] \times \{1 + [0.099 [1 + (75.4 / 4378)]]\}}$$

$= 0.982$

(b) for Al Kα:

$f(\chi)$ (Cu) – not required (Al Kα not measured on Cu standard)

$$f(\chi) \text{ (Al)} = \frac{1 + 0.192}{[1 + (546 / 3258)] \times \{1 + [0.192 [1 + (546 / 3258)]]\}}$$

$= 0.834$

$$f(\chi) \text{ (alloy)} = \frac{1 + 0.099}{[1 + (6979 / 3258)] \times \{1 + [0.099 [1 + (6979 / 3258)]]\}}$$

$= 0.267$

Step 5: Calculate the A-factors ($A = f(\chi)_{std} / f(\chi)_{sample}$):

(a) for Cu Kα: $A = 0.982 / 0.982 = 1.000$

(i.e. no effective absorption correction, reflecting the trivial difference in the absorption coefficients of Cu and Al for Cu Kα).

(b) for Al Kα: $A = 0.834 / 0.267 = 3.124$

(i.e. a substantial correction, reflecting the major difference in the absorption coefficients of Cu and Al for Al Kα; the calculated value of $f(\chi)$ for the alloy sample is very much smaller than 0.7, indicating that the correction is unlikely to be very accurate).

Step 6: Apply the A-corrections to the Z-corrected composition:

(a) Cu = 0.785 × 1.000 = 0.785, or 78.5 per cent
(b) Al = 0.069 × 3.127 = 0.216, or 21.6 per cent

The analysis total is now 100.1 per cent, apparently indicating a remarkable improvement. However, some caution is necessary; in view of the magnitude of the absorption correction for Al, at least some of the apparent improvement may be illusory. Careful thought should be given to the precision of the quoted results of such an analysis. It is also necessary to complete the ZAF procedure by applying a fluorescence correction, although in this case it should be relatively small.

If corrections are large, as in this example, simple iteration does not always lead to rapid convergence. Sometimes the corrected compositions do not converge at all, and sometimes successive corrections may oscillate about a mean value without converging on it. These problems are normally encountered only when large absorption corrections are involved, and particularly when a strongly absorbing element is determined by difference rather than by direct measurement.

In such cases it may be better to use a more sophisticated iteration procedure, and this is fairly common practice in computer-based correction programs. Unfortunately, the success of the improved iteration procedures may conceal the magnitude of the corrections made, and with them the associated error probabilities. Good correction programs will at least print corrected and uncorrected compositions side by side to alert the analyst to potential problems.

9.3.3 The fluorescence (F) factor

Fluoresence, meaning the excitation of one characteristic emission by one or more others of higher energy, will occur within a microprobe sample whenever one or more lines in the characteristic X-ray spectrum of element B have sufficient energy to excite the analytical characteristic line of element A, i.e. whenever the condition that $E_{\lambda(B)} > E_c(A)$. Some of the emitted $\lambda(B)$ will then be absorbed by atoms of A in photoelectric processes, some of which in turn result in the generation of additional $\lambda(A)$. As a result, the measured intensity of $\lambda(B)$ will be attenuated and that of $\lambda(A)$ will be enhanced. The attenuation of $\lambda(B)$ is compensated by the absorption correction, since its extent will be reflected in the mass absorption coefficient of A for $\lambda(B)$, but the enhancement of $\lambda(A)$

requires further correction. The magnitude of this correction will clearly depend on the concentrations of A and B in the sample and on the efficiency of the fluorescent excitation process.

Further, electron excitation of charateristic X-ray spectra is always accompanied by generation of the continuum, and all continuum X-rays having energies greater than $E_c(A)$ are capable of exciting additional $\lambda(A)$ and hence causing further enhancement. Since the continuum intensity is a function of atomic number (Ch. 2), the extent of continuum fluorescence enhancement is also composition-dependent.

CHARACTERISTIC FLUORESCENCE

In a simple binary system AB, the fluorescence-corrected composition of element A (c_A) will be given by

$$c_A = c_A \times [1 / (1 + \gamma_{charact.}(B) + \gamma_{continuum})] \qquad (9.31)$$

in which c_A is the measured concentration of A (corrected for stopping power, backscatter and absorption), $\gamma_{charact.}(B)$ is a factor representing the enhancement of $\lambda(A)$ by all the characteristic lines of element B that have sufficient energy to excite $\lambda(A)$, and $\gamma_{continuum}$ represents the total continuum enhancement of $\lambda(A)$. Corresponding γ-factors describe similar enhancement, where appropriate, of characteristic emissions from the calibration standards.

In compound systems containing three or more elements, the factor $\gamma_{charact.}(B)$ must be replaced by a factor $\Sigma\ (\gamma_{(j)})$, i.e. by the summation of $\gamma_{charact.}$ factors for all characteristic emissions of all elements j that have sufficient energy to excite $\lambda(A)$.

The magnitude of the fluorescence correction will depend on many variables, including

(a) The (weight) concentration of element A in the sample.
(b) The number of elements (j) emitting characteristic lines that are sufficiently energetic to excite $\lambda(A)$.
(c) The intensities of all of the exciting emissions from elements j, which in turn will depend in each case on

(i) the weight concentrations of the elements j;
(ii) the fluorescence efficiencies ω_j;
(iii) the atomic weights A_j;
(iv) the electron beam overvoltages U_j $(U = E_0 / E_c)$;
(v) the spectral series (K, L, . . .) of the exciting emissions; etc.

(d) The mass absorption coefficients for each of the potentially exciting emissions in pure A and in the compound target, which determine the proportion of each emission actually absorbed by element A.

THE 'SIMPLIFIED THEORY' APPROACH

(e) The jump ratio r_A of absorption coefficients on each immediate side of the relevant absorption edge of element A, which is a measure of the fraction of absorbed radiation which will produce ionizations in the appropriate shell of A.
(f) The fluorescence efficiency ω for the appropriate shell of A.
(g) The depth distribution of generation of each of the exciting emissions.
(h) The extent of possible 'third-element' effects, in which A is enhanced by B and C, but B is also enhanced by C, leading to yet further enhancement of A by the enhanced B.
(i) The atomic number and overvoltage dependence of the continuum intensity, which must be summed over the complete energy range from E_0 to E_c, taking into account any absorption edges which lie within that range and which therefore can modify the effective intensities of portions of the continuum.
(j) The absorption of emerging fluorescence-generated $\lambda_{(A)}$, calculated for the depth distribution of B(. . . j) radiation.
(k) The relative efficiencies of enhancement of AK by BK, of AL by BK, of AK by BL, of AL by BL, etc.

A comprehensive fluorescence correction procedure will obviously be complex and clearly not well suited to routine manual or direct on-line calculation. However, the corrections involved are usually relatively small (typically of the order of 10 per cent or less, and mostly less than 3 to 4 per cent), and again they often tend to cancel when sample : standard intensity ratios are taken – unless compositional differences are extreme or the fluorescence effects are unusually severe. This does not mean that the fluorescence correction can be neglected, although *continuum* fluorescence is often ignored because it is of relatively small magnitude and typically does not vary greatly from sample to standard; its correction calculation is also rather cumbersome (see below). Rather, microprobe analysts have again been encouraged to simplify the theory in order to develop correction procedures that are reasonably accurate but still suited to routine applications.

Once again we find that in most modern microprobe laboratories *ZAF* corrections are applied by on-line computer data processing, and it is common for many of the users to be almost totally ignorant of the fluorescence correction procedures (if any) that are being followed. However, these vary considerably from one laboratory, or even one application, to another, and it always remains the operator's responsibility to assess the probable magnitude of errors, and to ensure that adequate correction procedures are being followed.

K–K fluorescence (i.e. the excitation of fluorescent K radiation from one element by K radiation from another) can only be excited in the first element if the second is of higher atomic number. Up to $Z = 21$ (Sc), K radiation of an element of atomic number Z is most effectively excited by the Kα line of the element $(Z + 1)$, e.g. Sc K $(E_c = 4.488$ keV, $Z = 21)$ is excited by Ti Kα $(E = 4.508$ keV, $Z = 22)$. For $Z > 21$, K–K fluorescence is most effectively excited by the Kα line of element $(Z + 2)$; it is also excited by the Kβ line of $(Z + 1)$, but much less effectively because of the lower Kβ intensity. Kβ intensity for a given element in the normal working range for K emissions is typically about 10–15 per cent of the total intensity of the K spectrum; for this reason fluorescence excited by Kβ lines is often ignored.

Relatively severe K–K fluorescence effects can therefore be expected if two or more elements in a sample have atomic numbers differing by one or two (depending on the Z range) up to about five or six. Thus Cu $(Z = 29)$ produces moderately strong fluorescence of Fe K $(Z = 26)$; however, Fe K is not excited by Co Kα $(Z = 27)$, although some relatively minor Fe K fluorescence is produced by Co Kβ.

Fluorescence involving L lines is rather more complicated because of the differences between the L_I, L_{II} and L_{III} sub-shells. Each sub-shell, for example, has a different fluorescent yield (ω), but the yields are not known with great accuracy, and it is not uncommon to assume a mean value $\omega(L) = \omega(L_{III})$. For K–L and L–L fluorescence the accuracy of most correction procedures is thus significantly poorer than for K–K or L–K fluorescence. Experience suggests that the approximation is usually adequate, unless there are significant absorption edges of other elements between the relevant L edges.

In predicting the probable magnitude of fluorescence effects involving L lines or edges, the atomic number relationships are not as simple as those of the K–K type. For example, Na K $(Z = 12)$ is excited by the Lα_1 emissions of Ga $(Z = 31)$ or heavier elements, i.e. by $(Z + 19)$ or greater. Ca K $(Z = 20)$, however, is only excited by the Lα_1 emissions of Cs $(Z = 55)$ or heavier elements, i.e. by $(Z + 35)$ or greater. In assessing the probable magnitude of K–L or L–K fluorescence effects, there is really no alternative to careful study of tables of emission and absorption wavelengths/energies.

Continuum fluorescence is always present, and will be most pronounced for those spectral series whose absorption edges lie in the region of maximum continuum intensity. At any given wavelength/energy the continuum intensity varies approximately as Z^4, and continuum-excited fluorescence intensity typically increases from about 2 per cent of the primary (electron-excited) intensity for $Z = 20$ up to about 15 per cent for $Z = 35$. When the corresponding enhancement of standard intensities is also considered, the net error produced by continuum enhancement is usually found to be much smaller, and this is why some correction procedures ignore it altogether.

THE 'SIMPLIFIED THEORY' APPROACH

Characteristic fluorescence is most likely to be a problem when the energy of an exciting radiation (B) is only slightly higher than the absorption edge of the measured radiation (A), and when the concentration of the exciting element (B) is high. Continuum fluorescence is most significant when the absorption edge of A lies within the maximum intensity region of the continuum, and/or when compositional differences result in a relatively large difference in mean atomic number between samples and standards.

K–K fluorescence correction In those cases in which K radiation of element A is enhanced by the absorption of Kα radiation from element B, the correction parameter $\gamma_{A(B)}$ is usually calculated from an expression of a form similar to that given by Reed (1975):

$$\gamma_{A(B)} = 0.5 \times c_B \times \frac{\mu_B(A)}{\mu_B} \times \frac{(r_A - 1)}{r_A} \times \omega_K(B) \times \frac{A_A}{A_B} \times \ldots$$

$$\times \left(\frac{U_B - 1}{U_A - 1}\right)^{1.67} \times \left(\frac{\ln(1 + u)}{u} + \frac{\ln(1 + v)}{v}\right) \quad (9.32)$$

in which:

(a) the factor 0.5 is derived from a simplifying assumption that the exciting radiation is generated at the surface, so that half of it is emitted from the sample without producing any fluorescence;

(b) c_B is the weight concentration of element B;

(c) $\mu_B(A)$ is the mass absorption coefficient *of pure element A* for the K radiation of element B (usually taken as B Kα unless only B Kβ is effective, in which case the correction factor must be weighted down to compensate for the relatively low intensity of B Kβ);

(d) μ_B is the mass absorption coefficient *of the sample* for the K radiation of element B (again usually B Kα);

(e) r_A is the jump ratio at the A (K or L) absorption edge, which may be obtained from tabulated data such as those of Heinrich (1966) or Reed (1975), or calculated from expressions fitted to experimental data, such as those of Colby (1968):

$$(r_A - 1) / r_A = 0.954 - 0.002\,77 \times Z \quad \text{(K edge)} \quad (9.33)$$

$$(r_A - 1) / r_A = 0.996 - 0.004\,98 \times Z \quad \text{(L edge)} \quad (9.34)$$

(f) $\omega_K(B)$ is the K-fluorescence efficiency (fluorescent yield) for element B, which may also be obtained from tables such as those of Burhop and Asaad (1972, summarized by Reed 1975), or by calculation using any of several empirically based modifications of the expression

$$\omega = Z^4 / (a + Z^4) \qquad (9.35)$$

in which $a = 10^6$ for K and 10^8 for L excitation.

(g) A_A and A_B are the atomic weights of A and B respectively;
(h) U_A and U_B are the overvoltage factors (E_0 / E_c) for A and B;
(i) u is a factor allowing for absorption in the sample, calculated as

$$u = \chi_A / \mu_B = (\mu_A / \mu_B) \operatorname{cosec}(\psi) \qquad (9.36)$$

(j) v is a factor allowing for the depth distribution of excitation of the radiation from element B, based on the Lenard coefficient σ

$$v = \sigma / \mu_B \qquad (9.37)$$

where σ, as in the absorption correction, is given by

$$\sigma = 4.5 \times 10^5 / (E_0^{1.65} - E_c^{1.65})$$

A $\gamma_{A(j)}$ parameter must be calculated for each spectral line generated in the sample (or standard) that has sufficient energy to cause significant fluorescence of A. The net correction factor γ will be the sum of all of the individual γ terms, and the measured intensity of A is then corrected for characteristic fluorescent enhancement by multiplying it by the factor $1 / (1 + \gamma)$. Similar corrections must be applied to the measured standard intensities if the standards also contain elements capable of generating fluorescent enhancement.

Reed (1975) has suggested that the manual calculation of the γ parameters can be simplified without serious error if some of the terms of Equation 9.32 are combined into a single parameter $J(A)$, such that

$$J(A) = 0.5 \times \frac{r_A - 1}{r_A} \times \omega_K(B) \times \frac{A_A}{A_B} \qquad (9.38)$$

which is assumed to be constant for a given element A, regardless of element B, and is calculated for the element (B) of lowest atomic number capable of exciting fluorescence from A. Reed has presented conveniently tabulated values of $J(A)$ for K–K and L–K fluorescence of elements (A) from $Z = 11$ (Na) to $Z = 33$ (As), i.e. for the range of elements normally analysed by measurement of K radiation.

K–L, L–K and L–L fluorescence correction Although these cases are theoretically more complex, satisfactory results are usually obtained by assuming a single L absorption edge and/or emission line (normally the

THE 'SIMPLIFIED THEORY' APPROACH

L_{III} edge and/or the $L\alpha_1$ line). This approximation is only likely to introduce significant error if the sample (or standard) has an absorption edge of appreciable magnitude between the L lines of the exciting element (B), in which case it may be necessary to calculate separate corrections for the exciting L lines above and below the edge.

When the fluorescence-excited line of element A is an L line, the absorption edge jump ratio r_A is normally taken as the combined ratio for the L_I, L_{II} and L_{III} edges, i.e. as the product of the jump ratios for each of the three sub-shell edges. Values of $(r_A - 1) / r_A$ for an appropriate range of elements, calculated from the data of Heinrich (1966), have been given by Reed (1975). Alternatively, they may be calculated from expressions such as that of Colby (Eqn 9.34). Although each sub-shell has a different fluorescent yield, it is also often assumed, where appropriate, that $\omega(L_{III})$ may be used as an effective weighted mean fluorescent yield ω_L for the L shell. The $\omega_K(B)$ term in Equation 9.32 remains unchanged for K–L fluorescence, but must be replaced by $\omega_L(B)$ for L–K or L–L fluorescence. ω_L may be calculated from Equation 9.35, or the calculation may be simplified by using the $J(A)$ parameter (Eqn 9.38), for which Reed (1975) has presented tabulated values.

Finally, in K–L or L–K fluorescence allowance must be made for the relative intensities of K and L spectra. Reed (1975) has suggested the introduction of an empirically derived factor of 0.25 into Equation 9.33 for K–L fluorescence, and the inverse factor of 4.0 for L–K fluorescence. Note that if Reed's $J(A)$ parameters are used, these factors are already incorporated in the tabulated values.

The appropriate parameters to use in Equation 9.32 for the various possible combinations of K–L excitation/fluorescence are summarized in Table 9.1. For example, consider the analysis of an alloy of composition $Fe_{0.75}Ni_{0.25}$ at 25 kV, with a take-off angle ψ of 40°, using Kα radiation for the measurement of each element, and pure element standards. To calculate the characteristic fluorescence correction factor γ we will assume that the true composition is known; in practice it would of course be derived from the uncorrected data by ZAF iteration.

Table 9.1 Selection of parameters for characteristic fluorescence (excitation *of* element A *by* element B).

	For $\omega(B)$, use	For $(r_A - 1) / r_A$, use
$K_B \rightarrow K_A$	$\omega_K(B)$	K edge
$K_B \rightarrow L_A$	$\omega_K(B)$	L edge
$L_B \rightarrow K_A$	$\omega_L(B)$	K edge
$L_B \rightarrow L_A$	$\omega_L(B)$	L edge

COMPOSITION-DEPENDENT ERRORS IN EPA

From wavelength/energy tables (e.g. Table A.1),

	E_c (keV)	$E(K\alpha)$ (keV)	$E(K\beta)$ (keV)
Fe	7.110	6.398	7.057
Ni	8.330	7.471	8.263

There will be no characteristic fluorescence in the pure element standards, so that it is not necessary to calculate γ factors for correction of the standard intensities. In the alloy, both Ni Kα and Ni Kβ have energies higher than that of the Fe K absorption edge, so that Fe Kα intensities will require correction for Ni K-excited fluorescent enhancement. However, both Fe K lines have energies lower than that of the Ni K edge and therefore cannot excite fluorescent Ni K radiation; hence data for Ni in the alloy will not need correction for characteristic fluorescence.

For K–K fluorescence, no relative intensity correction factor is required, so that the γ factor for Fe Kα enhancement is obtained by substitution in Equation 9.32:

$$\gamma_{A(B)} = 0.5 \times c_B \times \frac{\mu_B(A)}{\mu_B} \times \frac{(r_A - 1)}{r_A} \times \omega_K(B) \times \frac{A_A}{A_B} \times \ldots$$
$$\times \left(\frac{U_B - 1}{U_A - 1}\right)^{1.67} \times \left(\frac{\ln(1 + u)}{u} + \frac{\ln(1 + v)}{v}\right)$$
(Eqn 9.32)

in which:

(a) c_B = weight concentration of exciting element (Ni) = 0.25

(b) $\mu_B(A) = (\mu/\varrho)$ for Ni Kα in Fe = 380

  $\mu_B = (\mu/\varrho)$ for Ni Kα in the alloy = $(0.75 \times 380) + (0.25 \times 59)$

  = 300

  $\mu_B(A) / \mu_B = 380 / 300 = 1.267$

(c) $(r_A - 1)/r_A$ (for Fe K edge) = 0.877 (Reed)

  or = 0.882 (Eqn 9.35)

(d) $\omega_K(B)$ (for Ni K) = 0.389 (Reed)

  or = 0.381 (Eqn 9.36)

THE 'SIMPLIFIED THEORY' APPROACH

(e) $\qquad A_A / A_B = 55.85 / 58.71 = 0.951$

(f) $\qquad U_A = 25.0 / 7.11 = 3.52$

$$U_B = 25.0 / 8.33 = 3.00$$

$$((U_B - 1) / (U_A - 1))^{1.67} = 0.680$$

(g) To find u:

$$\mu_A(\text{sample}) = (0.75 \times 71) + (0.25 \times 90) = 75.8$$

$$\mu_B = 300$$

$$\text{cosec}(\psi) = 1.5557$$

$$u = (\mu_A(\text{sample}) / \mu_B) \times \text{cosec}(\psi)$$

$$= 0.393$$

(h) To find v:

$$\sigma_B = 4.5 \times 10^5 / (25.0^{1.65} - 8.33^{1.65}) = 2654$$

$$v = \sigma_B / \mu_B = 2654 / 300 = 8.85$$

(i) $\dfrac{\ln(1+u)}{u} + \dfrac{\ln(1+v)}{v} = \dfrac{\ln(1.393)}{0.393} + \dfrac{\ln(9.85)}{8.85}$

$$= 1.10$$

(j) Substituting in Equation 9.32,

$$\gamma = 0.5 \times 0.25 \times 1.267 \times 0.877 \times 0.389 \times 0.951 \times 0.68 \times 1.10$$

$$= 0.038$$

The characteristic fluorescence correction factor for Fe Kα in an alloy of this composition is then $1 / (1 + \gamma) = 1 / 1.038 = 0.963$. The measured Fe K$\alpha$ intensities must be multiplied by this factor to compensate for characteristic fluorescence enhancement of almost 4 per cent.

CONTINUUM FLUORESCENCE

Continuum-excited fluorescence may contribute as much as 20 per cent or more of the intensity of a characteristic emission. The ratio of continuum-generated to electron-excited K series X-ray emission is given approximately by the expression

$$I_{\text{continuum}} / I_{\text{direct}} = Z^4 / 10^7 \qquad (9.39)$$

which indicates a ratio of 1.6 per cent for $Z = 20$ (Ca), rising to 11.8 per cent for $Z = 33$ (As). Because of its dependence on Z, continuum-excited fluorescence may be sensitive to composition, but the associated analytical error will be small if the samples and standards do not vary greatly in mean atomic number. Silicates, for example, may show only minor variation in Z_{mean} over quite substantial ranges of composition, and continuum fluorescence is then not a source of significant analytical error.

Furthermore, some of the fluorescent radiation is generated at greater depths in the sample than the primary radiation (Fig. 9.5) and hence is more heavily absorbed, particularly at low take-off angles.

For these reasons continuum fluorescence is often ignored in routine *ZAF* correction procedures. The omission is unlikely to lead to serious error unless there is a significant variation in Z_{mean} between samples and standards, or unless the samples/standards have one or more pronounced absorption edges between E_0 and E_c for the measured radiation.

For cases in which continuum fluorescence correction is deemed to be necessary, several different approaches to the problem have been suggested. The calculations involved are relatively complex, since they must take account of the effects of all continuum X-rays having sufficient energy to excite the measured radiation, including assessment of their relative intensities. The latter in turn require evaluation of the magnitude of all absorption edges between E_0 and E_c.

The distribution of generated continuum intensities in a single element target is described by an expression due to Kramers (Ch. 2), which can be rewritten in terms of energies rather than wavelengths as

$$dn = (k \times Z \times (E_0 - E) / E) \, dE \tag{9.40}$$

This expression describes the number of X-ray photons (dn) with energies between E and $(E + dE)$ generated by an incident electron of energy E_0. Continuum photons capable of exciting fluorescence have energies between E_0 and E_c, and the total number of such photons will therefore be given by

$$n = kZ \int_{E_c}^{E_0} [(E_0 - E) / E] \, dE \tag{9.41}$$

Substituting $U = E / E_c$,

$$n = kZE_c \int_1^{U_0} \left(\frac{U_0}{U} - 1\right) dU$$

$$= kZE_c \, (U_0 \ln U_0 - U_0 + 1) \tag{9.42}$$

THE 'SIMPLIFIED THEORY' APPROACH

If it is assumed that the continuum is generated at negligibly shallow depths, which is not strictly true but which simplifies the theory without introducing serious error, then approximately half of the continuum X-rays will leave the sample without producing any fluorescence. Of the remainder, a fraction $(r - 1) / r$ will produce K shell ionizations (where r is the K absorption edge jump ratio), and of these a fraction ω_K will produce fluorescent K photons. In a pure element sample, the K fluorescence intensity will then be given by

$$I_f = 0.5 \times \frac{(r - 1)}{r} \times \omega_K \times k \times Z \times E_c \times (U_0 \ln U_0 - U_0 + 1)$$

(9.43)

For compound targets, Springer (1967) derived an expression for continuum fluorescent excitation of K radiation for element A:

$$\gamma = \frac{I_f}{I_A}$$

$$= 4.34 \times 10^{-6} \times \frac{(r_A - 1)}{r_A} \times A \times Z_{mean} \times E_c \times \frac{\mu_K(A)}{\mu_K} \times h(u, U_0)$$

(9.44)

in which:

(a) $\mu_K(A)$ and μ_K are the mass absorption coefficients immediately on the high side of the K absorption edge for pure A and for the compound target respectively. Values of $\mu_K(A)$ and $\mu_L(A)$ were calculated by Springer from the data of Heinrich (1966), and have been summarized by Reed (1975).
(b) Z_{mean} is the mean atomic number, calculated as $\Sigma\, c_j\, Z_j$.
(c) $h(u, U_0)$ is a term to correct for absorption of the fluorescent radiation, calculated from an expression which Springer showed could be approximated by

$$h(u, U_0) = \frac{\ln(1 + u \times U_0)}{u \times U_0}$$

(9.45)

in which $u = \chi / \mu_K$ and $\chi = (\mu / \varrho) \operatorname{cosec}(\psi)$ for pure A.

For continuum excitation of L radiation, the constant in Equation 9.44 is divided by the empirically derived factor of 4.0 to become 1.08×10^{-6}, and it is used in conjunction with r_A values for the combined L edges and mass absorption coefficients for the high-energy side of the L_I edge.

If the target has one or more significant absorption edges between E_c and E_0 then the continuum intensity will be a discontinuous function of wavelength (Fig. 9.12), and it will be necessary to modify Equation 9.44 so that the integrated fluorescence effect (Eqn 9.41) can be evaluated separately for each of the discontinuous energy bands.

If we put

$$G = 4.34 \times 10^{-6} \times \frac{(r_A - 1)}{r_A} \times A \times Z_{mean} \quad (9.46)$$

then the fluorescence components for the successive energy bands, each lying between $E(n)$ and $E(n + 1)$, can be summed by rewriting Equation 9.45 as

$$\gamma = G \times V / (U_0 \ln U_0 - U_0 + 1) \quad (9.47)$$

in which

$$V = \sum_{E(n) = E_c}^{E(n+1) = E_0} \left[\left\{ \frac{\mu_K^A(n)}{\mu_K(n)} \left[E_c(n) \left[U_0(n) \ln U_0(n) - U_0(n) + 1 \right] \times h\left[u, U_0(n)\right] \right. \right.\right.$$

$$- E_c(n + 1) \left[U_0(n + 1) \ln U_0(n + 1) - U_0(n + 1) + 1 \right]$$

$$\left.\left.\left. \times h\left[u, U_0(n + 1)\right] \right] \right\} \right]$$

$$(9.48)$$

The summation of Equation 9.48 is then performed over each of the edge-bounded continuum energy bands between $E_c(A)$ and E_0. In each term of the summation, parameters identified as (n) refer to the lower energy band, and the $(n + 1)$ parameters refer to the upper. For the last (highest) energy band, $E_c(n + 1) = E_0$, which is not an absorption edge but the high-energy limit of the continuum; the term in $(n + 1)$ then becomes zero for this band. Note that Equation 9.48 reduces to Equation 9.45 if there are no intervening absorption edges between E_c and E_0 (i.e. the summation is made over only one band, and the terms in $(n + 1)$ are all zero).

Consider the example of the $Fe_{0.75}Ni_{0.25}$ alloy previously used to demonstrate the calculation of characteristic fluorescence corrections:

THE 'SIMPLIFIED THEORY' APPROACH

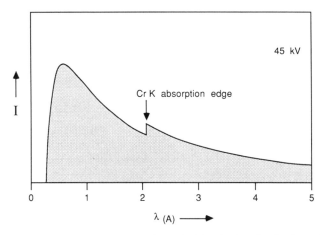

Figure 9.12 Absorption edges produce discontinuities in the distribution of continuum intensity as a function of wavelength, since continuum wavelengths generated within the sample and shorter than the absorption edge are more strongly absorbed by the sample than those of longer wavelength. Corrections for continuum-excited fluorescence depend on continuum intensity and must therefore be separately evaluated for each of the discontinuous wavelength bands (although the magnitude of a discontinuity is proportional to the concentration of the absorbing element, and it may not be significant if the latter is present only in low concentrations).

$$E_0 = 25 \text{ keV}$$
$$\psi = 40°$$
$$\text{Fe K}\alpha = 6.398 \text{ keV}, \quad E_c = 7.110 \text{ keV}$$
$$\text{Ni K}\alpha = 7.471 \text{ keV}, \quad E_c = 8.330 \text{ keV}$$

There are no absorption edges between $E_c(\text{Ni})$ and E_0 in the Ni standard, between $E_c(\text{Fe})$ and E_0 in the Fe standard, or between $E_c(\text{Ni})$ and E_0 in the alloy. Continuum fluorescence corrections for the standard intensities and for Ni in the alloy can therefore be made directly using Equation 9.45. However, for Fe in the alloy the Ni K absorption edge (8.330 keV) lies between $E_c(\text{Fe})$ (7.110 keV) and E_0 (25.0 keV). The summation of Equation 9.49 must therefore be performed over the two edge-bounded continuum energy bands, viz. 7.110–8.330 and 8.330–25.0 keV.

The simpler calculations (for Fe in the Fe standard, Ni in the Ni standard, and Ni in the alloy) may be illustrated by the example of Ni in the alloy. The required data (e.g. from the appropriate tables in the Appendix) are as follows:

$(r_A - 1) / r_A$ for Ni Kα = 0.871

$Z(Ni) = 28 \quad Z(Fe) = 26$

$A(Ni) = 58.7$

μ_K (pure Ni) for high-energy side of Ni K edge = 336

μ_K (pure Fe) for high-energy side of Ni K edge = 280

$\mu_{Ni\ K\alpha}$ (pure Ni) = 59, $\mu_{Fe\ K\alpha}$ (pure Ni) = 90

$\mu_{Ni\ K\alpha}$ (pure Fe) = 380, $\mu_{Fe\ K\alpha}$ (pure Fe) = 71

For the alloy,

$Z_{mean} = (0.75 \times 26) + (0.25 \times 28) = 26.5$

μ_K(alloy) for high-energy side of Ni K edge = (0.75×280)
$+ (0.25 \times 336)$
$= 294$

$\chi = \mu_{Ni\ K\alpha} \operatorname{cosec}(\psi) = [(0.75 \times 380) + (0.25 \times 59)] \times \operatorname{cosec}(40)$
$= 466$

$u = \chi / \mu_K = 466 / 294$
$= 1.585$

$U_0 = 25 / 8.33$
$= 3.00$

$h(u, U_0) = [\ln(1 + u \times U_0)] / (u \times U_0)$
$= 0.37$

Substituting in Equation 9.46,

$\gamma_{cont} = 4.34 \times 10^{-6} \times 0.871 \times 58.7 \times 26.5 \times 8.33 \times (336 / 294) \times 0.37$
$= 0.021$

There is no characteristic fluorescence of Ni Kα by Fe, so that in this instance $\gamma_{char} = 0$, and the total fluorescence correction factor is

$F = 1 / (1 + \gamma_{cont} + \gamma_{char}) = 1 / (1 + 0.021 + 0) = 0.980$

Similar calculations yield γ_{cont} values of 0.028 for Fe Kα in the (pure) Fe standard and 0.039 for Ni Kα in the (pure) Ni standard. Again no

THE 'SIMPLIFIED THEORY' APPROACH

characteristic fluorescence correction is required in either case, so that the net fluorescence correction factors (F) become

$$1 / (1 + 0.028) = 0.973$$

and

$$1 / (1 + 0.039) = 0.962$$

respectively.

For correction of measured Fe Kα intensity in the alloy, the V term of Equation 9.48 must now be evaluated and summed for the two energy bands 7.11–8.33 and 8.33–25.0 keV; since 25.0 keV is E_0 and not an absorption edge, the ($n + 1$) terms for the second band will be zero.

Step 1:   Evaluate the u, U_0 and h parameters at each absorption edge:

(a) at 7.110 keV:

$$U_0 = 25 / 7.110 = 3.52$$

This is the Fe K absorption edge, and the mass absorption coefficients of Fe and Ni for Fe Kα are 71 and 90 respectively.

$$\chi_{\text{Fe K}\alpha} = \mu_{\text{Fe K}\alpha} \text{ (alloy)} \times \text{cosec}(\psi)$$
$$= (0.75 \times 71) + (0.25 \times 90) \times \text{cosec}(40)$$
$$= 118$$

On the high-energy side of the Fe K absorption edge,

$$\mu_K \text{ (pure Fe)} = 421 \text{ and } \mu_K \text{ (pure Ni)} = 68$$

so that

$$\mu_K \text{ (alloy)} = (0.75 \times 421) + (0.25 \times 68)$$
$$= 333$$

At this edge,

$$u = \chi / \mu_K$$
$$= 118 / 333$$
$$= 0.354$$
$$h(u, U_0) = [\ln(1 + u \times U_0)] / (u \times U_0)$$

$$= [\ln(1 + 0.354 \times 3.52)] / (0.354 \times 3.52)$$
$$= 0.65$$

(b) at 8.330 keV:

$$U_0 = 25 / 8.33$$
$$= 3.00$$

This is the Ni K absorption edge, and the mass absorption coefficients of Fe and Ni for Ni Kα are 380 and 59 respectively.

$$\chi_{Ni\ K\alpha} = [(0.75 \times 380) + (0.25 \times 59)] \times \text{cosec}(40)$$
$$= 466$$

On the high-energy side of the Ni K absorption edge,

$$\mu_K \text{ (pure Fe)} = 280 \qquad \mu_K \text{ (pure Ni)} = 336$$
$$\mu_K \text{ (alloy)} = (0.75 \times 280) + (0.25 \times 336)$$
$$= 294$$

At this edge,

$$u = \chi / \mu_K$$
$$= 466 / 294$$
$$= 1.585$$
$$h(u, U_0) = [\ln(1 + u \times U_0)] / (u \times U_0)$$
$$= [\ln(1 + 1.585 \times 3.00)] / (1.585 \times 3.00)$$
$$= 0.37$$

For the two edges, we have thus derived the following values:

	Edge 1 (7.110 keV)	Edge 2 (8.330 keV)
E_c	7.110	8.330
$\mu_K(A)$	421	280
μ_K	333	294
U_0	3.52	3.00
$h(u, U_0)$	0.65	0.37

THE 'SIMPLIFIED THEORY' APPROACH

Step 2: Now substitute the appropriate values in Equation 9.48 for each of the two energy bands.

(a) 7.110–8.330 keV:

$$V = (421 / 333) \times [(7.110 \times (3.52 \times 1.258 - 3.52 + 1) \times 0.65)$$
$$- (8.33 \times (3.00 \times 1.099 - 3.00 + 1) \times 0.37)]$$
$$= 6.11$$

(b) 8.330–25.0 keV:

$$V = (280 / 294) \times [(8.33 \times (3.00 \times 1.099 - 3.00 + 1) \times 0.37)$$
$$- (0.0)]$$
$$= 3.80$$

Step 3: For the two bands,

$$\Sigma (V) = 6.11 + 3.80 = 9.91$$

Step 4: For Fe K,

$$U_0 \ln(U_0) - U_0 + 1 = 3.52 \times \ln(3.52) - 3.52 + 1$$
$$= 1.91$$

Step 5: For Fe K,

$$(r_A - 1) / r_A = 0.877$$
$$A = 55.85$$
$$Z_{mean} \text{ (alloy)} = (0.75 \times 26) + (0.25 \times 28)$$
$$= 26.5$$

Substituting in Equation 9.46,

$$G = 4.34 \times 10^{-6} \times 0.877 \times 55.85 \times 26.5 = 0.0056$$

Step 6: Substituting in Equation 9.47,

$$\gamma_{cont} = G \times V / (U_0 \ln(U_0) - U_0 + 1)$$
$$= 0.0056 \times 9.91 / 1.91$$
$$= 0.029$$

Since γ_{char} for Fe in the alloy has previously been calculated as 0.038, the total fluorescence correction factor F will therefore be given by

$$F = 1 / (1 + \gamma_{char} + \gamma_{cont})$$
$$= 1 / (1 + 0.038 + 0.029)$$
$$= 0.937$$

These calculations suggest that for an alloy of this composition, and under the specified analytical conditions, approximately 6 per cent of the measured Fe Kα intensity arises from characteristic and continuum fluorescence, compared to less than 3 per cent for the pure Fe standard. Slightly less than half of the fluorescent contribution is due to continuum fluorescence. It is the analyst's responsibility to determine whether an error of this magnitude is tolerable or whether all or part of the fluorescence correction must be made.

To complete the ZAF correction procedure, the apparent composition (derived from $I_{sample}/I_{standard}$ and already corrected for atomic number and absorption effects) is multiplied by the correction factor $F_{sample}/F_{standard}$. In the $Fe_{0.75}Ni_{0.25}$ example,

$$c_{Fe} = c'_{(Fe)} \times (F_{Fe} \text{ (alloy)} / F_{Fe} \text{ (Fe standard)})$$
$$= c'_{(Fe)} \times (0.937 / 0.973)$$
$$= c'_{(Fe)} \times 0.963$$

$$c_{Ni} = c'_{(Ni)} \times (F_{Ni} \text{ (alloy)} / F_{Ni} \text{ (Ni standard)})$$
$$= c'_{(Ni)} \times (0.980 / 0.962)$$
$$= c'_{(Ni)} \times 1.019$$

9.4 The empirical (α-factor) approach

The ZAF procedure of calculating correction factors by detailed evaluation of atomic number (Z), absorption (A) and fluorescence (F) effects is fundamentally complex and requires some simplification to reduce it to a practical level, even for on-line computer calculation. There is also still some degree of uncertainty associated with some of the parameters required for the calculations, such as the mass absorption coefficients for long-wavelength (low-energy) emissions.

Faced with these difficulties (and particularly in the earlier stages of microprobe evolution, when many of the empirical relationships used in the ZAF procedure had not yet been derived), some analysts adopted an

THE EMPIRICAL (α-FACTOR) APPROACH

alternative strategy based essentially on the analysis of a more or less comprehensive library of materials of known composition and the empirical determination of the magnitude of the resulting errors as functions of compositional variation. Once determined in this way, the compositional error functions could then be used (subject to some obvious limiting assumptions) to correct the raw intensity data collected from other samples of unknown but broadly similar composition. A similar procedure based on empirically determined 'influence coefficients' is used in XRF analysis, and has been described in Chapter 8.

The most successful attempts to implement this approach have been based on a linear relationship, often observed in simple (e.g. binary) systems such as a system AB, between the correction factor F_A required for an element A in such a system and its concentration c_A. If this relationship is indeed linear, then it can be expressed as

$$F_A = \alpha_{AB} + (1 - \alpha_{AB}) \times c_A \qquad (9.49)$$

in which α_{AB} is the limiting value for F_A as $c_A \to 0$; it can be determined by analysing a series of AB samples of known composition, using a standard of pure A ($F_A = 1$ when $c_A = 100$ per cent) and graphically or analytically deriving the value of α_{AB} required for a linear relationship between F_A and c_A.

If it is further assumed that the linear dependence of F_j on c_j in individual binary systems is maintained in more complex systems containing the same elements (e.g. Ziebold & Ogilvie 1966), then for a multi-element compound (ABCD . . .),

$$F_{A(ABCD...)} = \alpha_{AA} c_A + \alpha_{AB} c_B + \alpha_{AC} c_C \cdots$$
$$= \Sigma \, \alpha_{Aj} \, c_j \qquad (9.50)$$

By definition, $\alpha_{AA} = 1$; if the binary factors α_{AB}, α_{AC} . . . can be established independently then the calculation of overall factors F_j for each of the j elements in a multi-element sample becomes feasible. These factors are analogous to ZAF factors, but are much easier to calculate, since they require only simple substitution in Equation 9.50 and iteration to convergence. This is certainly a much easier procedure to follow for hand calculation.

The binary factors can be established either empirically, i.e. by analysis of appropriate binary samples of known composition, or they can be calculated for each pair of elements in a multi-component system by using the ZAF procedures described above (or any other suitable matrix correction procedure). Once the α-factors are known, they are much easier to apply in routine hand calculations than the ZAF procedures which may have been used to calculate them.

In one of the best known applications of this approach, Bence and Albee (1968) derived the necessary α-factors for the major elements most commonly determined in oxide and silicate minerals, using a combination of both calculation and empirical measurement procedures. The Norrish and Hutton XRF procedure for silicate analysis by sample fusion (Ch. 8) is analogous in that correction coefficients are calculated from (simplified) first principles and empirically refined in the light of analytical experience.

However, it must not be forgotten that the α-factor approach is based on the assumption of a linear relationship between F_A and c_A. A review of the principles of ZAF correction procedures will show that this assumption is not unreasonable for the absorption correction, which is strongly dependent on the term $(1 + \chi / \sigma)$ in the Philibert Equation, and this term in turn is a linear function of composition. On the other hand, the composition–fluorescence and composition–atomic number functions are hyperbolic rather than linear (although they may be approximately linear over limited but useful compositional ranges).

This means that the α-factor approach is potentially useful in systems in which absorption is the predominant source of error, and/or if there is only limited variation in composition between samples and standards. Otherwise it may be expected to yield relatively poor results. It is also important to note that the α-factors derived for a given set of experimental conditions (accelerating voltage, microprobe take-off angle etc.) cannot be used if any of those parameters is significantly changed, e.g. on another microprobe with a different take-off angle.

In more recent years the advent of inexpensive but powerful microcomputers that can be easily adapted to on-line data collection and reduction, and the ready availability of numerous ZAF-based correction programs, have eliminated much of the need for the α-factor approach, and it is now largely confined to a relatively small group of specialized applications. However, in such a context (for example in routine process control) it can still prove to be very useful because of the simplicity and speed of its calculations and because of the facility with which its correction coefficients can be progressively refined in the light of analytical experience.

APPENDIX A
Tables of X-ray analysis parameters

This appendix contains tables of various parameters useful in the derivation of analytical strategy and in the correction of the raw data obtained in X-ray fluorescence and electron microprobe analysis. In their original form, these parameters are all available in various literature sources which have been quoted, where appropriate, in the text. The data presented in the appendix tables are condensed and in some respects approximate. They are intended to serve principally for convenient reference.

Analysts have traditionally looked up the data needed for analytical applications in various 'standard' sources, among which the wavelength tables of White and Johnson (1972) and the mass absorption coefficients of Heinrich (1966) have been particularly widely used. These data can then be incorporated directly into whatever calculations are being undertaken. However, the proliferation of small computers into almost every analytical laboratory has had a marked impact on this process, and the analyst now has several additional options. For example, the data can be looked up, as before, and entered manually into the computer as required whenever a correction program (for example) is running – although this procedure is wasteful of time and effort and encourages transcription errors. Alternatively, the source data can be copied into disk or memory files to be accessed by a computer 'look-up' procedure when required – but this may be inefficient and wasteful of limited disk or memory storage.

Another alternative, which has been used over the years with varying degrees of success in data reduction procedures such as the well known electron microprobe correction programs MAGIC IV (Colby 1968) or FRAME (Yakowitz et al. 1973), is to fit the data to relatively simple mathematical expressions which can be evaluated whenever a numerical value for a parameter is required. Most commonly such expressions are derived by some form of linear or curvilinear regression performed on the available data and resulting in a set of coefficients for a polynomial expression of the form of

$$y = a + bx + cx^2 + dx^3 + \ldots$$

in which y is the desired parameter, x is the independent variable on which y is dependent, and $a, b, c, d \ldots$ are the coefficients derived from

APPENDIX A

the regression. It is then necessary to store only the coefficients in computer-accessible form, to be used by an operating program as desired.

As a guide to the practical utility of this technique, almost all of the tables in this appendix have been prepared in this way. Many of the parameters used in X-ray spectrometry are functions of the atomic number Z, which has thus been used as the independent variable on which the various dependent parameters (e.g. X-ray emission wavelengths, absorption edges, mass absorption coefficients) have been regressed. Since many of the Z-functions are exponential (e.g. Moseley's Law), the regressions were performed after logarithmic transformation of both Z and the dependent parameter, and the resulting polynomial coefficients are summarized in Table A.1. When using these coefficients, it is thus *essential* to use $\log_e(Z)$ as the independent variable (x in the polynomial), and to transform the resulting value of y to its natural exponent (e^y) to obtain the desired parameter.

For example, mass absorption coefficients are functions of λ, the wavelength of the X-rays being absorbed, and of Z, the atomic number of the absorbing element(s). They can be calculated from an expression of the general form of

$$(\mu/\varrho) = C^j \lambda^i$$

in which the constant C and the exponents i and j change values at each absorption edge of the absorbing medium (and may also change slightly in the interval between successive absorption edges, although this is not usually a severe effect). Therefore to calculate a mass absorption coefficient for a particular absorber and a specified spectral emission from a particular element, it is necessary to know

(a) the wavelength (energy) of the spectral emission;
(b) the wavelengths (energies) of each of the absorption edges of the absorber, in order to determine the appropriate pair of bounding absorption edges; and
(c) the values of C, i and j appropriate to the absorption region bounded by those edges.

Tables A.1 and A.2 include polynomial coefficients for the evaluation of each of the necessary parameters. The Table A.1 polynomials for emission wavelengths (K, L and M α and β series) have been fitted by the author to data compiled from several sources, but principally from Heinrich (1966), White and Johnson (1972), Reed (1975) and Tertian and Claisse (1982). The mass absorption coefficient data of Table A.2 have been kindly provided by Dr Keith Norrish, of the Commonwealth Scientific and Industrial Research Organization.

It is necessary to issue a few words of caution about the use either of

USEFUL PARAMETERS FOR XRS DATA CORRECTION

the polynomial coefficients or of the tabulated data produced with them. The most obvious is that the polynomials have been fitted to published numbers which themselves have been derived in part by extrapolation or by fitting a limited amount of experimental data to similar empirical expressions of sometimes uncertain validity. The polynomials can at best be no more accurate than the data on which they are based, and they may sometimes be significantly worse. It is often difficult or impossible to assess their accuracy because there may be no accepted standard against which they can be judged. This is particularly true of data pertaining to the long-wavelength (low-energy) regions of the X-ray spectrum, where experimental support is very limited. Published data for L series parameters show significantly more inconsistencies than those for the K series, and data for the M series are worse again.

Secondly, the question of mathematical precision is very significant to regression analysis, but often overlooked. In theory, a curve can be fitted closely to almost any set of x–y data by calculating a polynomial of sufficiently high order, i.e. with sufficient terms in x^a. In practice the x^a terms rapidly become very large as a increases (consider, for example, the numerical value of x^9 where x is the atomic number of, say, tin ($Z = 50$) – or even its logarithmic transformation). A small rounding error in the polynomial coefficient for a large term can result in significant error when the polynomial is evaluated. Fitting data to a regression curve therefore involves some compromise between 'goodness of fit' and minimizing precision errors associated with the higher terms.

For the data presented in this appendix, the maximum polynomial order has been limited to five, and it has been kept smaller whenever possible, consistent with reasonably good agreement with the source data. There are therefore some slight discrepancies between the data in the tables and the sources from which they were derived. The tabulated data have been checked in a fairly wide range of test calculations and have not yet been found to produce serious errors, but the possibility cannot be excluded and users are warned to exercise some care, particularly in long-wavelength regions, in the immediate vicinity of absorption edges, and particularly with M series parameters. Even slight errors, of the order of 1 per cent or less, will occasionally result in a calculated emission being placed on the wrong side of a calculated absorption edge. This is most likely to happen in the bands between the L_I and L_{III} and the M_I and M_V absorption edges, particularly when the edges are at long wavelengths. In the mass absorption coefficient tables (A.6 to A.8), identified errant values that do not agree with those of Heinrich (1966) have been italicized as a warning, but there may also be a few others which so far have escaped detection. In cases where emission wavelengths and absorption edges could not be estimated with sufficient precision for reliable assessment of their relative magnitudes, Heinrich very properly did not calculate mass absorption coefficients; however, their omission can be frustrating when an application calls for at least an order of magnitude estimate, so that the computed values for these cases have been included in Tables A.6 to A.8 regardless of their uncertainty. The user is specifically warned of the possibility of relatively serious error whenever an emission wavelength lies close to an absorption edge, particularly in the long wavelength region.

Notwithstanding these comments, data for regions beyond the N_I absorption edge are so fragmentary that confidence in the computed results is minimal; this has been indicated by following Heinrich's practice of tabulating absorption coefficients in these regions as zeroes.

There are also some discrepancies with Heinrich's data for oxygen absorption, some of which appear to be in error.

The regressions from which the coefficients of Table A.1 were derived were performed on an Apple Macintosh computer, with a program written in Macintosh Pascal (cf. Table A.15) and using double or extended precision throughout – that is, the calculations were performed with approximately 12–16 digit precision. The coefficients are reproduced in Table A.1 with about the same precision, and rounding or truncating them to shorter numbers may be convenient for some purposes but may also introduce significant errors, particularly in the higher-order terms.

In a few cases, values in the tables will not agree precisely with those used in worked examples in the text; this is because the text examples have used data from the original sources. No significant error is likely to arise from the differences, which are most apparent in the tables of emission and absorption edge energies/wavelengths. The reader is reminded that the numbers presented in this appendix are computed approximations which have been found to serve adequately for most practical purposes but should not be used on faith when absolute accuracy is mandatory.

Table A.1 Polynomial coefficients for calculation of X-ray emission and absorption wavelengths.

Parameter	A	B	C	D	E	F
Kα	9.068019359128	-4.522524060217	1.586321229666	-0.543135476264	0.095623759498	-0.006815040278
Kβ	-3.146296575168	15.804656840079	-11.636458440841	3.670946765013	-0.566341361019	0.034308904518
K edge	-2.359446425531	14.546361072520	-10.862614093087	3.440411792856	-0.532809528250	0.032362309762
Lα	-44.203492478029	73.353272820071	-41.002502059789	10.917977697082	-1.435389945541	0.074822289023
Lβ	-8.489546670833	22.764855200893	-12.409218877559	2.847980008737	-0.296473069854	0.010359965978
L_I edge	27.805924435315	-44.894464345371	32.807535451714	-11.335114078557	1.834009626933	-0.113685898371
L_{II} edge	-101.081705049676	123.781502152109	-54.702614695261	11.220500247413	-1.058205280387	0.034028621480
L_{III} edge	-100.790835352308	125.825882930448	-57.193277311 9493	12.291092922952	-1.260648152897	0.048438689903
Mα	76.886565822907	-44.851099915612	9.337875524501	-0.688397842401	0.0	0.0
Mβ	40.081035836506	-19.368715380172	3.472384519834	-0.239981403426	0.0	0.0
M_I edge	92.580456582022	-59.192355848285	13.427760136541	-1.067189363467	0.0	0.0
M_{II} edge	104.932007824875	-67.096805665675	15.140468077835	-1.192208758828	0.0	0.0
M_{III} edge	102.706985200956	-64.930371783528	14.456917292511	-1.120924744826	0.0	0.0
M_{IV} edge	176.111573903636	-114.928049419388	25.889762735507	-1.996453200321	0.0	0.0
M_V edge	172.570383069957	-112.247590576062	25.209388919051	-1.938232604284	0.0	0.0
N edge	-195.589167069815	132.954374871369	-28.788861523955	2.002861573284	0.0	0.0

$$\text{Parameter} = \exp(A + Bx + Cx^2 + Dx^3 + Ex^4 + Fx^5)$$

$$x = \log_e(Z)$$

Table A.2 Polynomial coefficients for calculation of mass absorption coefficients (courtesy of Dr Keith Norrish).

For λ :		Z range	A	B	C	n
<K edge		6–32	−0.12	0.215	−0.0013	$2.92 - (0.04\sqrt{Z})$
>K	<L_I	12–70	−0.21	0.1	−0.0002	$2.73 - 0.003(Z - 40)$ (if Z>40)
>L_I	<L_{II}	29–74	−0.12	0.092	−0.000159	"
>L_{II}	<L_{III}	30–76	−0.09	0.082	−0.00014	"
>L_{III}	<M_I	33–92	0.16	0.041	0.000112	2.6
>M_I	<M_{II}	55–92	0.75	0.0227	0.000216	2.6
>M_{II}	<M_{III}	56–92	0.77	0.019	0.000215	2.6
>M_{III}	<M_{IV}	58–92	0.86	0.015	0.00023	2.6
>M_{IV}	<M_V		0.9	0.013	0.00022	2.6
>M_V	<N_I	63–92	1.52	0.0047	0.00015	2.22

$$\mu \; (= \mu/\rho) \; = \; (A + BZ + CZ^2)^3 \cdot \lambda^n$$

USEFUL PARAMETERS FOR XRS DATA CORRECTION

Table A.3 Wavelengths and energies: K series.

Z	Element	Kα Å	Kα keV	Kβ Å	Kβ keV	K edge Å	K edge keV
6	C	44.386	0.279			41.558	0.298
7	N	31.602	0.392			30.814	0.402
8	O	23.626	0.525			23.315	0.532
9	F	18.316	0.677			18.057	0.686
10	Ne	14.604	0.849			14.303	0.867
11	Na	11.909	1.041	11.621	1.067	11.561	1.072
12	Mg	9.891	1.253	9.561	1.297	9.515	1.303
13	Al	8.342	1.486	7.990	1.551	7.955	1.558
14	Si	7.128	1.739	6.771	1.831	6.742	1.839
15	P	6.159	2.013	5.807	2.135	5.783	2.143
16	S	5.373	2.307	5.034	2.463	5.013	2.473
17	Cl	4.727	2.622	4.404	2.814	4.386	2.826
18	Ar	4.191	2.958	3.886	3.190	3.869	3.204
19	K	3.740	3.315	3.453	3.590	3.438	3.606
20	Ca	3.357	3.692	3.089	4.013	3.074	4.033
21	Sc	3.030	4.091	2.779	4.460	2.765	4.483
22	Ti	2.748	4.511	2.514	4.931	2.500	4.959
23	V	2.504	4.951	2.285	5.426	2.271	5.459
24	Cr	2.290	5.413	2.085	5.945	2.072	5.984
25	Mn	2.102	5.897	1.911	6.488	1.897	6.534
26	Fe	1.937	6.401	1.757	7.055	1.744	7.109
27	Co	1.789	6.927	1.621	7.647	1.608	7.710
28	Ni	1.658	7.475	1.500	8.264	1.487	8.337
29	Cu	1.541	8.044	1.392	8.905	1.379	8.990
30	Zn	1.436	8.635	1.295	9.572	1.282	9.669
31	Ga	1.340	9.248	1.208	10.263	1.195	10.375
32	Ge	1.254	9.883	1.129	10.981	1.116	11.108
33	As	1.176	10.541	1.057	11.724	1.044	11.868
34	Se	1.105	11.220	0.992	12.493	0.979	12.656
35	Br	1.040	11.922	0.933	13.289	0.920	13.471
36	Kr	0.980	12.647	0.878	14.110	0.866	14.315
37	Rb	0.925	13.394	0.829	14.959	0.816	15.187
38	Sr	0.875	14.164	0.783	15.834	0.771	16.087
39	Y	0.829	14.958	0.741	16.737	0.728	17.016
40	Zr	0.786	15.775	0.702	17.666	0.690	17.975
41	Nb	0.746	16.615	0.666	18.624	0.654	18.963
42	Mo	0.709	17.479	0.632	19.608	0.620	19.981
43	Tc	0.675	18.368	0.601	20.621	0.589	21.028
44	Ru	0.643	19.280	0.572	21.661	0.561	22.106
45	Rh	0.613	20.217	0.545	22.730	0.534	23.213
46	Pd	0.585	21.179	0.520	23.826	0.509	24.351
47	Ag	0.559	22.165	0.497	24.951	0.486	25.520
48	Cd	0.535	23.177	0.475	26.105	0.464	26.720
49	In	0.512	24.214	0.454	27.286	0.444	27.950
50	Sn	0.490	25.277	0.435	28.496	0.424	29.212

Table A.4 Wavelengths and energies: L series.

Z	Elem	Lα Å	Lα keV	Lβ Å	Lβ keV	L_I edge Å	L_I edge keV	L_II edge Å	L_II edge keV	L_III edge Å	L_III edge keV
25	Mn	19.449	0.637	19.108	0.649	16.156	0.767	19.064	0.650	19.401	0.639
26	Fe	17.589	0.705	17.267	0.718	14.704	0.843	17.255	0.718	17.573	0.705
27	Co	15.976	0.776	15.669	0.791	13.412	0.924	15.646	0.792	15.949	0.777
28	Ni	14.569	0.851	14.274	0.868	12.259	1.011	14.218	0.872	14.507	0.854
29	Cu	13.335	0.930	13.051	0.950	11.230	1.104	12.948	0.957	13.227	0.937
30	Zn	12.248	1.012	11.973	1.035	10.310	1.202	11.819	1.049	12.088	1.025
31	Ga	11.287	1.098	11.019	1.125	9.486	1.307	10.815	1.146	11.075	1.119
32	Ge	10.433	1.188	10.171	1.219	8.747	1.417	9.920	1.250	10.172	1.219
33	As	9.670	1.282	9.414	1.317	8.082	1.534	9.121	1.359	9.366	1.324
34	Se	8.988	1.379	8.735	1.419	7.484	1.656	8.406	1.475	8.644	1.434
35	Br	8.374	1.480	8.126	1.526	6.944	1.785	7.765	1.596	7.996	1.550
36	Kr	7.820	1.585	7.575	1.636	6.455	1.920	7.189	1.724	7.415	1.672
37	Rb	7.319	1.694	7.078	1.751	6.012	2.062	6.670	1.859	6.890	1.799
38	Sr	6.864	1.806	6.626	1.871	5.610	2.210	6.202	1.999	6.417	1.932
39	Y	6.450	1.922	6.214	1.995	5.245	2.364	5.778	2.145	5.988	2.070
40	Zr	6.072	2.042	5.838	2.123	4.911	2.524	5.394	2.298	5.600	2.214
41	Nb	5.726	2.165	5.495	2.256	4.606	2.691	5.045	2.457	5.246	2.363
42	Mo	5.409	2.292	5.179	2.393	4.328	2.864	4.727	2.622	4.924	2.517
43	Tc	5.116	2.423	4.889	2.535	4.073	3.044	4.437	2.794	4.630	2.677
44	Ru	4.847	2.557	4.622	2.682	3.838	3.230	4.171	2.972	4.361	2.842
45	Rh	4.599	2.696	4.375	2.833	3.622	3.423	3.928	3.156	4.114	3.013
46	Pd	4.368	2.838	4.147	2.989	3.423	3.622	3.705	3.346	3.888	3.189
47	Ag	4.155	2.984	3.935	3.150	3.239	3.827	3.499	3.543	3.679	3.370
48	Cd	3.956	3.133	3.738	3.316	3.069	4.040	3.309	3.746	3.486	3.556
49	In	3.772	3.287	3.555	3.487	2.911	4.258	3.134	3.955	3.308	3.747

50	Sn	3.599	3.444	3.384	3.663	2.765	4.484	2.972	4.171	3.143	3.944
51	Sb	3.438	3.605	3.225	3.844	2.629	4.716	2.821	4.394	2.990	4.145
52	Te	3.288	3.770	3.076	4.030	2.502	4.955	2.681	4.623	2.848	4.352
53	I	3.147	3.939	2.936	4.222	2.384	5.200	2.551	4.859	2.716	4.564
54	Xe	3.015	4.111	2.805	4.419	2.273	5.452	2.430	5.101	2.593	4.780
55	Cs	2.891	4.288	2.682	4.621	2.170	5.711	2.317	5.350	2.478	5.002
56	Ba	2.774	4.468	2.567	4.829	2.074	5.978	2.211	5.607	2.370	5.229
57	La	2.664	4.653	2.458	5.043	1.983	6.251	2.112	5.870	2.270	5.462
58	Ce	2.561	4.841	2.355	5.263	1.898	6.532	2.019	6.140	2.175	5.699
59	Pr	2.463	5.033	2.259	5.488	1.818	6.819	1.932	6.417	2.086	5.941
60	Nd	2.370	5.230	2.167	5.720	1.742	7.114	1.850	6.702	2.003	6.189
61	Pm	2.283	5.430	2.081	5.958	1.671	7.417	1.772	6.994	1.924	6.441
62	Sm	2.200	5.634	1.999	6.202	1.604	7.727	1.699	7.294	1.850	6.699
63	Eu	2.122	5.842	1.921	6.452	1.541	8.045	1.631	7.602	1.781	6.962
64	Gd	2.048	6.054	1.848	6.709	1.481	8.371	1.566	7.917	1.714	7.230
65	Tb	1.977	6.270	1.778	6.973	1.424	8.705	1.504	8.241	1.652	7.504
66	Dy	1.910	6.490	1.711	7.243	1.370	9.048	1.446	8.573	1.593	7.783
67	Ho	1.846	6.714	1.648	7.520	1.319	9.398	1.391	8.913	1.537	8.067
68	Er	1.786	6.943	1.588	7.804	1.270	9.758	1.338	9.262	1.483	8.357
69	Tm	1.728	7.175	1.531	8.096	1.224	10.127	1.289	9.620	1.433	8.652
70	Yb	1.673	7.411	1.477	8.395	1.180	10.504	1.241	9.987	1.385	8.953
71	Lu	1.620	7.651	1.425	8.701	1.138	10.891	1.196	10.363	1.339	9.260
72	Hf	1.570	7.895	1.375	9.015	1.098	11.288	1.153	10.749	1.295	9.572
73	Ta	1.522	8.143	1.328	9.337	1.060	11.695	1.112	11.145	1.253	9.890
74	W	1.477	8.395	1.282	9.667	1.023	12.112	1.073	11.550	1.214	10.214
75	Re	1.433	8.652	1.239	10.005	0.989	12.540	1.036	11.966	1.176	10.544
76	Os	1.391	8.912	1.197	10.352	0.955	12.978	1.000	12.394	1.139	10.881
77	Ir	1.351	9.176	1.158	10.707	0.923	13.427	0.966	12.832	1.105	11.223
78	Pt	1.313	9.444	1.120	11.071	0.893	13.889	0.933	13.280	1.071	11.572
79	Au	1.276	9.716	1.083	11.444	0.863	14.362	0.902	13.741	1.039	11.927

Table A.4 – *continued*.

Z	Elem	Lα Å	Lα keV	Lβ Å	Lβ keV	L$_I$ edge Å	L$_I$ edge keV	L$_{II}$ edge Å	L$_{II}$ edge keV	L$_{III}$ edge Å	L$_{III}$ edge keV
80	Hg	1.241	9.992	1.048	11.826	0.835	14.847	0.872	14.214	1.009	12.288
81	Tl	1.207	10.272	1.015	12.217	0.808	15.346	0.843	14.699	0.979	12.656
82	Pb	1.174	10.556	0.982	12.618	0.782	15.857	0.816	15.196	0.951	13.032
83	Bi	1.143	10.843	0.952	13.028	0.757	16.382	0.789	15.706	0.924	13.414
84	Po	1.113	11.134	0.922	13.448	0.733	16.921	0.764	16.230	0.898	13.803
85	At	1.085	11.430	0.893	13.879	0.709	17.475	0.739	16.768	0.873	14.199
86	Rn	1.057	11.729	0.866	14.320	0.687	18.044	0.716	17.320	0.849	14.603
87	Fr	1.030	12.032	0.839	14.771	0.665	18.628	0.693	17.887	0.826	15.014
88	Ra	1.005	12.339	0.814	15.234	0.645	19.230	0.671	18.469	0.803	15.432
89	Ac	0.980	12.649	0.789	15.707	0.625	19.847	0.650	19.066	0.782	15.858
90	Th	0.956	12.963	0.766	16.192	0.605	20.482	0.630	19.679	0.761	16.293
91	Pa	0.933	13.280	0.743	16.688	0.586	21.136	0.610	20.309	0.741	16.735
92	U	0.911	13.602	0.721	17.196	0.568	21.807	0.592	20.957	0.721	17.187

USEFUL PARAMETERS FOR XRS DATA CORRECTION

Table A.5 Wavelengths and energies: M series.

Z	Elem	Mα Å	Mα keV	Mβ Å	Mβ keV	M_I edge Å	M_I edge keV	M_{II} edge Å	M_{II} edge keV
54	Xe	17.567	0.706	17.113	0.724	10.893	1.138	12.507	0.991
55	Cs	16.566	0.748	16.163	0.767	10.310	1.202	11.793	1.051
56	Ba	15.650	0.792	15.288	0.811	9.772	1.269	11.138	1.113
57	La	14.809	0.837	14.480	0.856	9.274	1.337	10.534	1.177
58	Ce	14.036	0.883	13.732	0.903	8.812	1.407	9.978	1.242
59	Pr	13.322	0.930	13.039	0.951	8.382	1.479	9.462	1.310
60	Nd	12.663	0.979	12.395	1.000	7.981	1.553	8.985	1.380
61	Pm	12.052	1.029	11.796	1.051	7.608	1.629	8.541	1.451
63	Eu	10.957	1.131	10.719	1.156	6.931	1.789	7.742	1.601
64	Gd	10.466	1.184	10.233	1.211	6.623	1.872	7.381	1.679
65	Tb	10.007	1.239	9.778	1.268	6.334	1.957	7.043	1.760
66	Dy	9.579	1.294	9.352	1.325	6.062	2.045	6.726	1.843
67	Ho	9.177	1.351	8.953	1.385	5.806	2.135	6.429	1.928
68	Er	8.801	1.409	8.577	1.445	5.564	2.228	6.149	2.016
69	Tm	8.447	1.467	8.225	1.507	5.335	2.324	5.885	2.106
70	Yb	8.115	1.528	7.893	1.571	5.118	2.422	5.636	2.199
71	Lu	7.801	1.589	7.580	1.635	4.913	2.523	5.401	2.295
72	Hf	7.506	1.652	7.284	1.702	4.718	2.628	5.178	2.394
73	Ta	7.227	1.715	7.005	1.770	4.533	2.735	4.967	2.496
74	W	6.963	1.780	6.742	1.839	4.357	2.845	4.767	2.600
75	Re	6.713	1.846	6.492	1.909	4.190	2.959	4.577	2.708
76	Os	6.477	1.914	6.256	1.982	4.030	3.076	4.397	2.819
77	Ir	6.252	1.983	6.032	2.055	3.878	3.196	4.226	2.934
78	Pt	6.039	2.053	5.819	2.130	3.733	3.321	4.062	3.051
79	Au	5.836	2.124	5.617	2.207	3.595	3.448	3.907	3.173
80	Hg	5.643	2.197	5.425	2.285	3.462	3.580	3.759	3.298
81	Tl	5.459	2.271	5.242	2.365	3.336	3.716	3.617	3.427
82	Pb	5.284	2.346	5.068	2.446	3.215	3.856	3.482	3.560
83	Bi	5.117	2.423	4.903	2.528	3.099	4.000	3.352	3.698
84	Po	4.957	2.501	4.745	2.613	2.988	4.148	3.229	3.839
85	At	4.804	2.580	4.594	2.698	2.882	4.302	3.111	3.985
86	Rn	4.658	2.661	4.450	2.786	2.780	4.459	2.997	4.136
87	Fr	4.518	2.744	4.312	2.874	2.682	4.622	2.889	4.291
88	Ra	4.384	2.827	4.181	2.965	2.588	4.790	2.785	4.451
89	Ac	4.256	2.913	4.055	3.057	2.498	4.963	2.685	4.616
90	Th	4.132	3.000	3.935	3.151	2.411	5.142	2.589	4.787
91	Pa	4.014	3.088	3.819	3.246	2.327	5.326	2.498	4.963
92	U	3.900	3.178	3.709	3.342	2.247	5.516	2.409	5.145

APPENDIX A

Table A.5 – *Continued.*

		M_{III} edge		M_{IV} edge		M_V edge	
Z	Elem	Å	keV	Å	keV	Å	keV
54	Xe	13.344	0.929	18.227	0.680	18.603	0.666
55	Cs	12.610	0.983	17.047	0.727	17.409	0.712
56	Ba	11.936	1.039	15.983	0.776	16.332	0.759
57	La	11.316	1.095	15.019	0.825	15.356	0.807
58	Ce	10.743	1.154	14.143	0.876	14.469	0.857
59	Pr	10.214	1.214	13.344	0.929	13.660	0.907
60	Nd	9.723	1.275	12.613	0.983	12.920	0.959
61	Pm	9.266	1.338	11.324	1.095	11.615	1.067
63	Eu	8.445	1.468	10.755	1.153	11.037	1.123
64	Gd	8.074	1.535	10.227	1.212	10.503	1.180
65	Tb	7.726	1.604	9.738	1.273	10.007	1.239
66	Dy	7.400	1.675	9.283	1.335	9.546	1.299
67	Ho	7.094	1.747	8.859	1.399	9.116	1.360
68	Er	6.805	1.822	8.462	1.465	8.714	1.423
69	Tm	6.533	1.898	8.091	1.532	8.338	1.487
70	Yb	6.276	1.975	7.743	1.601	7.985	1.552
71	Lu	6.033	2.055	7.416	1.672	7.654	1.620
72	Hf	5.802	2.136	7.108	1.744	7.341	1.689
73	Ta	5.584	2.220	6.817	1.818	7.047	1.759
74	W	5.377	2.305	6.543	1.895	6.768	1.831
75	Re	5.180	2.393	6.283	1.973	6.505	1.906
76	Os	4.993	2.483	6.037	2.053	6.255	1.982
77	Ir	4.814	2.575	5.804	2.136	6.019	2.060
78	Pt	4.644	2.669	5.582	2.221	5.793	2.140
79	Au	4.482	2.766	5.371	2.308	5.579	2.222
80	Hg	4.327	2.865	5.170	2.398	5.375	2.306
81	Tl	4.179	2.966	4.978	2.490	5.180	2.393
82	Pb	4.037	3.071	4.795	2.585	4.994	2.482
83	Bi	3.901	3.177	4.620	2.683	4.817	2.573
84	Po	3.771	3.287	4.453	2.784	4.647	2.668
85	At	3.646	3.400	4.293	2.888	4.484	2.764
86	Rn	3.527	3.515	4.139	2.995	4.328	2.864
87	Fr	3.412	3.633	3.992	3.105	4.178	2.967
88	Ra	3.301	3.755	3.851	3.219	4.035	3.072
89	Ac	3.195	3.880	3.716	3.336	3.897	3.181
90	Th	3.093	4.008	3.585	3.457	3.764	3.293
91	Pa	2.995	4.139	3.460	3.582	3.636	3.409
92	U	2.900	4.274	3.340	3.712	3.514	3.528

Table A.6 Mass absorption coefficients for Kα emissions.

Absorber		11 Na	12 Mg	13 Al	14 Si	15 P	16 S	17 Cl	18 Ar	19 K	20 Ca	21 Sc	22 Ti	23 V	24 Cr	25 Mn	26 Fe
8	O	3653	2169	1345	865	574	391	273	195	141	105	78	60	46	36	28	22
9	F	5143	3058	1898	1222	812	554	387	276	201	148	111	85	65	51	40	32
10	Ne	6946	4135	2570	1656	1101	752	526	376	273	202	152	116	89	69	55	43
11	Na	562	5409	3365	2170	1444	987	691	494	359	266	200	152	117	92	72	57
12	Mg	768	463	4287	2768	1843	1261	883	632	460	341	256	195	151	118	93	74
13	Al	1019	614	386	3451	2300	1575	1104	790	576	427	321	245	189	148	116	93
14	Si	1319	794	499	325	2816	1929	1353	969	707	524	395	301	233	182	143	114
15	P	1670	1006	632	411	276	2325	1632	1169	854	633	477	364	281	220	174	138
16	S	2076	1251	786	511	343	236	1940	1391	1016	754	568	434	336	262	207	165
17	Cl	2542	1531	962	626	420	289	204	1634	1194	887	669	511	395	309	244	195
18	Ar	3070	1849	1162	756	507	350	246	177	1388	1032	778	595	460	360	285	227
19	K	3663	2207	1386	902	605	417	294	212	155	1188	896	686	531	415	329	262
20	Ca	4326	2606	1637	1066	715	493	347	250	183	136	1024	783	607	475	376	300
21	Sc	5061	3049	1915	1247	836	576	406	292	214	160	121	888	688	539	426	341
22	Ti	5871	3537	2222	1446	970	668	471	339	249	185	140	107	83	607	480	384
23	V	6759	4072	2558	1665	1117	770	543	390	286	213	161	123	96	75	538	430
24	Cr	7728	4655	2924	1903	1277	880	620	446	327	244	184	141	109	86	68	478
25	Mn	8780	5289	3323	2163	1451	1000	705	507	372	277	209	160	124	97	77	62

Table A.6 – continued.

Absorber		11 Na	12 Mg	13 Al	14 Si	15 P	16 S	17 Cl	18 Ar	19 K	20 Ca	21 Sc	22 Ti	23 V	24 Cr	25 Mn	26 Fe
26	Fe	9918	5975	3753	2443	1639	1129	796	573	420	313	236	181	140	110	87	70
27	Co	11146	6714	4218	2745	1842	1269	895	644	472	351	266	204	158	124	98	78
28	Ni	12464	7509	4717	3070	2060	1419	1001	720	528	393	297	228	176	138	110	88
29	Cu	12175	8359	5251	3418	2293	1580	1114	802	587	438	331	253	196	154	122	97
30	Zn	9777	9267	5821	3789	2542	1751	1235	889	651	485	367	281	218	171	135	108
31	Ga	2284	8943	6429	4184	2808	1934	1364	981	719	536	405	310	240	189	149	119
32	Ge	2505	9823	7074	4604	3089	2128	1501	1080	791	590	446	341	265	207	164	131
33	As	2740	1691	6756	5049	3388	2334	1646	1184	868	647	489	374	290	227	180	144
34	Se	2991	1846	7376	5520	3704	2552	1799	1295	949	707	534	409	317	249	197	157
35	Br	3259	2011	1292	5226	4037	2781	1961	1411	1034	770	582	446	346	271	215	172
36	Kr	3543	2186	1404	5673	4388	3023	2131	1534	1124	837	633	485	376	295	233	186
37	Rb	3845	2373	1524	1012	4122	3277	2311	1663	1219	908	686	526	407	319	253	202
38	Sr	4164	2570	1651	1096	4454	3544	2499	1798	1318	982	742	568	441	345	273	219
39	Y	4503	2779	1785	1186	811	3308	2696	1940	1422	1059	801	613	475	373	295	236
40	Zr	4861	3000	1927	1280	875	3558	2902	2088	1531	1140	862	660	512	401	318	254
41	Nb	5239	3233	2077	1380	943	662	2681	2234	1638	1220	923	707	548	430	340	272
42	Mo	5639	3480	2235	1485	1015	712	2064	2385	1749	1303	986	756	586	460	364	291
43	Tc	6060	3740	2402	1596	1091	765	549	2193	1864	1390	1051	806	625	491	389	311
44	Ru	6503	4013	2578	1712	1171	821	589	1682	1983	1479	1119	858	666	523	414	331
45	Rh	6970	4301	2763	1835	1255	880	631	461	1816	1571	1189	912	708	556	441	353

Z	El																
46	Pd	7460	4604	2957	1964	1343	942	675	494	1388	1666	1262	968	752	590	468	375
47	Ag	7976	4922	3161	2100	1436	1007	722	528	393	1521	1337	1026	797	626	497	397
48	Cd	8517	5256	3376	2243	1534	1075	771	564	419	1159	1414	1086	844	663	526	421
49	In	9085	5607	3601	2392	1636	1147	822	601	447	338	1287	1147	891	700	556	445
50	Sn	9679	5974	3837	2549	1743	1222	876	641	476	360	978	1210	941	739	587	470
51	Sb	10302	6358	4083	2713	1855	1301	933	682	507	383	293	1099	992	780	619	496
52	Te	10954	6761	4342	2884	1972	1383	992	725	539	407	312	833	899	821	652	522
53	I	10517	7181	4612	3064	2095	1469	1053	770	573	433	331	257	946	863	686	550
54	Xe	11092	7621	4895	3252	2224	1559	1118	817	608	459	352	273	715	781	720	578
55	Cs	9395	8081	5190	3448	2358	1653	1185	866	644	487	373	289	227	820	756	606
56	Ba	9894	7604	5498	3652	2498	1751	1256	918	683	516	395	306	240	619	683	636
57	La	9360	8009	5819	3866	2643	1854	1329	971	723	546	418	324	255	202	716	666
58	Ce	9840	8433	6154	4088	2796	1961	1405	1027	764	577	442	343	269	213	539	601
59	Pr	10343	7115	6503	4320	2954	2072	1485	1086	808	610	467	363	284	226	181	453
60	Nd	10868	6707	5996	4562	3119	2188	1568	1146	853	644	493	383	300	238	191	473
61	Pm	11416	7046	6305	4813	3291	2308	1655	1210	900	680	521	404	317	251	201	163
62	Sm	3331	7399	5307	5075	3470	2434	1745	1275	949	717	549	426	334	265	212	171
63	Eu	3430	7768	5575	4626	3656	2564	1838	1344	999	755	578	449	352	279	224	181
64	Gd	3532	8154	5237	4859	3850	2700	1935	1415	1052	795	609	472	371	294	235	190
65	Tb	3638	7402	5495	4082	4050	2840	2036	1489	1107	836	641	497	390	309	248	200
66	Dy	3747	2481	5764	4284	3662	2987	2141	1565	1164	879	674	523	410	325	260	210
67	Ho	3860	2556	6045	4016	3842	3139	2250	1645	1223	924	708	549	431	342	274	221
68	Er	3976	2633	6338	4210	3223	3296	2363	1727	1285	971	743	577	453	359	287	232
69	Tm	4096	2713	1859	4413	3378	2962	2480	1813	1348	1019	780	605	475	377	302	244
70	Yb	4220	2795	1915	4624	3540	3105	2602	1902	1414	1068	818	635	498	395	316	256

Table A.6 – continued.

		Emitter														
		11 Na	12 Mg	13 Al	14 Si	15 P	16 S	17 Cl	18 Ar	19 K	20 Ca	21 Sc	22 Ti	23 V	24 Cr	25 Mn	26 Fe
Absorber																	
71	Lu	4348	2879	1973	4844	3313	3253	2727	1994	1483	1120	858	666	522	414	332	268
72	Hf	4480	2967	2033	4353	3470	2724	2442	2089	1554	1174	899	697	547	434	348	281
73	Ta	4616	3057	2095	1477	3633	2852	2557	2188	1627	1229	942	730	573	455	364	294
74	W	4757	3150	2158	1522	3802	2985	2676	2290	1703	1287	985	765	600	476	381	308
75	Re	4902	3246	2224	1569	3979	2790	2239	2047	1782	1346	1031	800	628	498	398	322
76	Os	5051	3345	2292	1616	3558	2919	2342	2141	1863	1407	1078	836	656	520	417	337
77	Ir	5205	3447	2362	1666	1204	3053	2450	2238	1947	1471	1127	874	686	544	435	352
78	Pt	5364	3553	2434	1717	1241	3193	2289	1872	1740	1536	1177	913	716	568	455	367
79	Au	5529	3661	2509	1769	1279	2845	2393	1956	1818	1604	1229	953	748	593	475	384
80	Hg	5698	3773	2585	1823	1318	2971	2501	2044	1900	1674	1283	995	781	619	496	400
81	Tl	5872	3889	2665	1879	1358	1003	2613	1910	1588	1498	1338	1038	815	646	517	418
82	Pb	6052	4008	2746	1937	1400	1034	2729	1995	1658	1564	1395	1082	849	674	539	436
83	Bi	6237	4131	2830	1996	1443	1066	2422	2084	1731	1308	1454	1128	885	702	562	454
84	Po	6428	4257	2917	2057	1487	1098	827	2175	1807	1365	1305	1175	922	731	586	473
85	At	6625	4388	3006	2120	1533	1132	852	2271	1689	1424	1361	1224	961	762	610	493
86	Rn	0	4522	3098	2185	1580	1167	878	2009	1762	1485	1138	1274	1000	793	635	513
87	Fr	0	4660	3193	2252	1628	1202	905	693	1839	1549	1187	1148	1041	825	661	534
88	Ra	0	4803	3291	2321	1678	1239	933	714	1918	1449	1237	1196	1083	859	687	555
89	Ac	0	4950	3392	2392	1729	1277	961	736	1693	1511	1289	1000	978	893	715	577
90	Th	0	0	3495	2465	1782	1316	991	758	1764	1575	1343	1042	1019	928	743	600
91	Pa	0	0	3602	2540	1836	1356	1021	781	607	1642	1258	1086	852	964	772	624
92	U	0	0	3712	2618	1893	1398	1052	805	625	1445	1310	1131	887	876	802	648

Absorber		Emitter															
		27 Co	28 Ni	29 Cu	30 Zn	31 Ga	32 Ge	33 As	34 Se	35 Br	36 Kr	37 Rb	38 Sr	39 Y	40 Zr	41 Nb	42 Mo
8	O	18	14	12	10	8	/	6	5	4	3	3	2	2	2	2	1
9	F	25	21	17	14	11	9	8	7	6	5	4	3	3	3	2	2
10	Ne	35	28	23	19	16	13	11	9	8	6	6	5	4	3	3	3
11	Na	46	37	30	25	21	17	14	12	10	9	7	6	5	5	4	3
12	Mg	59	48	39	32	27	22	18	15	13	11	9	8	7	6	5	5
13	Al	74	60	49	40	33	28	23	20	16	14	12	10	9	8	7	6
14	Si	92	74	61	50	41	34	29	24	20	17	15	13	11	9	8	7
15	P	111	90	74	60	50	42	35	29	25	21	18	15	13	11	10	9
16	S	133	108	88	72	60	50	42	35	30	25	22	18	16	14	12	10
17	Cl	157	127	104	85	71	59	49	41	35	30	25	22	19	16	14	12
18	Ar	183	148	121	100	83	69	58	49	41	35	30	26	22	19	16	14
19	K	211	171	140	115	95	80	67	56	48	40	35	30	26	22	19	17
20	Ca	242	196	160	132	109	91	76	64	55	46	40	34	29	25	22	19
21	Sc	274	223	182	150	124	104	87	73	62	53	45	39	33	29	25	22
22	Ti	309	251	206	169	140	117	98	83	70	60	51	44	38	33	28	25
23	V	346	281	230	190	157	131	110	93	79	67	57	49	42	37	32	28
24	Cr	386	313	257	212	176	146	123	104	88	75	64	55	47	41	36	31
25	Mn	427	347	284	234	195	162	136	115	97	83	71	61	53	46	40	34

Table A.6 – continued.

Absorber		27 Co	28 Ni	29 Cu	30 Zn	31 Ga	32 Ge	33 As	34 Se	35 Br	36 Kr	37 Rb	38 Sr	39 Y	40 Zr	41 Nb	42 Mo
26	Fe	56	382	313	258	214	179	150	127	108	92	78	67	58	50	44	38
27	Co	63	51	344	283	235	197	165	139	118	101	86	74	64	55	48	42
28	Ni	71	57	47	310	257	215	180	152	129	110	94	81	70	61	53	46
29	Cu	79	64	52	43	280	234	196	166	141	120	103	88	76	66	57	50
30	Zn	87	71	58	48	40	253	213	180	153	130	111	96	83	72	62	54
31	Ga	96	78	64	53	44	36	230	194	165	141	121	104	89	78	67	59
32	Ge	106	86	70	58	48	40	34	209	178	152	130	112	96	84	73	63
33	As	116	94	77	64	53	44	37	31	191	163	140	120	104	90	78	68
34	Se	127	103	84	69	58	48	40	34	29	25	149	129	111	96	84	73
35	Br	138	112	92	76	63	52	44	37	31	27	23	137	119	103	89	78
36	Kr	150	122	100	82	68	57	48	40	34	29	25	21	126	110	95	83
37	Rb	163	132	108	89	74	62	52	44	37	31	27	23	20	116	101	88
38	Sr	176	143	117	97	80	67	56	47	40	34	29	25	22	19	107	94
39	Y	190	154	126	104	86	72	60	51	43	37	31	27	23	20	17	99
40	Zr	205	166	136	112	93	78	65	55	46	40	34	29	25	22	19	16
41	Nb	219	178	146	120	100	83	70	59	50	42	36	31	27	23	20	18
42	Mo	235	191	156	129	107	89	75	63	54	46	39	33	29	25	22	19
43	Tc	251	204	167	138	114	95	80	68	57	49	42	36	31	27	23	20
44	Ru	267	217	178	147	122	102	85	72	61	52	45	38	33	29	25	22
45	Rh	285	232	190	156	130	108	91	77	65	56	48	41	35	30	26	23

Z	El																
46	Pd	302	246	202	166	138	115	97	82	69	59	51	43	38	32	28	25
47	Ag	321	261	214	177	147	123	103	87	74	63	54	46	40	35	30	26
48	Cd	340	277	227	187	156	130	109	92	78	67	57	49	42	37	32	28
49	In	360	293	240	198	165	138	116	98	83	71	61	52	45	39	34	29
50	Sn	380	309	254	210	174	146	122	103	88	75	64	55	48	41	36	31
51	Sb	401	326	268	221	184	154	129	109	93	79	68	58	50	44	38	33
52	Te	422	344	282	233	194	162	136	115	98	83	71	61	53	46	40	35
53	I	444	362	297	246	204	171	144	121	103	88	75	65	56	49	42	37
54	Xe	467	381	313	258	215	180	151	128	109	93	79	68	59	51	44	39
55	Cs	491	400	328	271	226	189	159	134	114	97	84	72	62	54	47	41
56	Ba	514	420	345	285	237	198	167	141	120	102	88	76	65	57	49	43
57	La	539	440	361	299	249	208	175	148	126	107	92	79	69	59	52	45
58	Ce	564	460	378	313	260	218	183	155	132	113	97	83	72	62	54	47
59	Pr	590	481	396	327	272	228	192	163	138	118	101	87	75	65	57	50
60	Nd	532	503	413	342	285	239	201	170	145	124	106	91	79	68	59	52
61	Pm	400	525	432	357	298	249	210	178	151	129	111	95	83	72	62	54
62	Sm	417	473	450	373	311	260	219	186	158	135	116	100	86	75	65	57
63	Eu	147	355	406	389	324	271	229	194	165	141	121	104	90	78	68	60
64	Gd	155	370	423	405	337	283	238	202	172	147	126	109	94	82	71	62
65	Tb	163	134	317	365	351	295	248	210	179	153	131	113	98	85	74	65
66	Dy	171	141	329	379	366	307	258	219	186	159	137	118	102	89	77	68
67	Ho	180	148	122	284	329	319	269	228	194	166	142	123	106	92	80	70
68	Er	189	155	128	295	246	331	279	237	202	172	148	128	111	96	84	73
69	Tm	198	163	135	112	255	298	290	246	209	179	154	133	115	100	87	76
70	Yb	208	171	141	117	265	222	301	255	217	186	160	138	119	104	91	79

Table A.6 – continued.

		Emitter															
Absorber		27 Co	28 Ni	29 Cu	30 Zn	31 Ga	32 Ge	33 As	34 Se	35 Br	36 Kr	37 Rb	38 Sr	39 Y	40 Zr	41 Nb	42 Mo
71	Lu	218	179	148	123	103	231	271	265	226	193	166	143	124	108	94	82
72	Hf	229	188	155	129	108	239	202	238	234	200	172	148	129	112	98	85
73	Ta	239	196	162	135	113	95	209	247	242	207	178	154	133	116	101	89
74	W	251	206	170	141	118	99	217	184	218	215	185	159	138	120	105	92
75	Re	262	215	178	148	124	104	88	190	162	222	191	165	143	124	109	95
76	Os	274	225	186	155	129	109	92	197	168	200	198	171	148	129	112	98
77	Ir	286	235	194	162	135	114	96	82	174	149	178	177	153	133	116	102
78	Pt	299	246	203	169	141	119	100	85	180	154	184	183	158	138	120	105
79	Au	312	256	212	176	147	124	105	89	76	159	137	165	164	142	124	109
80	Hg	326	268	221	184	154	129	109	93	79	164	141	122	169	147	128	113
81	Tl	340	279	231	192	160	135	114	97	83	71	146	126	152	152	133	116
82	Pb	355	291	240	200	167	141	119	101	86	74	151	130	113	137	137	120
83	Bi	370	303	251	208	174	147	124	106	90	77	67	134	117	141	141	124
84	Po	385	316	261	217	182	153	129	110	94	81	69	139	120	105	127	128
85	At	401	329	272	226	189	159	135	115	98	84	72	62	124	108	95	131
86	Rn	418	343	283	236	197	166	140	119	102	87	75	65	128	111	97	119
87	Fr	435	357	295	245	205	173	146	124	106	91	78	68	59	115	100	88
88	Ra	452	371	307	255	213	179	152	129	110	95	81	70	61	118	103	91
89	Ac	470	386	319	265	222	187	158	134	115	98	85	73	64	55	106	93
90	Th	489	401	331	276	231	194	164	140	119	102	88	76	66	58	110	96
91	Pa	508	417	344	286	240	202	171	145	124	106	91	79	69	60	52	99
92	U	528	433	358	297	249	209	177	151	129	110	95	82	71	62	54	102

Table A.6 – continued.

Absorber		Emitter							
		43 Tc	44 Ru	45 Rh	46 Pd	47 Ag	48 Cd	49 In	50 Sn
8	O	1	1	1	1	1	1	1	0
9	F	2	1	1	1	1	1	1	1
10	Ne	2	2	2	2	1	1	1	1
11	Na	3	3	2	2	2	2	1	1
12	Mg	4	3	3	3	2	2	2	2
13	Al	5	4	4	3	3	3	2	2
14	Si	6	5	5	4	4	3	3	3
15	P	8	7	6	5	4	4	3	3
16	S	9	8	7	6	5	5	4	4
17	Cl	11	9	8	7	6	6	5	4
18	Ar	13	11	10	8	7	7	6	5
19	K	15	13	11	10	9	8	7	6
20	Ca	17	15	13	11	10	9	8	7
21	Sc	19	17	15	13	11	10	9	8
22	Ti	22	19	17	15	13	11	10	9
23	V	24	21	19	16	15	13	11	10
24	Cr	27	24	21	18	16	14	13	11
25	Mn	30	26	23	20	18	16	14	13
26	Fe	33	29	26	23	20	18	16	14
27	Co	37	32	28	25	22	19	17	15
28	Ni	40	35	31	27	24	21	19	17
29	Cu	44	38	34	30	26	23	21	18
30	Zn	48	42	37	32	29	25	23	20
31	Ga	51	45	40	35	31	27	24	22
32	Ge	55	49	43	38	33	30	26	23
33	As	60	52	46	41	36	32	28	25
34	Se	64	56	49	44	39	34	30	27
35	Br	68	60	53	47	41	37	33	29
36	Kr	73	64	56	50	44	39	35	31
37	Rb	77	68	60	53	47	42	37	33
38	Sr	82	72	64	56	50	44	39	35
39	Y	87	76	67	59	53	47	42	37
40	Zr	92	81	71	63	56	49	44	39
41	Nb	15	85	75	66	59	52	46	41
42	Mo	16	14	79	70	62	55	49	44
43	Tc	18	15	14	73	65	58	51	46
44	Ru	19	17	15	13	68	60	54	48
45	Rh	20	18	16	14	12	11	56	50
46	Pd	21	19	17	15	13	11	10	52
47	Ag	23	20	18	16	14	12	11	10
48	Cd	24	21	19	17	15	13	12	10
49	In	26	23	20	18	16	14	12	11
50	Sn	27	24	21	19	16	15	13	12

APPENDIX A

Table A.6 – *continued.*

		\multicolumn{7}{c}{Emitter}						
		43 Tc	44 Ru	45 Rh	46 Pd	47 Ag	48 Cd	49 In	50 Sn
Absorber									
51	Sb	29	25	22	20	17	15	14	12
52	Te	31	27	24	21	18	16	15	13
53	I	32	28	25	22	19	17	15	14
54	Xe	34	30	26	23	21	18	16	14
55	Cs	36	31	28	24	22	19	17	15
56	Ba	38	33	29	26	23	20	18	16
57	La	40	35	31	27	24	21	19	17
58	Ce	42	36	32	28	25	22	20	18
59	Pr	44	38	34	30	26	23	21	19
60	Nd	46	40	35	31	28	25	22	19
61	Pm	48	42	37	33	29	26	23	20
62	Sm	50	44	39	34	30	27	24	21
63	Eu	52	46	40	36	32	28	25	22
64	Gd	54	48	42	37	33	29	26	23
65	Tb	57	50	44	39	35	31	27	24
66	Dy	59	52	46	41	36	32	28	25
67	Ho	62	54	48	42	38	33	30	26
68	Er	64	57	50	44	39	35	31	28
69	Tm	67	59	52	46	41	36	32	29
70	Yb	69	61	54	48	42	38	33	30
71	Lu	72	63	56	50	44	39	35	31
72	Hf	75	66	58	51	46	41	36	32
73	Ta	78	68	60	53	47	42	38	34
74	W	81	71	63	55	49	44	39	35
75	Re	83	73	65	57	51	45	40	36
76	Os	86	76	67	60	53	47	42	37
77	Ir	89	79	70	62	55	49	43	39
78	Pt	93	82	72	64	57	50	45	40
79	Au	96	84	74	66	59	52	46	42
80	Hg	99	87	77	68	61	54	48	43
81	Tl	102	90	80	70	63	56	50	44
82	Pb	105	93	82	73	65	58	51	46
83	Bi	109	96	85	75	67	59	53	47
84	Po	112	99	87	77	69	61	55	49
85	At	116	102	90	80	71	63	56	50
86	Rn	119	105	93	82	73	65	58	52
87	Fr	108	108	96	85	75	67	60	54
88	Ra	80	111	99	87	78	69	62	55
89	Ac	82	101	101	90	80	71	64	57
90	Th	85	75	92	93	82	73	66	59
91	Pa	87	77	68	95	85	75	67	60
92	U	90	79	70	86	87	78	69	62

Table A.7 Mass absorption coefficients for Lα emissions.

Absorber		31 Ga	32 Ge	33 As	34 Se	35 Br	36 Kr	37 Rb	38 Sr	39 Y	40 Zr	41 Nb	42 Mo	43 Tc	44 Ru	45 Rh	46 Pd
8	O	3142	2519	2036	1658	1359	1122	931	778	653	551	468	398	341	293	253	219
9	F	4426	3550	2871	2339	1918	1584	1316	1100	924	780	662	564	483	415	358	310
10	Ne	5979	4799	3882	3164	2597	2145	1783	1490	1253	1058	898	766	656	564	487	422
11	Na	7815	6275	5079	4141	3400	2810	2336	1954	1643	1388	1179	1005	861	741	640	554
12	Mg	664	535	435	5274	4332	3582	2979	2492	2096	1772	1505	1284	1101	947	818	709
13	Al	881	710	577	473	390	4463	3714	3108	2615	2211	1879	1604	1375	1183	1022	886
14	Si	1139	919	747	612	504	418	349	293	3200	2707	2301	1965	1685	1450	1253	1087
15	P	1442	1163	946	774	638	530	442	371	313	265	2772	2368	2031	1749	1512	1312
16	S	1793	1447	1176	963	794	659	550	461	389	330	281	241	207	2079	1797	1560
17	Cl	2196	1771	1440	1179	972	806	673	565	477	404	344	295	253	218	189	1832
18	Ar	2652	2139	1739	1424	1174	974	813	682	576	488	416	356	306	264	229	199
19	K	3165	2553	2075	1699	1401	1162	970	814	687	582	496	425	365	315	273	237
20	Ca	3737	3014	2450	2006	1654	1372	1145	961	811	688	586	502	431	372	322	280
21	Sc	4372	3526	2867	2347	1935	1605	1340	1125	949	805	686	587	504	435	377	327
22	Ti	5072	4091	3325	2723	2245	1862	1554	1305	1101	933	795	681	585	505	437	380
23	V	5839	4709	3828	3135	2584	2144	1789	1502	1267	1075	916	784	673	581	503	437
24	Cr	6675	5385	4377	3584	2955	2451	2046	1717	1449	1229	1047	896	770	664	575	500
25	Mn	7584	6118	4973	4072	3357	2785	2324	1951	1646	1396	1189	1018	875	755	654	568

Table A.7 – continued

Absorber		31 Ga	32 Ge	33 As	34 Se	35 Br	36 Kr	37 Rb	38 Sr	Emitter 39 Y	40 Zr	41 Nb	42 Mo	43 Tc	44 Ru	45 Rh	46 Pd
26	Fe	8568	6911	5618	4600	3792	3146	2626	2204	1860	1577	1344	1150	988	853	738	642
27	Co	9628	7766	6313	5169	4261	3535	2951	2477	2090	1772	1510	1292	1110	958	830	721
28	Ni	10767	8685	7060	5780	4765	3954	3300	2770	2337	1982	1688	1445	1242	1071	928	806
29	Cu	10517	9668	7859	6435	5305	4401	3673	3083	2602	2206	1880	1609	1382	1193	1033	898
30	Zn	11635	9385	8713	7134	5881	4879	4073	3418	2884	2446	2084	1783	1533	1322	1145	995
31	Ga	1987	10344	8408	7879	6495	5389	4498	3775	3185	2701	2301	1969	1692	1460	1265	1099
32	Ge	2179	1775	9236	7562	7147	5930	4949	4154	3505	2972	2532	2167	1862	1607	1392	1210
33	As	2384	1942	1595	8280	6826	6503	5427	4555	3844	3259	2777	2377	2042	1762	1526	1326
34	Se	2602	2121	1741	1439	7451	6182	5933	4980	4202	3563	3036	2598	2233	1926	1668	1450
35	Br	2835	2310	1896	1568	1304	4871	5617	5428	4580	3884	3309	2832	2434	2100	1818	1581
36	Kr	3082	2511	2062	1704	1418	1187	4412	5118	4978	4221	3597	3078	2645	2282	1977	1718
37	Rb	3344	2725	2237	1849	1539	1288	1084	4009	4677	3966	3899	3337	2868	2474	2143	1862
38	Sr	3622	2952	2423	2003	1667	1395	1174	994	846	4285	3651	3608	3101	2676	2317	2014
39	Y	3917	3192	2620	2166	1802	1509	1270	1075	914	781	3935	3368	3346	2887	2500	2173
40	Zr	4228	3446	2829	2338	1946	1629	1371	1160	987	844	724	2617	3113	3107	2691	2339
41	Nb	4558	3714	3049	2520	2097	1755	1478	1251	1064	909	781	673	2402	2870	2878	2501
42	Mo	4905	3997	3281	2713	2257	1889	1590	1346	1145	979	840	724	627	2210	2652	2305
43	Tc	5271	4296	3526	2915	2425	2030	1709	1446	1230	1052	903	778	674	585	2037	2455
44	Ru	5657	4610	3785	3128	2603	2179	1834	1552	1320	1129	969	835	723	628	548	479
45	Rh	6063	4941	4056	3353	2790	2335	1966	1664	1415	1210	1038	895	775	673	587	514

Z	El																
46	Pd	6490	5289	4342	3589	2986	2499	2104	1781	1515	1295	1112	958	830	721	629	550
47	Ag	6938	5654	4642	3837	3192	2672	2250	1904	1620	1384	1188	1025	887	771	672	588
48	Cd	7409	6038	4957	4097	3409	2853	2402	2033	1729	1478	1269	1094	947	823	718	628
49	In	7903	6440	5287	4370	3636	3044	2562	2169	1845	1577	1354	1167	1010	878	765	670
50	Sn	8420	6862	5633	4656	3874	3243	2730	2311	1965	1680	1442	1243	1076	935	815	714
51	Sb	8962	7303	5996	4956	4124	3452	2906	2459	2092	1788	1535	1323	1146	995	868	759
52	Te	9529	7765	6375	5270	4384	3670	3090	2615	2224	1901	1632	1407	1218	1058	923	807
53	I	10122	8249	6772	5598	4657	3898	3282	2778	2363	2019	1734	1495	1294	1124	980	858
54	Xe	9649	8754	7187	5941	4943	4137	3483	2948	2507	2143	1840	1586	1373	1193	1040	910
55	Cs	10171	8289	7620	6299	5241	4387	3693	3125	2659	2272	1951	1682	1456	1265	1103	965
56	Ba	8606	8734	8072	6673	5552	4647	3912	3311	2816	2407	2067	1782	1542	1340	1169	1022
57	La	9060	9200	7553	7063	5876	4919	4141	3504	2981	2548	2187	1886	1632	1418	1237	1082
58	Ce	8560	7769	7952	6573	6214	5202	4379	3706	3153	2694	2313	1994	1726	1500	1308	1144
59	Pr	8997	7332	6709	6918	6567	5497	4627	3916	3331	2847	2444	2108	1824	1585	1382	1209
60	Nd	9454	7704	7055	5832	6055	5804	4886	4135	3518	3007	2581	2225	1926	1674	1459	1277
61	Pm	9931	8093	6644	6130	6367	5329	5155	4363	3712	3172	2723	2348	2032	1766	1540	1347
62	Sm	10429	8499	6977	5768	5359	5602	4716	4601	3913	3345	2871	2476	2143	1862	1624	1421
63	Eu	3045	8923	7325	6055	5630	4712	4955	4847	4123	3524	3025	2608	2258	1962	1711	1497
64	Gd	3136	8114	7689	6356	5288	4948	5205	4405	4341	3710	3185	2746	2377	2065	1801	1576
65	Tb	3230	2712	8068	6669	5549	4645	4373	4625	3934	3904	3352	2890	2501	2173	1895	1658
66	Dy	3326	2793	2360	6996	5821	4872	4589	3884	4129	3529	3524	3038	2630	2285	1993	1744
67	Ho	3426	2877	2431	6332	6104	5110	4302	4074	3465	3703	3703	3193	2764	2401	2094	1832
68	Er	3530	2964	2504	2129	6400	5357	4510	3817	3634	3883	3334	3353	2902	2522	2199	1924
69	Tm	3636	3053	2580	2193	1874	5615	4727	4001	3810	3256	3495	3013	3046	2647	2308	2020
70	Yb	3746	3146	2658	2259	1931	5060	4953	4192	3566	3412	2929	3158	3195	2776	2421	2118

Table A.7 – continued

Absorber	31 Ga	32 Ge	33 As	34 Se	35 Br	36 Kr	37 Rb	38 Sr	39 Y	40 Zr	41 Nb	42 Mo	43 Tc	44 Ru	45 Rh	46 Pd
71 Lu	3860	3241	2739	2328	1989	1709	5189	4392	3736	3193	3069	2646	2864	2911	2538	2221
72 Hf	3977	3339	2822	2398	2050	1761	4663	4600	3913	3344	3214	2771	3000	2606	2659	2327
73 Ta	4098	3441	2907	2471	2112	1815	1566	4128	4097	3501	3006	2901	2512	2729	2379	2437
74 W	4223	3546	2996	2547	2176	1870	1614	1400	4288	3665	3146	2713	2629	2284	2490	2179
75 Re	4352	3654	3087	2624	2243	1927	1663	1443	3839	3835	3293	2839	2751	2390	2084	2280
76 Os	4484	3765	3182	2704	2311	1986	1714	1487	1295	3429	3445	2970	2571	2500	2180	2385
77 Ir	4621	3880	3279	2787	2382	2046	1766	1532	1334	1167	3603	3106	2689	2336	2280	1995
78 Pt	4762	3999	3379	2872	2455	2109	1820	1579	1375	1203	3214	3248	2812	2443	2383	2085
79 Au	4908	4121	3482	2960	2530	2173	1876	1627	1417	1239	1088	2894	2939	2554	2227	2179
80 Hg	5058	4247	3589	3050	2607	2240	1934	1677	1461	1277	1121	988	3072	2669	2327	2036
81 Tl	5213	4377	3699	3144	2687	2308	1993	1728	1505	1316	1156	1018	2731	2789	2432	2128
82 Pb	5373	4511	3812	3240	2769	2379	2054	1781	1551	1357	1191	1049	928	2477	2540	2223
83 Bi	5537	4649	3929	3339	2854	2452	2117	1836	1599	1398	1227	1081	956	848	2653	2321
84 Po	5707	4792	4049	3441	2941	2527	2181	1892	1648	1441	1265	1115	985	874	2352	2423
85 At	5882	4939	4173	3547	3031	2604	2248	1950	1698	1485	1304	1149	1016	901	801	2146
86 Rn	6062	5090	4301	3656	3124	2684	2317	2010	1750	1531	1344	1184	1047	928	826	737
87 Fr	0	5246	4432	3767	3220	2766	2388	2071	1804	1578	1385	1220	1079	957	851	759
88 Ra	0	0	4568	3883	3318	2851	2461	2134	1859	1626	1427	1257	1112	986	877	783
89 Ac	0	0	4708	4002	3420	2938	2536	2200	1916	1676	1471	1296	1146	1016	904	807
90 Th	0	0	0	4124	3525	3028	2614	2267	1975	1727	1516	1336	1181	1047	932	831
91 Pa	0	0	0	0	3632	3121	2694	2336	2035	1780	1562	1376	1217	1079	960	857
92 U	0	0	0	0	3743	3216	2776	2408	2097	1834	1610	1418	1254	1112	989	883

	Emitter															
Absorber	47 Ag	48 Cd	49 In	50 Sn	51 Sb	52 Te	53 I	54 Xe	55 Cs	56 Ba	57 La	58 Ce	59 Pr	60 Nd	61 Pm	62 Sm
8 O	190	166	145	127	112	99	87	77	69	61	55	49	44	39	35	32
9 F	270	235	206	180	159	140	124	110	98	87	78	70	62	56	50	45
10 Ne	367	320	280	246	216	191	169	150	133	119	106	95	85	76	69	62
11 Na	482	421	368	323	284	251	222	197	175	156	140	125	112	101	91	82
12 Mg	617	538	471	414	364	322	285	253	225	201	179	160	144	129	117	105
13 Al	771	673	590	518	456	403	357	317	282	251	225	201	181	162	146	132
14 Si	946	826	724	636	560	495	438	389	346	309	276	248	222	200	180	163
15 P	1142	997	874	768	677	598	530	470	419	374	334	299	269	242	218	197
16 S	1358	1187	1040	914	806	712	631	561	499	445	398	357	321	289	260	235
17 Cl	1595	1394	1222	1074	947	837	742	660	587	524	469	420	378	340	307	277
18 Ar	173	152	1421	1249	1102	974	864	767	684	610	546	490	440	396	357	323
19 K	207	181	159	140	123	1122	995	884	788	703	630	565	507	457	412	372
20 Ca	244	214	187	165	146	129	114	1010	900	804	719	645	580	522	471	426
21 Sc	286	250	219	193	170	151	134	119	106	95	815	732	658	592	534	483
22 Ti	331	290	254	224	198	175	155	138	123	110	98	88	740	667	602	544
23 V	381	334	293	258	228	201	179	159	142	127	113	102	91	82	74	609
24 Cr	436	382	335	295	260	230	204	182	162	145	130	116	105	94	85	77
25 Mn	495	433	380	335	296	262	232	206	184	164	147	132	119	107	97	87

Table A.7 – continued

Absorber		Emitter															
		47 Ag	48 Cd	49 In	50 Sn	51 Sb	52 Te	53 I	54 Xe	55 Cs	56 Ba	57 La	58 Ce	59 Pr	60 Nd	61 Pm	62 Sm
26	Fe	560	490	430	378	334	296	262	233	208	186	166	149	134	121	109	99
27	Co	629	550	483	425	375	332	295	262	234	209	187	168	151	136	123	111
28	Ni	703	615	540	475	420	371	329	293	261	233	209	188	169	152	137	124
29	Cu	783	685	601	529	467	413	367	326	291	260	233	209	188	169	153	138
30	Zn	868	759	667	587	518	458	407	362	322	288	258	232	208	188	169	153
31	Ga	959	839	736	648	572	506	449	399	356	318	285	256	230	207	187	169
32	Ge	1055	923	810	713	629	557	494	440	392	350	314	281	253	228	206	186
33	As	1157	1012	888	782	690	611	542	482	430	384	344	309	277	250	226	204
34	Se	1265	1106	971	855	754	668	592	527	470	420	376	337	303	273	247	223
35	Br	1378	1206	1058	932	822	728	646	574	512	458	410	368	331	298	269	243
36	Kr	1498	1311	1150	1013	894	791	702	624	557	497	445	400	359	324	292	264
37	Rb	1624	1421	1247	1098	969	858	761	677	603	539	483	433	390	351	317	286
38	Sr	1756	1537	1349	1187	1048	927	823	732	653	583	522	469	421	380	343	310
39	Y	1895	1658	1455	1281	1130	1001	888	790	704	629	563	506	455	409	370	334
40	Zr	2040	1785	1566	1379	1217	1077	956	850	758	677	606	544	489	441	398	360
41	Nb	2182	1909	1676	1475	1302	1153	1023	910	811	725	649	583	524	472	426	385
42	Mo	2329	2038	1790	1576	1391	1232	1093	972	867	775	694	623	560	505	456	412
43	Tc	2142	2172	1907	1679	1483	1313	1166	1037	925	827	741	665	598	539	487	440
44	Ru	2277	1993	2029	1787	1578	1397	1241	1104	985	880	789	708	637	574	518	469
45	Rh	451	1526	1858	1898	1677	1485	1318	1173	1047	936	839	753	677	611	551	499

46	Pd	483	425	1421	1735	1533	1575	1399	1245	1111	993	890	799	719	648	586	530		
47	Ag	516	454	401	1324	1623	1438	1481	1319	1177	1052	943	847	762	687	621	562		
48	Cd	551	485	429	380	1237	1520	1350	1395	1245	1113	998	896	807	728	657	595		
49	In	588	518	457	405	359	1157	1028	1270	1315	1176	1055	947	853	769	695	629		
50	Sn	626	551	487	431	383	341	304	965	1195	1070	1113	1000	900	812	733	664		
51	Sb	667	587	518	459	408	363	324	290	907	1127	1010	1054	948	856	773	700		
52	Te	709	624	551	488	433	386	344	308	276	854	1063	956	998	901	814	737		
53	I	753	663	585	518	460	410	366	327	293	263	805	724	905	947	856	775		
54	Xe	799	704	621	550	489	435	388	347	311	280	252	760	684	857	775	814		
55	Cs	847	746	659	583	518	461	412	368	330	296	267	241	718	648	813	737		
56	Ba	898	790	698	618	549	489	436	390	350	314	283	255	230	629	614	772		
57	La	950	836	739	654	581	517	461	413	370	332	299	270	244	221	200	582		
58	Ce	1005	885	781	692	614	547	488	436	391	351	316	285	258	234	212	192		
59	Pr	1062	935	826	731	649	578	516	461	413	371	334	302	273	247	224	203		
60	Nd	1121	987	872	772	685	610	544	487	437	392	353	318	288	261	236	215		
61	Pm	1183	1041	920	814	723	644	574	514	461	414	373	336	304	275	249	227		
62	Sm	1247	1098	970	859	762	679	606	542	486	436	393	354	320	290	263	239		
63	Eu	1314	1157	1022	905	803	715	638	571	512	460	414	373	337	305	277	252		
64	Gd	1383	1218	1076	953	846	753	672	601	539	484	436	393	355	322	292	265		
65	Tb	1456	1282	1132	1002	890	792	707	632	567	509	458	414	374	338	307	279		
66	Dy	1531	1348	1190	1054	936	833	743	665	596	535	482	435	393	356	323	293		
67	Ho	1608	1416	1251	1108	983	875	781	699	626	563	507	457	413	374	339	308		
68	Er	1689	1487	1313	1163	1033	919	820	734	658	591	532	480	434	393	356	323		
69	Tm	1773	1561	1378	1221	1084	965	861	770	690	620	558	504	455	412	374	340		
70	Yb	1860	1637	1446	1281	1137	1012	903	808	724	651	586	528	477	432	392	356		

Table A.7 – *continued*

		Emitter														
		47 Ag	48 Cd	49 In	50 Sn	51 Sb	52 Te	53 I	54 Xe	55 Cs	56 Ba	57 La	58 Ce	59 Pr	60 Nd	61 Pm	62 Sm
Absorber																	
71	Lu	1950	1717	1516	1342	1192	1061	947	847	759	682	614	554	501	453	411	373
72	Hf	2043	1799	1588	1407	1249	1112	992	887	796	715	643	580	524	475	431	391
73	Ta	2139	1884	1663	1473	1308	1164	1039	929	833	748	674	608	549	497	451	410
74	W	2239	1972	1741	1542	1369	1219	1088	973	872	783	705	636	575	520	472	429
75	Re	2342	2062	1821	1613	1432	1275	1138	1018	912	820	738	665	601	544	494	449
76	Os	2093	2157	1904	1687	1498	1333	1190	1064	954	857	771	696	629	569	516	469
77	Ir	2189	1927	1990	1763	1565	1393	1243	1112	997	896	806	727	657	595	540	490
78	Pt	1831	2014	1779	1841	1635	1455	1299	1162	1041	936	842	760	686	622	564	512
79	Au	1913	1685	1859	1646	1707	1520	1356	1213	1087	977	879	793	717	649	589	535
80	Hg	1999	1760	1554	1720	1782	1586	1415	1266	1135	1019	918	828	748	677	614	558
81	Tl	2088	1838	1623	1796	1595	1655	1476	1321	1184	1064	957	864	780	707	641	582
82	Pb	1951	1920	1695	1501	1665	1482	1540	1377	1234	1109	998	901	814	737	668	607
83	Bi	2037	1794	1770	1567	1392	1547	1380	1435	1287	1156	1041	939	848	768	697	633
84	Po	2127	1873	1654	1636	1452	1293	1440	1288	1341	1204	1084	978	884	800	726	659
85	At	2220	1955	1727	1707	1515	1349	1204	1344	1204	1254	1129	1019	920	833	756	687
86	Rn	1965	2040	1802	1596	1581	1407	1256	1123	1256	*1306*	1176	1060	958	868	787	715
87	Fr	2049	2129	1880	1665	1478	1467	1309	1171	1050	1176	1223	1103	997	903	819	744
88	Ra	700	*1880*	1961	1736	1542	1530	1365	1221	1095	*1226*	1104	1148	1037	939	852	774
89	Ac	722	647	1731	1811	1608	1431	1423	1273	1141	1025	1150	1037	1079	977	886	805
90	Th	744	667	600	1597	1676	1492	1331	1326	1189	1068	961	1080	976	1015	921	837
91	Pa	766	688	618	1663	1747	1555	1388	1241	1238	1112	1001	903	1017	921	957	869
92	U	790	709	637	574	1538	1621	1446	1293	1290	1159	1043	941	850	958	869	903

	Absorber		Emitter													
			63 Eu	64 Gd	65 Tb	66 Dy	67 Ho	68 Er	69 Tm	70 Yb	71 Lu	72 Hf	73 Ta	74 W	75 Re	76 Os	77 Ir	78 Pt
8	O		29	26	24	21	20	18	16	15	14	12	11	10	10	9	8	7
9	F		41	37	34	31	28	25	23	21	19	18	16	15	14	13	12	11
10	Ne		56	51	46	42	38	35	32	29	26	24	22	20	19	17	16	15
11	Na		74	67	61	55	50	46	42	38	35	32	29	27	25	23	21	19
12	Mg		95	86	78	71	65	59	54	49	45	41	38	35	32	29	27	25
13	Al		119	108	98	89	81	74	68	62	56	52	48	44	40	37	34	31
14	Si		147	133	121	110	100	91	83	76	70	64	59	54	50	46	42	39
15	P		178	161	146	133	121	111	101	92	84	77	71	65	60	55	51	47
16	S		213	193	175	159	145	132	121	110	101	93	85	78	72	66	61	56
17	Cl		251	227	206	187	171	156	142	130	119	109	100	92	85	78	72	67
18	Ar		292	265	240	219	199	182	166	152	139	128	117	108	99	91	84	78
19	K		337	306	278	252	230	210	192	175	161	147	135	125	115	106	98	90
20	Ca		385	350	318	289	263	240	219	201	184	169	155	143	131	121	112	103
21	Sc		437	397	360	328	299	273	249	228	209	192	176	162	149	138	127	118
22	Ti		493	447	406	370	337	307	281	257	236	216	199	183	169	155	144	133
23	V		551	500	455	414	377	344	315	288	264	242	223	205	189	174	161	149
24	Cr		70	557	506	461	420	383	350	321	294	270	248	228	210	194	179	166
25	Mn		79	72	65	59	465	424	388	355	326	299	275	253	233	215	199	184

Table A.7 – continued

		Emitter														
		63 Eu	64 Gd	65 Tb	66 Dy	67 Ho	68 Er	69 Tm	70 Yb	71 Lu	72 Hf	73 Ta	74 W	75 Re	76 Os	77 Ir	78 Pt
Absorber																	
26	Fe	89	81	74	67	61	56	427	391	359	330	303	279	257	237	219	203
27	Co	100	91	83	75	69	63	57	52	48	361	332	306	282	260	240	222
28	Ni	112	102	93	84	77	70	64	59	54	49	45	334	308	284	263	243
29	Cu	125	113	103	94	86	78	71	65	60	55	51	46	43	39	286	264
30	Zn	139	126	114	104	95	87	79	72	66	61	56	52	47	44	40	37
31	Ga	153	139	126	115	105	96	87	80	73	67	62	57	52	48	45	41
32	Ge	168	153	139	126	115	105	96	88	81	74	68	63	58	53	49	45
33	As	185	168	152	139	126	115	105	97	88	81	75	69	63	58	54	50
34	Se	202	183	167	152	138	126	115	106	97	89	82	75	69	64	59	54
35	Br	220	200	181	165	151	137	126	115	105	97	89	82	75	70	64	59
36	Kr	239	217	197	180	164	149	137	125	115	105	97	89	82	76	70	64
37	Rb	259	235	214	195	177	162	148	136	124	114	105	96	89	82	76	70
38	Sr	281	254	231	210	192	175	160	147	134	123	113	104	96	89	82	76
39	Y	303	275	249	227	207	189	173	158	145	133	122	112	104	96	88	82
40	Zr	326	296	269	244	223	203	186	170	156	143	132	121	112	103	95	88
41	Nb	349	317	288	262	239	218	199	183	167	154	141	130	120	110	102	94
42	Mo	373	339	308	280	256	233	213	195	179	164	151	139	128	118	109	101
43	Tc	399	362	329	299	273	249	228	209	191	176	162	149	137	126	117	108
44	Ru	425	386	351	319	291	266	243	223	204	187	172	159	146	135	125	115
45	Rh	452	410	373	340	310	283	259	237	217	200	183	169	156	144	133	123

Z	El															
46	Pd	480	436	396	361	329	301	275	252	231	212	195	180	166	153	141	131
47	Ag	509	462	420	383	349	319	292	267	245	225	207	191	176	162	150	139
48	Cd	539	490	445	406	370	338	309	283	260	239	219	202	186	172	159	147
49	In	570	518	471	429	391	358	327	300	275	253	232	214	197	182	168	156
50	Sn	602	547	497	453	413	378	346	317	291	267	246	226	208	192	178	165
51	Sb	635	576	524	478	436	398	365	334	307	282	259	239	220	203	188	174
52	Te	668	607	552	503	459	420	384	352	323	297	273	252	232	214	198	183
53	I	703	639	581	530	483	442	404	371	340	313	288	265	244	226	209	193
54	Xe	739	671	611	557	508	464	425	390	358	329	302	279	257	237	219	203
55	Cs	775	704	641	584	533	488	446	409	376	345	318	293	270	249	231	213
56	Ba	701	738	672	613	559	511	468	429	394	362	333	307	283	262	242	224
57	La	528	667	704	642	586	536	491	450	413	380	350	322	297	275	254	235
58	Ce	553	502	636	579	613	561	514	471	432	398	366	337	311	288	266	246
59	Pr	185	525	478	606	553	586	537	493	452	416	383	353	326	301	278	258
60	Nd	195	178	500	456	578	529	561	515	473	435	400	369	340	315	291	269
61	Pm	206	188	172	476	434	397	506	464	493	454	418	385	356	329	304	281
62	Sm	217	198	181	165	453	414	380	484	445	473	436	402	371	343	317	294
63	Eu	229	209	191	174	159	146	396	363	464	426	454	419	387	357	331	306
64	Gd	241	220	201	183	168	154	141	378	347	320	409	436	403	372	345	319
65	Tb	254	231	211	193	177	162	149	137	362	333	306	393	363	388	359	332
66	Dy	267	243	222	203	186	170	156	144	132	346	319	294	378	349	373	346
67	Ho	280	255	233	213	195	179	164	151	139	128	331	306	282	261	336	359
68	Er	294	268	245	224	205	188	173	159	146	135	124	318	293	271	251	324
69	Tm	309	282	257	235	215	197	181	166	153	141	130	120	111	282	261	242
70	Yb	324	295	270	247	226	207	190	175	161	148	137	126	117	108	271	251

Table A.7 – continued

Absorber		63 Eu	64 Gd	65 Tb	66 Dy	67 Ho	68 Er	69 Tm	70 Yb	71 Lu	72 Hf	73 Ta	74 W	75 Re	76 Os	77 Ir	78 Pt
71	Lu	340	310	283	258	237	217	199	183	168	155	143	132	122	113	105	260
72	Hf	356	324	296	271	248	227	209	192	177	163	150	139	128	119	110	102
73	Ta	373	340	310	284	260	238	219	201	185	170	157	145	134	124	115	107
74	W	390	356	325	297	272	249	229	210	194	178	165	152	141	130	121	112
75	Re	408	372	340	311	284	261	239	220	202	187	172	159	147	136	126	117
76	Os	427	389	355	325	297	273	250	230	212	195	180	166	154	142	132	122
77	Ir	446	407	371	339	311	285	261	240	221	204	188	174	161	149	138	128
78	Pt	466	425	388	354	325	298	273	251	231	213	197	182	168	155	144	134
79	Au	487	444	405	370	339	311	285	262	241	222	205	190	175	162	150	140
80	Hg	508	463	423	386	354	324	298	274	252	232	214	198	183	169	157	146
81	Tl	530	483	441	403	369	338	311	285	263	242	223	206	191	177	164	152
82	Pb	552	504	460	420	385	353	324	298	274	252	233	215	199	184	171	158
83	Bi	576	525	479	438	401	368	338	310	286	263	243	224	207	192	178	165
84	Po	600	547	499	456	418	383	352	323	298	274	253	234	216	200	185	172
85	At	625	570	520	475	435	399	366	337	310	286	263	243	225	208	193	179
86	Rn	651	593	541	495	453	415	381	350	323	297	274	253	234	217	201	187
87	Fr	677	617	563	515	471	432	397	365	336	309	285	264	244	226	209	194
88	Ra	704	642	586	536	490	450	413	379	349	322	297	274	254	235	218	202
89	Ac	732	667	609	557	510	468	429	395	363	335	309	285	264	244	226	210
90	Th	761	694	633	579	530	486	446	410	378	348	321	297	274	254	235	218
91	Pa	791	721	658	602	551	505	464	426	392	362	334	308	285	264	245	227
92	U	822	749	684	625	572	525	482	443	407	376	347	320	296	274	254	236

		Emitter													
Absorber		79 Au	80 Hg	81 Tl	82 Pb	83 Bi	84 Po	85 At	86 Rn	87 Fr	88 Ra	89 Ac	90 Th	91 Pa	92 U
8	O	7	6	6	5	5	5	4	4	4	4	3	3	3	3
9	F	10	9	8	8	7	7	6	6	5	5	5	4	4	4
10	Ne	14	13	12	11	10	9	9	8	7	7	6	6	6	5
11	Na	18	17	15	14	13	12	11	11	10	9	9	8	8	7
12	Mg	23	21	20	18	17	16	15	14	13	12	11	10	10	9
13	Al	29	27	25	23	21	20	19	17	16	15	14	13	12	11
14	Si	36	33	31	29	27	25	23	21	20	19	17	16	15	14
15	P	44	40	37	35	32	30	28	26	24	23	21	20	18	17
16	S	52	48	45	42	39	36	33	31	29	27	25	24	22	21
17	Cl	62	57	53	49	46	42	39	37	34	32	30	28	26	24
18	Ar	72	67	62	57	53	50	46	43	40	37	35	33	31	29
19	K	83	77	72	66	62	57	53	50	46	43	40	38	35	33
20	Ca	96	89	82	76	71	66	61	57	53	50	46	43	41	38
21	Sc	109	101	93	87	81	75	70	65	61	57	53	49	46	43
22	Ti	123	114	105	98	91	85	79	73	68	64	60	56	52	49
23	V	138	128	118	110	102	95	88	82	77	72	67	63	59	55
24	Cr	153	142	132	122	114	106	99	92	86	80	75	70	65	61
25	Mn	170	158	146	136	126	117	109	102	95	89	83	78	73	68

Table A.7 – continued

Absorber		79 Au	80 Hg	81 Tl	82 Pb	83 Bi	84 Po	85 At	86 Rn	87 Fr	88 Ra	89 Ac	90 Th	91 Pa	92 U
26	Fe	188	174	161	150	139	130	121	112	105	98	92	86	80	75
27	Co	206	191	177	164	153	142	132	124	115	108	101	94	88	83
28	Ni	225	208	193	180	167	155	145	135	126	118	110	103	96	90
29	Cu	245	227	211	196	182	169	158	147	137	128	120	112	105	99
30	Zn	265	246	228	212	197	184	171	160	149	139	130	122	114	107
31	Ga	38	35	33	229	213	198	185	172	161	150	141	132	123	116
32	Ge	42	39	36	34	31	214	199	186	173	162	152	142	133	125
33	As	46	43	40	37	34	32	30	28	186	174	163	152	143	134
34	Se	50	47	43	40	37	35	32	30	28	26	25	163	153	143
35	Br	55	51	47	44	41	38	35	33	31	29	27	25	23	153
36	Kr	60	55	51	48	44	41	38	36	33	31	29	27	25	24
37	Rb	65	60	56	52	48	45	42	39	36	34	31	29	28	26
38	Sr	70	65	60	56	52	48	45	42	39	36	34	32	30	28
39	Y	75	70	65	60	56	52	48	45	42	39	37	34	32	30
40	Zr	81	75	70	65	60	56	52	49	45	42	40	37	35	32
41	Nb	87	81	75	70	65	60	56	52	49	45	42	40	37	35
42	Mo	93	87	80	75	69	64	60	56	52	49	46	43	40	37
43	Tc	100	93	86	80	74	69	64	60	56	52	49	46	43	40
44	Ru	107	99	92	85	79	74	69	64	60	56	52	49	46	43
45	Rh	114	105	98	91	84	78	73	68	64	59	56	52	49	46

Z	El														
46	Pd	121	112	104	97	90	84	78	73	68	63	59	55	52	49
47	Ag	128	119	110	103	95	89	83	77	72	67	63	59	55	52
48	Cd	136	126	117	109	101	94	88	82	76	71	67	62	58	55
49	In	144	134	124	115	107	100	93	87	81	76	71	66	62	58
50	Sn	152	141	131	122	113	105	98	92	86	80	75	70	66	61
51	Sb	161	149	139	129	120	111	104	97	90	84	79	74	69	65
52	Te	170	157	146	136	126	118	110	102	95	89	83	78	73	69
53	I	179	166	154	143	133	124	116	108	101	94	88	82	77	72
54	Xe	188	175	162	151	140	130	122	113	106	99	93	87	81	76
55	Cs	198	183	170	158	147	137	128	119	111	104	97	91	85	80
56	Ba	208	193	179	166	155	144	134	125	117	109	102	96	90	84
57	La	218	202	188	174	162	151	141	132	123	115	107	101	94	88
58	Ce	228	212	197	183	170	158	148	138	129	120	113	105	99	93
59	Pr	239	222	206	191	178	166	155	144	135	126	118	111	104	97
60	Nd	250	232	215	200	186	174	162	151	141	132	123	116	108	102
61	Pm	261	242	225	209	195	181	169	158	148	138	129	121	113	106
62	Sm	272	253	235	218	203	189	177	165	154	144	135	126	118	111
63	Eu	284	264	245	228	212	198	184	172	161	150	141	132	124	116
64	Gd	296	275	255	237	221	206	192	179	168	157	147	138	129	121
65	Tb	308	286	266	247	230	215	200	187	175	163	153	143	134	126
66	Dy	321	298	277	257	240	223	208	195	182	170	159	149	140	131
67	Ho	333	310	288	268	249	232	217	202	189	177	166	155	146	137
68	Er	300	322	299	278	259	242	225	210	197	184	172	162	152	142
69	Tm	312	290	310	289	269	251	234	219	204	191	179	168	157	148
70	Yb	233	301	279	300	279	260	243	227	212	199	186	174	163	153

Table A.7 – continued

Absorber	Emitter													
	79 Au	80 Hg	81 Tl	82 Pb	83 Bi	84 Po	85 At	86 Rn	87 Fr	88 Ra	89 Ac	90 Th	91 Pa	92 U
71 Lu	241	224	208	270	251	270	252	235	220	206	193	181	170	159
72 Hf	250	232	216	201	261	243	261	244	228	214	200	188	176	165
73 Ta	99	241	224	208	194	181	235	253	236	221	207	194	182	171
74 W	104	97	232	216	201	188	175	228	213	229	215	201	189	177
75 Re	109	101	94	224	208	194	181	169	220	206	222	208	196	184
76 Os	114	106	98	92	85	201	188	175	164	154	200	188	202	190
77 Ir	119	111	103	96	89	83	194	181	170	159	149	194	182	196
78 Pt	124	115	107	100	93	87	81	188	175	164	154	144	188	177
79 Au	130	121	112	105	97	91	85	79	181	170	159	149	140	132
80 Hg	135	126	117	109	102	95	89	83	78	175	164	154	145	136
81 Tl	141	131	122	114	106	99	93	87	81	76	71	159	149	140
82 Pb	147	137	127	119	111	103	96	90	84	79	74	70	154	145
83 Bi	153	143	133	124	115	108	101	94	88	82	77	73	68	149
84 Po	160	149	138	129	120	112	105	98	92	86	81	76	71	67
85 At	166	155	144	134	125	117	109	102	95	89	84	79	74	69
86 Rn	173	161	150	140	130	122	114	106	99	93	87	82	77	72
87 Fr	180	168	156	145	136	127	118	111	103	97	91	85	80	75
88 Ra	188	174	162	151	141	132	123	115	108	101	95	89	83	78
89 Ac	195	181	169	157	147	137	128	120	112	105	98	92	87	81
90 Th	203	189	176	164	152	142	133	124	116	109	102	96	90	85
91 Pa	211	196	182	170	158	148	138	129	121	113	106	100	94	88
92 U	219	204	189	176	165	154	144	134	126	118	110	103	97	91

Table A.8 Mass absorption coefficients for Mα emissions.

Absorber		56 Ba	57 La	58 Ce	59 Pr	60 Nd	61 Pm	62 Sm	Emitter 63 Eu	64 Gd	65 Tb	66 Dy	67 Ho	68 Er	69 Tm	70 Yb	71 Lu
8	O	7864	6735	5793	5004	4339	3777	3299	2891	2542	2241	1982	1758	1563	1393	1244	1114
9	F	11052	9468	8147	7040	6107	5317	4646	4073	3582	3160	2795	2479	2205	1966	1757	1573
10	Ne	833	716	10992	9501	8245	7181	6276	5504	4842	4272	3780	3354	2984	2661	2379	2131
11	Na	1184	1018	879	763	664	580	8201	7194	6331	5588	4946	4389	3906	3484	3115	2791
12	Mg	1620	1393	1203	1044	909	794	696	612	540	478	424	5590	4975	4439	3970	3558
13	Al	2149	1848	1597	1385	1205	1053	923	812	717	634	563	501	446	399	358	434
14	Si	2780	2391	2065	1791	1559	1362	1194	1050	927	820	728	647	577	516	463	416
15	P	3520	3027	2615	2268	1974	1725	1512	1330	1174	1038	921	820	731	654	586	526
16	S	4377	3765	3252	2820	2455	2145	1880	1654	1459	1291	1146	1019	909	813	729	654
17	Cl	5359	4609	3981	3452	3005	2626	2302	2025	1786	1581	1403	1248	1113	995	892	801
18	Ar	6472	5566	4808	4169	3630	3171	2780	2445	2158	1909	1694	1507	1344	1202	1077	967
19	K	7724	6643	5738	4976	4332	3785	3318	2918	2575	2278	2022	1799	1604	1434	1286	1155
20	Ca	9121	7844	6776	5876	5115	4469	3918	3446	3041	2691	2388	2124	1895	1694	1518	1363
21	Sc	10670	9177	7926	6874	5984	5228	4584	4032	3557	3148	2793	2485	2216	1982	1776	1595
22	Ti	12378	10645	9195	7974	6942	6065	5317	4677	4126	3651	3240	2883	2571	2299	2060	1850
23	V	14250	12256	10586	9180	7992	6982	6122	5384	4750	4204	3730	3319	2960	2647	2372	2130
24	Cr	16292	14012	12103	10496	9137	7983	6999	6156	5431	4806	4265	3794	3384	3026	2712	2435
25	Mn	18511	15920	13751	11925	10381	9070	7952	6994	6171	5461	4845	4311	3845	3438	3081	2767
26	Fe	18498	15910	15534	13471	11728	10246	8983	7901	6971	6169	5474	4870	4344	3884	3480	3126
27	Co	15078	17824	15396	15138	13179	11514	10095	8879	7834	6932	6151	5473	4881	4364	3911	3512
28	Ni	3469	3005	17170	14891	12963	12876	11289	9929	8760	7752	6878	6120	5459	4880	4374	3928
29	Cu	3835	3322	2889	2523	14396	12578	11027	11054	9752	8630	7658	6813	6077	5433	4869	4373
30	Zn	4227	3662	3185	2781	2437	10100	12199	10729	9466	9567	8489	7553	6737	6023	5398	4848

Table A.8 – continued.

		Emitter															
Absorber		56 Ba	57 La	58 Ce	59 Pr	60 Nd	61 Pm	62 Sm	63 Eu	64 Gd	65 Tb	66 Dy	67 Ho	68 Er	69 Tm	70 Yb	71 Lu
31	Ga	4647	4025	3501	3057	2679	2356	2078	8578	10434	9233	8192	8341	7440	6652	5961	5354
32	Ge	5096	4414	3840	3353	2938	2583	2279	2017	1790	7352	8999	8006	7141	7320	6559	5891
33	As	5576	4830	4201	3668	3214	2826	2494	2207	1959	1743	1556	6352	7819	6991	6265	6460
34	Se	6087	5273	4586	4004	3509	3086	2722	2409	2138	1903	1698	1519	1363	5527	6839	6142
35	Br	6631	5744	4996	4362	3823	3361	2965	2624	2329	2073	1850	1655	1484	1334	1202	4839
36	Kr	7209	6245	5431	4742	4156	3654	3224	2853	2532	2254	2011	1800	1614	1451	1307	1180
37	Rb	7823	6776	5894	5146	4510	3966	3498	3096	2748	2446	2183	1953	1751	1574	1418	1280
38	Sr	8473	7340	6384	5574	4885	4295	3789	3353	2976	2649	2364	2115	1897	1705	1536	1387
39	Y	9162	7937	6903	6027	5282	4645	4098	3626	3218	2865	2556	2287	2051	1844	1661	1499
40	Zr	9891	8568	7452	6507	5702	5014	4423	3915	3474	3092	2760	2469	2214	1990	1793	1618
41	Nb	10661	9235	8032	7013	6146	5404	4768	4219	3745	3333	2974	2661	2387	2145	1932	1744
42	Mo	11473	9939	8644	7548	6614	5816	5131	4541	4030	3587	3201	2864	2568	2309	2080	1877
43	Tc	12330	10681	9290	8111	7108	6250	5514	4880	4331	3855	3440	3078	2760	2481	2235	2018
44	Ru	13232	11462	9970	8705	7628	6708	5918	5237	4648	4137	3692	3303	2962	2663	2399	2165
45	Rh	14182	12285	10685	9330	8176	7189	6342	5613	4982	4434	3957	3540	3175	2854	2571	2321
46	Pd	15180	13150	11437	9986	8751	7695	6789	6008	5332	4746	4235	3789	3398	3055	2752	2484
47	Ag	16229	14058	12228	10676	9356	8227	7258	6423	5701	5074	4528	4051	3633	3266	2942	2656
48	Cd	17330	15012	13057	11401	9991	8785	7750	6859	6088	5418	4835	4326	3880	3487	3141	2836
49	In	18485	16013	13927	12160	10656	9371	8267	7316	6493	5779	5158	4614	4138	3720	3351	3025
50	Sn	19695	17061	14839	12957	11354	9984	8808	7795	6918	6158	5495	4916	4409	3963	3570	3223

Z	El															
51	Sb	15486	16643	14476	12640	12085	9375	8297	7364	6554	5849	5233	4693	4218	3800	3430
52	Te	14764	14148	15281	13342	11692	9968	8822	7830	6969	6219	5564	4990	4485	4040	3647
53	I	15544	13465	12974	11328	12337	10589	9371	8317	7403	6606	5910	5300	4764	4292	3874
54	Xe	16360	14172	12327	11938	10461	10093	8932	8826	7856	7011	6272	5625	5056	4555	4112
55	Cs	17215	14912	12971	11325	9924	10640	9416	8358	8330	7433	6650	5964	5361	4829	4359
56	Ba	18109	15687	13644	11913	10440	9180	9003	9922	8807	7838	7875	7045	6318	5679	5116	4618
57	La	5284	16497	14349	12529	10979	9654	8517	8387	9276	8256	7368	7457	6687	6011	5415	4888
58	Ce	5438	4811	15086	13172	11543	10150	8955	7924	7833	6972	7757	6940	7072	6357	5727	5170
59	Pr	5598	4952	4396	13845	12132	10668	9412	8329	7393	7334	6545	7304	6550	5888	6052	5463
60	Nd	5762	5098	4525	4030	11110	11210	9890	8752	7768	6914	6882	6157	6891	6194	5580	5768
61	Pm	5933	5248	4659	4149	3707	10247	9194	8160	7263	6481	6472	5805	6513	5867	5296
62	Sm	6109	5404	4797	4272	3817	3420	9479	9655	8569	7627	6807	6090	6099	5483	6167	5567
63	Eu	6290	5565	4940	4399	3930	3522	3164	8794	8997	8008	7146	6393	5734	5154	5188	4683
64	Gd	6478	5731	5087	4531	4048	3627	3259	8181	8405	7501	6711	6018	5410	4873	4918
65	Tb	6672	5902	5239	4666	4169	3735	3356	3024	2472	7870	7041	6315	5676	5113	4616
66	Dy	6872	6079	5396	4806	4294	3847	3457	3114	2813	2547	7387	6624	5954	5364	4842
67	Ho	7079	6262	5559	4951	4423	3963	3561	3208	2897	2623	2164	6947	6245	5625	5078
68	Er	7292	6451	5726	5100	4556	4082	3668	3305	2985	2702	2452	2230	2032	6547	5898	5324
69	Tm	7512	6646	5899	5254	4694	4206	3779	3405	3075	2784	2526	2297	2093	1911	5322	5580
70	Yb	7740	6847	6078	5413	4836	4333	3893	3508	3168	2868	2602	2366	2156	1969	1801	5028
71	Lu	7974	7054	6262	5577	4982	4464	4011	3614	3264	2955	2681	2438	2222	2028	1855	1700
72	Hf	8216	7268	6452	5746	5134	4600	4133	3724	3363	3045	2763	2512	2289	2090	1912	1752
73	Ta	8466	7489	6648	5921	5290	4740	4259	3837	3465	3137	2847	2589	2359	2153	1970	1805
74	W	8724	7717	6851	6101	5451	4884	4389	3954	3571	3233	2933	2667	2431	2219	2030	1860
75	Re	8990	7953	7060	6287	5617	5033	4522	4074	3680	3331	3023	2749	2505	2287	2092	1917
76	Os	9264	8195	7275	6479	5788	5187	4660	4198	3792	3433	3115	2833	2581	2356	2155	1975
77	Ir	9547	8446	7497	6677	5965	5345	4803	4327	3908	3538	3210	2919	2660	2428	2221	2035
78	Pt	9839	8704	7726	6881	6148	5508	4949	4459	4027	3646	3308	3008	2741	2503	2289	2098
79	Au	10140	8970	7963	7092	6336	5677	5101	4595	4150	3757	3409	3100	2825	2579	2359	2162
80	Hg	0	9244	8206	7309	6529	5850	5257	4736	4277	3872	3514	3195	2911	2658	2431	2228

Table A.8 – continued.

		Emitter															
		56 Ba	57 La	58 Ce	59 Pr	60 Nd	61 Pm	62 Sm	63 Eu	64 Gd	65 Tb	66 Dy	67 Ho	68 Er	69 Tm	70 Yb	71 Lu
Absorber																	
81	Tl	0	0	8457	7532	6729	6030	5418	4881	4408	3991	3621	3293	3001	2739	2506	2296
82	Pb	0	0	8716	7763	6935	6214	5584	5030	4543	4113	3732	3394	3092	2823	2583	2366
83	Bi	0	0	0	8001	7148	6405	5755	5184	4682	4239	3846	3498	3187	2910	2662	2439
84	Po	0	0	0	0	7367	6601	5931	5343	4826	4369	3964	3605	3285	2999	2743	2514
85	At	0	0	0	0	0	6803	6113	5507	4974	4503	4086	3715	3385	3091	2827	2591
86	Rn	0	0	0	0	0	0	6300	5675	5126	4641	4211	3829	3489	3185	2914	2670
87	Fr	0	0	0	0	0	0	0	0	5283	4783	4340	3946	3596	3283	3003	2752
88	Ra	0	0	0	0	0	0	0	0	0	4929	4473	4067	3706	3384	3095	2836
89	Ac	0	0	0	0	0	0	0	0	0	0	4609	4192	3819	3487	3190	2923
90	Th	0	0	0	0	0	0	0	0	0	0	0	4320	3936	3594	3287	3012
91	Pa	0	0	0	0	0	0	0	0	0	0	0	0	4056	3703	3388	3104
92	U	0	0	0	0	0	0	0	0	0	0	0	0	0	3817	3491	3199

	Absorber	72 Hf	73 Ta	74 W	75 Re	76 Os	77 Ir	78 Pt	79 Au	80 Hg	81 Tl	82 Pb	83 Bi	84 Po	85 At	86 Rn	87 Fr
8	O	1000	899	810	731	661	599	543	493	449	409	373	341	312	286	262	241
9	F	1412	1270	1144	1033	934	846	768	698	635	579	528	483	442	405	371	341
10	Ne	1913	1721	1551	1401	1267	1148	1042	947	862	786	718	656	600	550	505	463
11	Na	2506	2255	2033	1836	1661	1506	1367	1243	1132	1032	942	861	788	723	663	609
12	Mg	3196	2876	2593	2343	2120	1922	1745	1587	1445	1318	1204	1101	1008	924	848	779
13	Al	3983	3586	3234	2922	2645	2398	2178	1981	1804	1646	1503	1375	1259	1154	1059	973
14	Si	374	337	305	3575	3237	2935	2666	2426	2210	2016	1842	1685	1543	1415	1299	1194
15	P	474	427	386	349	317	288	262	238	2662	2429	2220	2031	1860	1706	1566	1440
16	S	589	531	480	434	394	358	325	296	270	247	226	207	2211	2028	1862	1712
17	Cl	721	650	587	532	482	438	398	363	331	302	276	253	232	213	196	180
18	Ar	871	785	709	642	582	529	481	438	400	365	334	306	280	258	237	218
19	K	1039	937	846	766	695	631	574	523	477	436	398	365	335	307	282	260
20	Ca	1227	1106	1000	905	820	745	678	617	563	514	471	431	395	363	334	307
21	Sc	1435	1294	1169	1058	960	871	793	722	659	602	551	504	462	425	390	359
22	Ti	1665	1501	1357	1228	1113	1011	920	838	764	698	639	585	536	492	453	417
23	V	1917	1729	1562	1413	1282	1164	1059	964	880	804	735	673	618	567	521	480
24	Cr	2192	1976	1786	1616	1465	1331	1210	1103	1006	919	841	770	706	648	596	548
25	Mn	2490	2245	2029	1836	1665	1512	1375	1253	1143	1044	955	875	802	736	677	623
26	Fe	2813	2537	2292	2074	1881	1708	1553	1415	1291	1179	1079	988	906	832	765	704
27	Co	3161	2850	2575	2331	2113	1919	1746	1590	1451	1325	1212	1110	1018	935	859	791
28	Ni	3535	3188	2880	2607	2363	2146	1952	1778	1622	1482	1356	1242	1139	1046	961	884
29	Cu	3935	3549	3206	2902	2631	2389	2173	1980	1806	1650	1509	1382	1268	1164	1070	984
30	Zn	4363	3934	3554	3217	2917	2649	2409	2195	2002	1829	1673	1533	1405	1290	1186	1091

Table A.8 – *continued*.

Absorber		72 Hf	73 Ta	74 W	75 Re	76 Os	77 Ir	78 Pt	79 Au	80 Hg	81 Tl	82 Pb	83 Bi	84 Po	85 At	86 Rn	87 Fr
31	Ga	4818	4345	3925	3553	3221	2925	2661	2424	2211	2020	1848	1693	1552	1425	1310	1205
32	Ge	5302	4781	4319	3909	3544	3219	2928	2667	2433	2223	2033	1862	1708	1568	1441	1326
33	As	5814	5243	4737	4287	3887	3530	3211	2925	2668	2438	2230	2042	1873	1720	1581	1454
34	Se	5528	5732	5178	4687	4249	3859	3510	3198	2917	2665	2438	2233	2048	1880	1728	1590
35	Br	6017	5426	4902	5108	4632	4206	3826	3485	3180	2905	2657	2434	2232	2049	1883	1733
36	Kr	1067	4262	5322	4817	4367	4572	4159	3788	3456	3157	2888	2645	2426	2227	2047	1884
37	Rb	1158	1049	952	3773	4730	4295	3907	4107	3747	3423	3131	2868	2630	2415	2219	2042
38	Sr	1254	1136	1032	938	855	3355	4221	3845	3508	3701	3386	3101	2844	2611	2400	2208
39	Y	1356	1229	1116	1015	924	843	770	2995	3782	3455	3160	3346	3068	2817	2589	2382
40	Zr	1464	1327	1204	1095	998	910	832	761	697	2684	3399	3114	2855	3032	2787	2565
41	Nb	1578	1430	1298	1180	1075	981	896	820	752	690	633	2402	3051	2801	2575	2743
42	Mo	1698	1539	1397	1270	1157	1056	965	883	809	742	682	627	577	2157	2746	2527
43	Tc	1825	1654	1501	1365	1244	1135	1037	949	869	797	733	674	620	572	528	1942
44	Ru	1958	1775	1611	1465	1335	1218	1113	1018	933	856	786	723	666	614	566	523
45	Rh	2099	1902	1727	1570	1430	1305	1192	1091	1000	917	843	775	714	658	607	561
46	Pd	2247	2036	1848	1681	1531	1397	1276	1168	1070	982	902	830	764	704	650	600
47	Ag	2402	2177	1976	1797	1637	1493	1365	1249	1144	1050	964	887	817	753	695	642
48	Cd	2565	2324	2110	1919	1748	1595	1457	1333	1222	1121	1030	947	872	804	742	685
49	In	2736	2479	2251	2047	1864	1701	1554	1422	1303	1196	1098	1010	930	858	791	731
50	Sn	2915	2642	2398	2181	1987	1812	1656	1515	1388	1274	1170	1076	991	914	843	779

51	Sb	3103	2812	2552	2321	2114	1929	1763	1613	1478	1356	1246	1146	1055	972	897	829
52	Te	3299	2990	2714	2468	2248	2051	1874	1715	1571	1442	1324	1218	1122	1034	954	882
53	I	3504	3176	2883	2622	2388	2179	1991	1822	1669	1531	1407	1294	1191	1098	1014	936
54	Xe	3719	3370	3060	2782	2534	2312	2113	1933	1771	1625	1493	1373	1264	1166	1076	994
55	Cs	3943	3573	3244	2950	2687	2452	2240	2050	1878	1723	1583	1456	1341	1236	1141	1054
56	Ba	4177	3785	3436	3125	2847	2597	2373	2171	1990	1825	1677	1542	1420	1309	1208	1116
57	La	4421	4007	3637	3308	3013	2749	2512	2298	2106	1932	1775	1632	1503	1386	1279	1181
58	Ce	4676	4237	3847	3498	3186	2907	2656	2431	2227	2043	1877	1726	1590	1466	1352	1249
59	Pr	4941	4478	4065	3697	3367	3072	2807	2568	2353	2159	1984	1824	1680	1549	1429	1320
60	Nd	5217	4728	4292	3903	3555	3244	2964	2712	2485	2280	2094	1926	1774	1635	1509	1394
61	Pm	5505	4988	4529	4118	3751	3423	3127	2861	2622	2406	2210	2033	1872	1725	1592	1471
62	Sm	5036	5260	4775	4342	3955	3609	3297	3017	2764	2536	2330	2143	1973	1819	1679	1551
63	Eu	5291	4795	4353	4575	4167	3802	3474	3179	2913	2672	2455	2258	2079	1917	1769	1634
64	Gd	4448	5036	4572	4158	4388	4003	3658	3347	3067	2814	2585	2377	2189	2018	1862	1720
65	Tb	4669	4231	4800	4365	3976	4212	3848	3521	3227	2960	2720	2501	2303	2123	1960	1810
66	Dy	4380	4440	4031	4582	4174	3808	4047	3703	3393	3113	2860	2630	2422	2233	2060	1903
67	Ho	4593	4162	4228	3845	3503	3995	3650	3340	3565	3271	3005	2764	2545	2346	2165	2000
68	Er	4816	4364	3962	4032	3673	3351	3828	3503	3210	3435	3156	2903	2673	2464	2274	2101
69	Tm	5047	4574	4152	3776	3851	3513	3210	3672	3365	3087	3312	3046	2805	2586	2387	2205
70	Yb	5289	4793	4351	3957	3604	3682	3364	3078	2820	3236	2972	3196	2943	2713	2503	2313
71	Lu	4759	5021	4558	4145	3776	3445	3148	3225	2955	2711	3114	2864	2638	2844	2624	2424
72	Hf	1608	4512	4774	4341	3955	3608	3297	3016	3095	2839	2608	3000	2762	2546	2750	2540
73	Ta	1657	1523	4284	4546	4141	3778	3452	3158	2894	2973	2731	2512	2892	2666	2460	2660
74	W	1707	1569	1445	4074	4334	3954	3613	3306	3029	2779	2858	2629	2421	2232	2575	2379
75	Re	1759	1617	1489	1373	3880	4138	3781	3459	3170	2908	2672	2751	2533	2335	2155	2489
76	Os	1813	1667	1535	1415	1307	3700	3381	3619	3316	3043	2795	2571	2650	2443	2254	2082
77	Ir	1868	1717	1581	1458	1347	1245	1153	3232	3469	3182	2924	2689	2476	2554	2357	2178
78	Pt	1925	1770	1630	1503	1388	1283	1188	1101	3094	3328	3057	2812	2589	2387	2203	2276
79	Au	1984	1824	1680	1549	1430	1322	1224	1135	1053	2965	3196	2939	2707	2495	2303	2127
80	Hg	2045	1880	1731	1596	1474	1363	1262	1170	1085	1009	2845	3072	2829	2608	2406	2223

Table A.8 – continued.

		Emitter															
		72 Hf	73 Ta	74 W	75 Re	76 Os	77 Ir	78 Pt	79 Au	80 Hg	81 Tl	82 Pb	83 Bi	84 Po	85 At	86 Rn	87 Fr
Absorber																	
81	Tl	2107	1937	1784	1645	1519	1405	1300	1205	1119	1039	967	2732	2956	2725	2514	2323
82	Pb	2172	1997	1839	1695	1566	1448	1340	1242	1153	1071	996	928	2625	2420	2626	2426
83	Bi	2239	2058	1895	1747	1613	1492	1381	1280	1188	1104	1027	956	891	2526	2331	2534
84	Po	2307	2121	1953	1801	1663	1538	1424	1320	1225	1138	1058	985	918	857	800	2246
85	At	2378	2186	2013	1856	1714	1585	1467	1360	1262	1173	1091	1016	947	883	824	771
86	Rn	2451	2253	2074	1913	1766	1633	1512	1402	1301	1209	1124	1047	975	910	850	794
87	Fr	2526	2322	2138	1971	1820	1683	1558	1445	1341	1246	1159	1079	1005	938	876	818
88	Ra	2603	2393	2203	2032	1876	1735	1606	1489	1382	1284	1194	1112	1036	967	903	843
89	Ac	2683	2466	2271	2094	1933	1788	1655	1534	1424	1323	1231	1146	1068	996	930	869
90	Th	2765	2542	2340	2158	1993	1842	1706	1581	1468	1363	1268	1181	1100	1027	959	896
91	Pa	2849	2619	2412	2224	2054	1899	1758	1630	1512	1405	1307	1217	1134	1058	988	923
92	U	2936	2699	2485	2292	2116	1957	1812	1679	1559	1448	1347	1254	1169	1090	1018	951

USEFUL PARAMETERS FOR XRS DATA CORRECTION

Table A.8 Mass absorption coefficients for Mα emissions (continued)

Absorber		Emitter 88 Ra	89 Ac	90 Th	91 Pa	92 U	Absorber		Emitter 88 Ra	89 Ac	90 Th	91 Pa	92 U
8	O	221	203	187	173	159	51	Sb	767	710	657	610	566
9	F	313	288	266	245	226	52	Te	815	754	699	648	601
10	Ne	426	392	361	333	307	53	I	866	801	742	688	639
							54	Xe	919	851	788	731	678
11	Na	560	515	475	438	404	55	Cs	974	902	835	775	719
12	Mg	716	659	608	560	517							
13	Al	895	824	760	701	647	56	Ba	1032	955	885	821	762
14	Si	1098	1011	932	860	794	57	La	1092	1011	937	869	806
15	P	1325	1220	1125	1038	959	58	Ce	1155	1069	991	919	852
							59	Pr	1221	1130	1047	971	901
16	S	1575	1451	1338	1235	1141	60	Nd	1289	1193	1105	1025	951
17	Cl	1850	1705	1572	1451	1341							
18	Ar	201	185	171	158	146	61	Pm	1360	1259	1166	1081	1004
19	K	239	221	204	188	174	62	Sm	1434	1327	1230	1140	1058
20	Ca	283	261	241	222	205	63	Eu	1511	1399	1296	1201	1115
							64	Gd	1591	1472	1364	1265	1174
21	Sc	331	305	281	260	240	65	Tb	1674	1549	1435	1331	1235
22	Ti	384	354	326	302	279							
23	V	442	407	376	347	321							
24	Cr	505	466	430	397	367	66	Dy	1760	1629	1509	1399	1299
25	Mn	574	529	488	451	417	67	Ho	1850	1712	1586	1471	1365
							68	Er	1942	1798	1666	1544	1433
							69	Tm	2039	1887	1748	1621	1504
26	Fe	648	598	551	509	471	70	Yb	2138	1979	1834	1700	1578
27	Co	728	672	620	572	529							
28	Ni	814	751	693	640	592	71	Lu	2242	2075	1922	1782	1654
29	Cu	907	836	772	713	659	72	Hf	2349	2174	2014	1868	1733
30	Zn	1005	927	855	790	730	73	Ta	2460	2277	2109	1956	1815
							74	W	2200	2383	2208	2047	1900
31	Ga	1110	1024	945	873	807	75	Re	2301	2130	2310	2142	1987
32	Ge	1222	1126	1039	960	888							
33	As	1340	1235	1140	1053	973	76	Os	2407	2228	2064	2239	2078
34	Se	1464	1350	1246	1151	1064	77	Ir	2014	1864	2158	2001	1857
35	Br	1596	1472	1358	1255	1160	78	Pt	2105	1948	1805	2092	1941
							79	Au	2200	2036	1886	1749	2028
36	Kr	1735	1600	1476	1364	1261	80	Hg	2056	2128	1971	1828	1696
37	Rb	1881	1734	1600	1478	1367							
38	Sr	2034	1875	1731	1599	1478	81	Tl	2148	1988	2059	1909	1771
39	Y	2194	2023	1867	1725	1595	82	Pb	2244	2077	1924	1993	1850
40	Zr	2362	2178	2010	1857	1717	83	Bi	2343	2169	2009	1863	1931
							84	Po	2446	2264	2098	1945	1805
41	Nb	2526	2329	2150	1986	1837	85	At	2167	2363	2189	2030	1884
42	Mo	2328	2487	2295	2121	1961							
43	Tc	2479	2286	2111	2260	2090	86	Rn	743	2091	2285	2119	1966
44	Ru	484	1754	2244	2074	1918	87	Fr	766	717	2020	1873	2051
45	Rh	519	480	445	1588	2036	88	Ra	789	739	692	1952	1812
							89	Ac	813	761	713	668	627
46	Pd	555	514	476	441	410	90	Th	838	784	735	689	646
47	Ag	594	549	509	472	438							
48	Cd	634	587	543	504	468	91	Pa	864	808	757	710	666
49	In	676	626	580	537	499	92	U	890	833	780	732	686
50	Sn	720	657	618	573	531							

Table A.9 Mass absorption coefficients for high side of K edge.

		Emitter															
		11 Na	12 Mg	13 Al	14 Si	15 P	16 S	17 Cl	18 Ar	19 K	20 Ca	21 Sc	22 Ti	23 V	24 Cr	25 Mn	26 Fe
Absorber																	
8	O	3361	1945	1177	740	481	322	221	156	112	82	61	46	35	27	21	17
9	F	4734	2743	1661	1046	681	456	314	221	159	116	86	65	50	38	30	24
10	Ne	6394	3711	2250	1417	923	620	427	300	216	158	118	89	68	52	41	32
11	Na	8355	4854	2947	1858	1212	814	561	395	284	208	155	117	89	69	54	43
12	Mg	709	6181	3756	2371	1548	1040	717	506	364	267	199	150	115	89	70	55
13	Al	940	552	4680	2957	1932	1299	897	633	456	334	249	188	144	112	88	69
14	Si	1216	715	438	3618	2366	1592	1100	777	560	411	306	232	177	138	108	85
15	P	1540	905	555	353	2850	1919	1326	938	676	496	370	280	215	167	131	103
16	S	1915	1125	690	439	289	2281	1577	1116	805	591	441	334	256	199	156	124
17	Cl	2344	1377	845	538	354	240	1853	1311	947	696	520	394	302	235	184	146
18	Ar	2831	1663	1020	650	427	289	201	1524	1101	809	605	458	352	273	215	170
19	K	3379	1985	1218	775	510	345	240	170	1267	933	697	529	406	316	248	197
20	Ca	3990	2344	1438	915	602	408	283	201	146	1065	796	604	464	361	284	225

Z	Sym																
21	Sc	4668	2743	1682	1071	704	477	331	235	170	125	903	685	527	410	322	256
22	Ti	5415	3181	1951	1242	817	553	384	273	197	146	109	771	593	461	363	288
23	V	6234	3663	2246	1430	941	637	442	314	227	168	125	95	664	516	406	323
24	Cr	7128	4188	2568	1635	1076	728	506	359	260	192	143	109	84	575	452	359
25	Mn	8098	4758	2918	1858	1222	827	574	408	295	218	163	124	95	74	500	398
26	Fe	9148	5375	3296	2099	1381	935	649	461	334	246	184	140	108	84	66	438
27	Co	10280	6040	3704	2358	1551	1050	729	518	375	276	207	157	121	94	74	59
28	Ni	11496	6754	4142	2637	1735	1175	816	579	419	309	231	176	135	105	83	66
29	Cu	11230	7519	4611	2936	1931	1308	908	645	467	344	258	196	150	117	92	73
30	Zn	12423	8336	5112	3255	2141	1450	1007	715	517	381	286	217	167	130	102	81
31	Ga	2115	8044	5646	3595	2365	1601	1112	789	571	421	315	240	184	143	113	90
32	Ge	2319	8836	6212	3955	2602	1762	1223	868	629	463	347	264	203	158	124	99
33	As	2537	1529	6813	4338	2854	1932	1341	952	690	508	381	289	222	173	136	108
34	Se	2770	1669	6477	4742	3120	2112	1466	1041	754	556	416	316	243	189	149	118
35	Br	3017	1818	5104	5169	3400	2302	1598	1135	822	606	453	344	265	206	162	129
36	Kr	3280	1977	1241	4874	3696	2502	1737	1233	893	658	493	374	288	224	176	140
37	Rb	3560	2145	1346	3817	4007	2713	1883	1337	968	714	534	406	312	243	191	152
38	Sr	3856	2323	1458	949	3751	2933	2037	1446	1047	772	578	439	338	263	207	164
39	Y	4169	2512	1577	1026	2922	3165	2197	1560	1130	833	623	473	364	283	223	177
40	Zr	4501	2712	1702	1107	743	2945	2365	1679	1216	896	671	510	392	305	240	191
41	Nb	4851	2923	1835	1194	801	3146	2530	1797	1301	960	719	546	420	327	257	204
42	Mo	5221	3146	1975	1285	862	595	2331	1918	1390	1025	768	584	449	350	275	219

Table A.9 – *continued*.

									Emitter							
Absorber		27 Co	28 Ni	29 Cu	30 Zn	31 Ga	32 Ge	33 As	34 Se	35 Br	36 Kr	37 Rb	38 Sr	39 Y	40 Zr	41 Nb	42 Mo
8	O	13	11	9	7	6	5	4	3	3	2	2	2	1	1	1	1
9	F	19	15	12	10	8	7	6	5	4	3	3	2	2	2	2	1
10	Ne	26	21	17	14	11	9	8	6	5	5	4	3	3	2	2	2
11	Na	34	27	22	18	15	12	10	9	7	6	5	4	4	3	3	2
12	Mg	44	35	29	23	19	16	13	11	9	8	7	6	5	4	4	3
13	Al	55	45	36	29	24	20	17	14	12	10	8	7	6	5	5	4
14	Si	68	55	45	36	30	25	21	17	15	12	10	9	8	7	6	5
15	P	83	67	54	44	36	30	25	21	18	15	13	11	9	8	7	6
16	S	99	80	65	53	44	36	30	25	21	18	15	13	11	10	8	7
17	Cl	117	94	76	63	51	43	36	30	25	21	18	15	13	11	10	8
18	Ar	136	110	89	73	60	50	42	35	29	25	21	18	15	13	11	10
19	K	157	127	103	84	70	58	48	40	34	29	24	21	18	15	13	12
20	Ca	180	145	118	97	80	66	55	46	39	33	28	24	21	18	15	13

Z	El	1	2	3	4	5	6	7	8	9	10	11	12	13	14	15	16
21	Sc	205	165	134	110	91	75	63	53	44	38	32	27	23	20	17	15
22	Ti	231	186	152	124	103	85	71	60	50	43	36	31	27	23	20	17
23	V	259	209	170	139	115	96	80	67	56	48	41	35	30	26	22	19
24	Cr	288	233	190	155	128	107	89	75	63	53	45	39	33	29	25	22
25	Mn	319	258	210	172	142	118	99	83	70	59	50	43	37	32	28	24
26	Fe	351	284	232	190	157	130	109	91	77	65	56	48	41	35	31	26
27	Co	385	312	254	209	172	143	120	101	85	72	61	52	45	39	34	29
28	Ni	53	340	278	228	188	157	131	110	93	79	67	57	49	42	37	32
29	Cu	59	47	302	248	205	170	142	120	101	86	73	63	54	46	40	35
30	Zn	65	53	43	269	222	185	155	130	110	93	79	68	58	50	44	38
31	Ga	72	58	47	39	240	200	167	140	119	101	86	74	63	55	47	41
32	Ge	79	64	52	43	35	215	180	151	128	109	93	79	68	59	51	44
33	As	87	70	57	47	39	32	193	162	137	117	100	85	73	63	55	48
34	Se	95	76	62	51	42	35	29	174	147	125	107	91	79	68	59	51
35	Br	103	83	68	56	46	38	32	27	157	133	114	98	84	72	63	55
36	Kr	112	91	74	60	50	41	35	29	24	142	121	104	89	77	67	58
37	Rb	122	98	80	66	54	45	37	31	27	22	129	111	95	82	71	62
38	Sr	132	106	86	71	58	49	41	34	29	24	21	117	101	87	75	66
39	Y	142	115	93	76	63	52	44	37	31	26	22	19	107	92	80	69
40	Zr	153	123	100	82	68	56	47	39	33	28	24	21	18	97	84	73
41	Nb	164	132	108	88	73	61	51	42	36	30	26	22	19	16	89	77
42	Mo	175	142	115	95	78	65	54	45	38	33	28	24	20	17	15	81

Table A.9 – continued.

		Emitter							
Absorber		43 Tc	44 Ru	45 Rh	46 Pd	47 Ag	48 Cd	49 In	50 Sn
8	O	1	1	1	1	0	0	0	0
9	F	1	1	1	1	1	1	1	0
10	Ne	2	1	1	1	1	1	1	1
11	Na	2	2	2	1	1	1	1	1
12	Mg	3	2	2	2	2	1	1	1
13	Al	3	3	3	2	2	2	2	2
14	Si	4	4	3	3	2	2	2	2
15	P	5	4	4	3	3	3	2	2
16	S	6	5	5	4	4	3	3	3
17	Cl	7	6	6	5	4	4	3	3
18	Ar	9	8	7	6	5	4	4	3
19	K	10	9	8	7	6	5	5	4
20	Ca	12	10	9	8	7	6	5	5

21	Sc	13	11	10	9	8	7	6	5	
22	Ti	15	13	11	10	9	8	7	6	
23	V	17	15	13	11	10	9	8	7	
24	Cr	19	16	14	13	11	10	9	8	
25	Mn	21	18	16	14	12	11	10	9	
26	Fe	23	20	18	15	14	12	11	9	
27	Co	25	22	19	17	15	13	12	10	
28	Ni	28	24	21	19	16	15	13	11	
29	Cu	30	26	23	20	18	16	14	12	
30	Zn	33	29	25	22	20	17	15	14	
31	Ga	36	31	27	24	21	19	17	15	
32	Ge	39	34	30	26	23	20	18	16	
33	As	41	36	32	28	25	22	19	17	
34	Se	44	39	34	30	26	23	21	18	
35	Br	48	42	36	32	28	25	22	20	
36	Kr	51	44	39	34	30	27	24	21	
37	Rb	54	47	41	36	32	28	25	22	
38	Sr	57	50	44	39	34	30	27	24	
39	Y	61	53	47	41	36	32	28	25	
40	Zr	64	56	49	43	38	34	30	27	
41	Nb	67	59	52	46	40	36	32	28	
42	Mo	71	62	55	48	42	38	33	30	

Table A.10 Mass absorption coefficients for high side of L_I edge.

Absorber		31 Ga	32 Ge	33 As	34 Se	35 Br	36 Kr	37 Rb	38 Sr	39 Y	40 Zr	41 Nb	42 Mo	43 Tc	44 Ru	45 Rh	46 Pd
30	Zn	8268	6625	5340	4328	3527	2890	2381	1971	1639	1370	1151	970	822	699	597	511
31	Ga	9131	7317	5897	4779	3895	3192	2629	2176	1811	1513	1271	1072	908	772	659	565
32	Ge	8764	8051	6489	5259	4286	3512	2893	2395	1992	1665	1398	1179	999	849	725	621
33	As	1517	7689	7116	5768	4701	3852	3172	2626	2185	1826	1533	1293	1095	931	795	681
34	Se	1656	1341	6765	6305	5139	4211	3468	2871	2389	1996	1676	1414	1197	1018	869	745
35	Br	1804	1461	1190	5969	5601	4589	3780	3129	2603	2176	1827	1541	1305	1110	948	812
36	Kr	1961	1588	1293	1059	5281	4988	4109	3402	2830	2365	1986	1675	1419	1206	1030	883
37	Rb	2128	1724	1403	1149	946	4687	4454	3688	3068	2564	2153	1816	1538	1308	1117	957
38	Sr	2305	1867	1520	1244	1024	847	4171	3988	3317	2773	2328	1964	1663	1414	1207	1035
39	Y	2493	2019	1644	1346	1107	916	762	3722	3579	2992	2512	2119	1794	1526	1303	1116
40	Zr	2691	2179	1774	1453	1196	989	822	687	3331	3220	2704	2281	1932	1643	1402	1202
41	Nb	2901	2349	1913	1566	1289	1066	886	740	2569	2974	2891	2439	2066	1757	1501	1286
42	Mo	3122	2528	2058	1685	1387	1147	954	797	669	2290	2664	2603	2206	1876	1603	1374
43	Tc	3355	2717	2212	1811	1490	1233	1025	856	718	606	2047	2394	2350	2000	1708	1465
44	Ru	3600	2915	2374	1943	1599	1323	1100	919	771	650	550	1836	2157	2127	1817	1559
45	Rh	3858	3125	2544	2083	1714	1418	1179	985	826	697	590	502	1651	1948	1930	1656
46	Pd	4130	3345	2723	2230	1835	1518	1262	1054	885	746	631	537	458	1489	1765	1756
47	Ag	4416	3576	2912	2384	1962	1623	1349	1127	946	797	675	574	490	420	1347	1603
48	Cd	4715	3818	3109	2545	2095	1733	1441	1203	1010	851	721	613	523	448	386	1222
49	In	5029	4073	3316	2715	2234	1848	1537	1283	1077	908	769	654	558	478	411	355

Z	El																
50	Sn	5359	4339	3533	2893	2381	1969	1637	1368	1148	968	819	697	595	510	438	378
51	Sb	5703	4619	3761	3079	2534	2096	1742	1456	1222	1030	872	741	633	542	467	403
52	Te	6064	4911	3999	3274	2694	2229	1853	1548	1299	1095	927	788	673	577	496	428
53	I	6442	5217	4248	3478	2862	2367	1968	1644	1380	1163	985	837	715	613	527	455
54	Xe	6836	5536	4508	3691	3037	2512	2089	1745	1464	1234	1045	889	759	650	559	483
55	Cs	7249	5870	4780	3913	3220	2664	2215	1850	1552	1309	1108	942	804	689	593	512
56	Ba	7679	6218	5063	4145	3411	2822	2346	1960	1645	1386	1174	998	852	730	628	542
57	La	7185	6582	5359	4387	3611	2987	2483	2074	1741	1467	1242	1056	902	773	665	574
58	Ce	7564	6961	5668	4640	3819	3159	2626	2194	1841	1552	1314	1117	954	817	703	607
59	Pr	6382	6447	5989	4903	4035	3338	2775	2318	1945	1640	1389	1181	1008	864	743	641
60	Nd	6711	6782	5522	5177	4261	3525	2930	2448	2054	1732	1466	1247	1064	912	785	677
61	Pm	6320	5713	5807	5463	4496	3719	3092	2583	2167	1827	1547	1315	1123	962	828	715
62	Sm	6637	6003	6104	4997	4740	3921	3260	2723	2285	1926	1631	1387	1184	1015	873	753
63	Eu	6968	5643	5135	5251	4321	4131	3434	2869	2408	2030	1719	1461	1247	1069	920	794
64	Gd	7314	5923	4823	4414	4539	4350	3616	3021	2535	2137	1809	1538	1313	1125	968	836
65	Tb	7675	6215	5061	4633	4765	3942	3805	3178	2667	2249	1904	1619	1382	1184	1019	879
66	Dy	6957	6520	5309	4346	4001	4137	4001	3342	2805	2364	2002	1702	1453	1245	1071	925
67	Ho	2330	6837	5567	4558	4198	3472	3608	3512	2947	2485	2104	1789	1527	1309	1126	972
68	Er	2400	2004	5837	4779	3933	3641	3785	3161	3095	2609	2209	1878	1604	1374	1182	1020
69	Tm	2472	2065	6118	5009	4122	3817	3173	3314	3248	2739	2319	1971	1683	1442	1241	1071
70	Yb	2547	2127	1785	5249	4319	3573	3326	3474	2915	2873	2432	2068	1765	1513	1301	1123
71	Lu	2624	2192	1839	4723	4525	3743	3484	2910	3054	2550	2168	1851	1586	1364	1178
72	Hf	2704	2258	1895	1597	4739	3920	3259	3048	2558	2697	2672	2272	1939	1662	1430	1234
73	Ta	2786	2327	1952	1646	4253	4105	3412	2850	2678	2823	2390	2379	2031	1740	1497	1292
74	W	2871	2398	2012	1696	1436	4296	3572	2984	2803	2363	2502	2490	2126	1822	1567	1353

Table A.10 – continued.

Absorber		31 Ga	32 Ge	33 As	34 Se	35 Br	36 Kr	37 Rb	38 Sr	39 Y	40 Zr	41 Nb	42 Mo	43 Tc	44 Ru	45 Rh	46 Pd
75	Re	2958	2471	2073	1748	1480	3846	3738	3122	2620	2473	2094	2226	2224	1906	1639	1415
76	Os	3049	2546	2136	1801	1525	1297	3910	3266	2741	2587	2190	2328	1987	1993	1714	1480
77	Ir	3142	2624	2202	1856	1572	1337	3492	3416	2867	2417	2290	1947	2078	2083	1791	1546
78	Pt	3238	2704	2269	1913	1620	1377	1176	3048	2998	2527	2394	2036	1738	1861	1871	1615
79	Au	3337	2787	2338	1971	1669	1420	1212	1040	3134	2642	2237	2128	1816	1945	1673	1687
80	Hg	3439	2872	2410	2032	1720	1463	1249	1072	2790	2761	2338	1988	1898	1626	1748	1760
81	Tl	3544	2960	2484	2094	1773	1508	1288	1104	951	2885	2443	2077	1982	1699	1461	1576
82	Pb	3653	3051	2560	2158	1827	1554	1327	1138	980	2563	2552	2170	1852	1774	1526	1645
83	Bi	3765	3144	2638	2224	1883	1601	1368	1173	1010	873	2665	2266	1934	1852	1593	1375
84	Po	3880	3240	2719	2292	1941	1651	1410	1209	1041	900	2363	2366	2020	1731	1662	1435
85	At	3999	3340	2802	2362	2000	1701	1453	1246	1073	927	804	2095	2108	1807	1734	1497
86	Rn	4121	3442	2888	2435	2061	1753	1497	1284	1106	956	829	2185	2200	1885	1622	1562
87	Fr	4247	3547	2977	2509	2125	1807	1543	1323	1139	985	854	744	1945	1967	1692	1461
88	Ra	4377	3656	3068	2586	2190	1862	1590	1364	1174	1015	881	767	670	2051	1765	1523
89	Ac	4511	3768	3161	2665	2257	1919	1639	1406	1210	1046	908	790	690	1811	1840	1589
90	Th	0	3883	3258	2747	2326	1978	1689	1449	1247	1078	935	814	711	624	1623	1656
91	Pa	0	4002	3358	2830	2397	2038	1741	1493	1285	1111	964	839	733	643	1690	1726
92	U	0	0	3460	2917	2470	2101	1794	1539	1325	1145	993	865	756	662	582	1520

		Emitter																	
		47 Ag	48 Cd	49 In	50 Sn	51 Sb	52 Te	53 I	54 Xe	55 Cs	56 Ba	57 La	58 Ce	59 Pr	60 Nd	61 Pm	62 Sm
Absorber																	
30	Zn	440	380	329	286	249	217	191	167	147	130	115	102	91	81	72	65
31	Ga	486	419	363	315	275	240	210	185	163	144	127	113	100	89	80	71
32	Ge	534	461	399	347	302	264	231	203	179	158	140	124	110	98	88	79
33	As	586	506	438	380	332	290	254	223	196	174	154	136	121	108	96	86
34	Se	641	553	479	416	362	317	278	244	215	190	168	149	132	118	105	94
35	Br	698	603	522	453	395	345	303	266	234	207	183	162	144	129	115	103
36	Kr	759	655	567	493	429	375	329	289	254	225	199	176	157	140	125	112
37	Rb	823	710	615	534	465	407	356	313	276	244	216	191	170	151	135	121
38	Sr	890	768	665	578	503	440	385	339	298	263	233	207	184	164	146	131
39	Y	960	829	718	623	543	475	416	365	322	284	252	223	198	177	158	141
40	Zr	1034	892	772	671	585	511	448	393	347	306	271	240	214	190	170	152
41	Nb	1106	955	827	719	626	547	480	422	371	328	290	258	229	204	182	163
42	Mo	1182	1020	884	768	669	585	513	451	397	351	311	276	245	218	195	174
43	Tc	1260	1088	943	819	714	624	547	481	424	375	332	294	262	233	208	186
44	Ru	1341	1158	1004	872	761	665	583	513	452	399	354	314	279	249	222	199
45	Rh	1425	1231	1067	927	809	707	620	545	481	425	376	334	297	265	236	212
46	Pd	1512	1306	1132	984	858	751	658	579	510	451	400	355	316	281	251	225
47	Ag	1601	1383	1199	1043	910	796	698	614	541	478	424	376	335	299	267	239
48	Cd	1459	1463	1269	1103	962	842	739	650	573	507	449	399	355	316	283	253
49	In	1111	1332	1340	1166	1017	890	781	687	606	536	475	422	375	335	299	268

Table A.10 – *continued*.

										Emitter							
Absorber		47 Ag	48 Cd	49 In	50 Sn	51 Sb	52 Te	53 I	54 Xe	55 Cs	56 Ba	57 La	58 Ce	59 Pr	60 Nd	61 Pm	62 Sm
50	Sn	328	1012	1218	1230	1073	939	824	725	640	566	501	445	396	354	316	283
51	Sb	349	303	925	1117	1131	990	869	765	675	597	529	470	418	373	333	299
52	Te	371	322	281	846	1026	1042	915	805	710	628	557	495	441	393	351	315
53	I	394	342	299	261	777	944	962	847	747	661	586	521	464	414	370	331
54	Xe	418	363	317	277	243	714	871	889	785	694	616	547	487	435	389	348
55	Cs	443	385	336	294	258	227	658	804	824	729	646	574	512	457	408	366
56	Ba	470	408	356	311	273	240	212	607	744	764	678	602	537	479	428	384
57	La	497	432	377	329	289	254	224	198	561	690	710	631	562	502	449	402
58	Ce	526	457	398	348	306	269	237	210	587	520	641	660	588	525	470	421
59	Pr	556	483	421	368	323	284	250	221	196	543	482	596	615	549	492	441
60	Nd	587	510	445	389	341	300	264	234	207	184	504	448	555	574	514	460
61	Pm	619	538	469	410	360	316	279	247	219	194	173	468	417	517	536	481
62	Sm	653	567	495	432	379	334	294	260	230	205	182	163	435	388	483	501
63	Eu	688	598	521	456	400	351	310	274	243	216	192	171	153	405	362	452
64	Gd	724	629	549	480	421	370	326	289	256	227	202	180	161	144	377	338
65	Tb	762	662	577	505	443	389	343	304	269	239	213	190	170	152	136	352
66	Dy	801	696	607	531	466	409	361	319	283	251	224	200	178	160	143	129
67	Ho	842	732	638	558	489	430	379	335	297	264	235	210	187	168	151	135
68	Er	884	768	670	586	514	452	398	352	312	277	247	220	197	176	158	142
69	Tm	928	806	703	615	539	474	418	370	328	291	259	231	207	185	166	149

70	Yb	973	846	737	645	566	497	439	388	344	305	272	243	217	194	174	157
71	Lu	1020	887	773	676	593	522	460	407	360	320	285	254	227	204	183	164
72	Hf	1069	929	810	708	621	546	482	426	378	335	299	266	238	213	191	172
73	Ta	1120	973	848	742	651	572	505	446	395	351	313	279	249	223	200	180
74	W	1172	1018	888	776	681	599	528	467	414	368	327	292	261	234	210	189
75	Re	1226	1065	929	812	712	627	553	489	433	385	342	305	273	245	219	197
76	Os	1282	1114	971	849	745	655	578	511	453	402	358	319	286	256	230	206
77	Ir	1340	1164	1015	888	779	685	604	534	473	420	374	334	298	267	240	216
78	Pt	1399	1216	1060	927	813	715	631	558	494	439	391	349	312	279	251	225
79	Au	1461	1270	1107	968	849	747	659	582	516	458	408	364	325	292	262	235
80	Hg	1525	1325	1156	1011	886	780	687	608	539	478	426	380	340	304	273	245
81	Tl	1591	1383	1206	1054	925	813	717	634	562	499	444	396	354	317	285	256
82	Pb	1425	1442	1257	1099	964	848	748	661	586	520	463	413	370	331	297	267
83	Bi	1487	1503	1310	1146	1005	884	779	689	611	542	483	431	385	345	310	278
84	Po	1243	1349	1365	1194	1047	921	812	718	636	565	503	449	401	359	323	290
85	At	1297	1407	1226	1243	1090	959	846	748	663	589	524	468	418	374	336	302
86	Rn	1353	1176	1279	1294	1135	998	880	778	690	613	546	487	435	390	350	314
87	Fr	1411	1226	1069	1166	1181	1039	916	810	718	638	568	507	453	406	364	327
88	Ra	1471	1278	1114	1215	1066	1081	953	843	747	663	591	527	471	422	379	340
89	Ac	1376	1332	1162	1016	1110	977	991	876	777	690	614	548	490	439	394	354
90	Th	1435	1388	1210	1058	928	1017	1030	911	807	717	638	570	509	456	409	368
91	Pa	1495	1300	1261	1103	967	851	934	947	839	745	663	592	529	474	425	382
92	U	1558	1354	1181	1148	1007	886	973	860	871	774	689	615	550	492	442	397

Table A.10 – *continued*.

Absorber		63 Eu	64 Gd	65 Tb	66 Dy	67 Ho	68 Er	69 Tm	70 Yb	71 Lu	72 Hf	73 Ta	74 W	75 Re	76 Os	77 Ir	78 Pt
30	Zn	58	52	47	42	38	262	237	215	195	177	161	146	133	121	111	101
31	Ga	64	57	52	46	42	38	34	232	211	191	174	158	144	131	120	109
32	Ge	70	63	57	51	46	42	38	34	31	206	187	170	155	141	129	118
33	As	77	69	62	56	50	46	41	37	34	31	28	183	167	152	139	127
34	Se	84	76	68	61	55	50	45	41	37	33	30	28	25	163	148	136
35	Br	92	82	74	67	60	54	49	44	40	36	33	30	27	25	23	145
36	Kr	100	90	81	72	65	59	53	48	44	40	36	33	30	27	25	22
37	Rb	108	97	87	79	71	64	58	52	47	43	39	35	32	29	27	24
38	Sr	117	105	94	85	77	69	62	57	51	46	42	38	35	32	29	26
39	Y	126	113	102	92	83	75	67	61	55	50	46	41	38	34	31	28
40	Zr	136	122	110	99	89	80	73	66	59	54	49	45	40	37	34	31
41	Nb	146	131	118	106	95	86	78	71	64	58	53	48	43	40	36	33
42	Mo	156	140	126	113	102	92	83	76	68	62	56	51	47	42	39	35
43	Tc	167	150	135	121	109	99	89	81	73	66	60	55	50	45	41	38
44	Ru	178	160	144	129	117	105	95	86	78	71	64	59	53	49	44	40
45	Rh	190	170	153	138	124	112	102	92	83	76	69	62	57	52	47	43
46	Pd	202	181	163	147	132	119	108	98	89	80	73	66	61	55	50	46
47	Ag	214	192	173	156	140	127	115	104	94	85	78	71	64	59	53	49
48	Cd	227	204	183	165	149	135	122	110	100	91	82	75	68	62	57	52
49	In	240	216	194	175	158	142	129	117	106	96	87	79	72	66	60	55

Z	El																
50	Sn	254	228	205	185	167	151	136	123	112	102	92	84	77	70	64	58
51	Sb	268	241	216	195	176	159	144	130	118	107	98	89	81	74	67	61
52	Te	282	254	228	206	186	168	152	138	125	113	103	94	85	78	71	65
53	I	297	267	240	217	196	177	160	145	132	119	109	99	90	82	75	68
54	Xe	313	281	253	228	206	186	168	153	138	126	114	104	95	86	79	72
55	Cs	328	295	266	239	216	196	177	160	146	132	120	109	100	91	83	76
56	Ba	344	310	279	251	227	205	186	168	153	139	126	115	105	96	87	80
57	La	361	325	292	264	238	215	195	177	160	146	133	121	110	100	92	84
58	Ce	378	340	306	276	249	226	204	185	168	153	139	127	115	105	96	88
59	Pr	396	356	320	289	261	236	214	194	176	160	146	132	121	110	101	92
60	Nd	413	372	335	302	273	247	224	203	184	167	152	139	126	115	105	96
61	Pm	432	388	350	316	285	258	234	212	192	175	159	145	132	121	110	101
62	Sm	450	405	365	329	298	269	244	221	201	183	166	151	138	126	115	105
63	Eu	469	422	381	343	310	281	254	231	210	191	173	158	144	131	120	110
64	Gd	423	440	396	358	323	293	265	241	219	199	181	165	150	137	125	115
65	Tb	317	396	413	372	337	305	276	251	228	207	188	172	157	143	131	119
66	Dy	329	296	371	387	350	317	287	261	237	215	196	179	163	149	136	124
67	Ho	122	308	278	349	364	330	299	271	246	224	204	186	170	155	142	129
68	Er	128	320	289	261	328	342	310	282	258	233	212	193	176	161	147	135
69	Tm	134	121	300	271	245	308	322	293	266	242	220	201	183	167	153	140
70	Yb	141	127	115	281	254	230	290	304	276	251	229	209	190	174	159	145
71	Lu	148	133	120	109	263	239	216	273	286	260	237	216	197	180	165	151
72	Hf	155	140	126	114	103	247	224	204	258	270	246	224	205	187	171	156
73	Ta	162	146	132	120	108	98	233	211	192	243	255	232	212	194	177	162
74	W	170	153	138	125	113	103	93	219	199	181	229	241	220	201	184	168

Table A.10 – *continued*.

Absorber		63 Eu	64 Gd	65 Tb	66 Dy	67 Ho	68 Er	69 Tm	70 Yb	71 Lu	72 Hf	73 Ta	74 W	75 Re	76 Os	77 Ir	78 Pt
75	Re	178	160	145	131	119	108	98	89	206	187	171	217	227	208	190	174
76	Os	186	168	151	137	124	112	102	93	213	194	177	161	205	215	197	180
77	Ir	194	175	158	143	130	118	107	97	88	201	183	167	152	194	203	186
78	Pt	203	183	165	149	135	123	112	101	92	84	189	172	157	144	183	192
79	Au	212	191	173	156	141	128	116	106	96	88	80	178	163	149	136	173
80	Hg	221	199	180	163	148	134	122	110	101	92	84	76	168	154	141	129
81	Tl	231	208	188	170	154	140	127	115	105	96	87	80	73	159	145	133
82	Pb	240	217	196	177	160	146	132	120	109	100	91	83	76	69	150	137
83	Bi	251	226	204	185	167	152	138	125	114	104	95	87	79	72	155	142
84	Po	261	235	213	192	174	158	144	131	119	108	99	90	82	75	69	146
85	At	272	245	222	200	182	165	149	136	124	113	103	94	86	78	72	66
86	Rn	283	255	231	209	189	171	156	142	129	117	107	98	89	82	75	68
87	Fr	295	266	240	217	197	178	162	147	134	122	111	102	93	85	78	71
88	Ra	306	276	250	226	205	186	168	153	139	127	116	106	97	88	81	74
89	Ac	319	287	260	235	213	193	175	159	145	132	120	110	101	92	84	77
90	Th	331	299	270	244	221	201	182	166	151	137	125	114	104	96	87	80
91	Pa	344	310	280	254	230	208	189	172	157	143	130	119	109	99	91	83
92	U	358	323	291	264	239	216	197	179	163	148	135	123	113	103	94	86

										Emitter					
		79 Au	80 Hg	81 Tl	82 Pb	83 Bi	84 Po	85 At	86 Rn	87 Fr	88 Ra	89 Ac	90 Th	91 Pa	92 U
Absorber															
30	Zn	92	84	77	71	65	59	54	50	46	42	39	35	33	30
31	Ga	100	91	84	76	70	64	59	54	50	45	42	38	35	32
32	Ge	108	98	90	82	75	69	63	58	53	49	45	41	38	35
33	As	116	106	97	89	81	74	68	63	57	53	48	44	41	38
34	Se	124	113	104	95	87	80	73	67	62	57	52	48	44	40
35	Br	132	121	111	101	93	85	78	72	66	60	56	51	47	43
36	Kr	141	129	118	108	99	91	83	76	70	64	59	54	50	46
37	Rb	22	20	125	115	105	97	89	81	75	69	63	58	53	49
38	Sr	24	22	20	18	112	102	94	86	79	73	67	61	56	52
39	Y	26	24	22	20	18	17	99	91	84	77	71	65	60	55
40	Zr	28	26	23	21	20	18	16	96	88	81	75	69	63	58
41	Nb	30	27	25	23	21	19	18	16	15	86	79	72	67	61
42	Mo	32	29	27	25	23	21	19	17	16	15	13	76	70	64
43	Tc	34	32	29	26	24	22	20	19	17	16	14	13	74	68
44	Ru	37	34	31	28	26	24	22	20	18	17	15	14	13	12
45	Rh	39	36	33	30	28	25	23	21	19	18	16	15	14	13
46	Pd	42	38	35	32	29	27	25	23	21	19	17	16	15	13
47	Ag	45	41	37	34	31	29	26	24	22	20	19	17	16	14
48	Cd	47	43	40	36	33	30	28	25	23	21	20	18	17	15
49	In	50	46	42	38	35	32	29	27	25	23	21	19	18	16

Table A.10 – *continued*.

Absorber	79 Au	80 Hg	81 Tl	82 Pb	83 Bi	84 Po	85 At	86 Rn	87 Fr	88 Ra	89 Ac	90 Th	91 Pa	92 U
50 Sn	53	49	44	41	37	34	31	29	26	24	22	20	19	17
51 Sb	56	51	47	43	39	36	33	30	28	26	23	22	20	18
52 Te	59	54	50	45	42	38	35	32	29	27	25	23	21	19
53 I	62	57	52	48	44	40	37	34	31	28	26	24	22	20
54 Xe	66	60	55	50	46	42	39	36	33	30	28	25	23	21
55 Cs	69	63	58	53	49	45	41	38	34	32	29	27	25	23
56 Ba	73	67	61	56	51	47	43	39	36	33	31	28	26	24
57 La	76	70	64	59	54	49	45	41	38	35	32	30	27	25
58 Ce	80	73	67	62	56	52	47	44	40	37	34	31	29	26
59 Pr	84	77	70	64	59	54	50	46	42	39	35	33	30	28
60 Nd	88	80	74	68	62	57	52	48	44	40	37	34	31	29
61 Pm	92	84	77	71	65	59	55	50	46	42	39	36	33	30
62 Sm	96	88	81	74	68	62	57	52	48	44	41	37	34	32
63 Eu	100	92	84	77	71	65	60	55	50	46	42	39	36	33
64 Gd	105	96	88	81	74	68	62	57	52	48	44	41	38	35
65 Tb	109	100	92	84	77	71	65	60	55	50	46	43	39	36
66 Dy	114	104	95	87	80	74	68	62	57	52	48	44	41	38
67 Ho	118	108	99	91	84	77	70	65	59	55	50	46	43	39
68 Er	123	113	103	95	87	80	73	67	62	57	52	48	44	41
69 Tm	128	117	107	99	90	83	76	70	64	59	54	50	46	42

Z	El														
70	Yb	133	122	112	102	94	86	79	73	67	62	57	52	48	44
71	Lu	138	126	116	106	98	90	82	76	70	64	59	54	50	46
72	Hf	143	131	120	110	101	93	85	78	72	66	61	56	52	48
73	Ta	148	136	125	114	105	96	89	81	75	69	63	58	54	49
74	W	154	141	129	119	109	100	92	84	78	71	66	60	56	51
75	Re	159	146	134	123	113	104	95	87	80	74	68	63	58	53
76	Os	165	151	138	127	117	107	98	91	83	77	71	65	60	55
77	Ir	170	156	143	131	121	111	102	94	86	79	73	67	62	57
78	Pt	176	161	148	136	125	115	105	97	89	82	76	70	64	59
79	Au	182	167	153	140	129	119	109	100	92	85	78	72	66	61
80	Hg	164	172	158	145	133	122	113	104	95	88	81	74	69	63
81	Tl	122	155	163	150	138	126	116	107	98	91	83	77	71	65
82	Pb	126	115	147	155	142	130	120	110	102	94	86	79	73	67
83	Bi	130	119	109	140	146	135	124	114	105	97	89	82	75	70
84	Po	134	123	113	103	132	139	128	117	108	100	92	84	78	72
85	At	138	127	116	107	98	125	132	121	111	103	95	87	80	74
86	Rn	63	130	120	110	101	93	119	125	115	106	97	90	83	76
87	Fr	65	60	123	113	104	96	88	113	118	109	100	93	85	79
88	Ra	68	62	57	117	107	99	91	84	107	112	103	95	88	81
89	Ac	71	65	59	55	110	102	93	86	79	102	106	98	90	83
90	Th	73	67	62	57	114	105	96	89	82	75	96	101	93	86
91	Pa	76	70	64	59	54	108	99	91	84	77	71	91	96	88
92	U	79	73	67	61	56	52	102	94	86	80	73	68	87	91

APPENDIX A

Table A.11 Absorption jump ratios and fluorescence yields for K fluorescence.

Z	Elem	At. Wt.	$(r_A-1)/r_A$	ω_K
8	O	15.999	0.957	0.008
9	F	18.998	0.950	0.012
10	Ne	20.183	0.944	0.016
11	Na	22.991	0.938	0.021
12	Mg	24.312	0.933	0.028
13	Al	26.982	0.928	0.037
14	Si	28.086	0.923	0.047
15	P	30.974	0.918	0.059
16	S	32.064	0.914	0.074
17	Cl	35.453	0.910	0.090
18	Ar	39.948	0.906	0.109
19	K	39.102	0.903	0.130
20	Ca	40.080	0.899	0.153
21	Sc	44.956	0.896	0.178
22	Ti	47.900	0.893	0.204
23	V	50.942	0.890	0.232
24	Cr	51.996	0.886	0.262
25	Mn	54.938	0.884	0.292
26	Fe	55.847	0.881	0.323
27	Co	58.933	0.878	0.355
28	Ni	58.710	0.875	0.387
29	Cu	63.540	0.872	0.419
30	Zn	65.370	0.870	0.451
31	Ga	69.720	0.867	0.483
32	Ge	72.590	0.864	0.515
33	As	74.922	0.862	0.547
34	Se	78.960	0.859	0.579
35	Br	79.909	0.857	0.611

The absorption jump ratios in this table and in Table A.12 were calculated from the polynomials of Tables A.1 and A.2. The fluorescence yields were calculated by fitting similar polynomials to the data of Reed (1975), after Burhop and Asaad (1972).

Table A.12 Absorption jump ratios and fluorescence yields for L fluorescence.

Z	Elem	At. Wt.	$(r_A-1)/r_A$	ω_L	Z	Elem	At. Wt.	$(r_A-1)/r_A$	ω_L
29	Cu	63.540	0.866	0.007	61	Pm	147.000	0.795	0.122
30	Zn	65.370	0.865	0.008	62	Sm	150.350	0.792	0.130
31	Ga	69.720	0.864	0.009	63	Eu	151.960	0.790	0.137
32	Ge	72.590	0.863	0.010	64	Gd	157.250	0.787	0.144
33	As	74.922	0.862	0.012	65	Tb	158.924	0.784	0.152
34	Se	78.960	0.861	0.013	66	Dy	162.500	0.781	0.160
35	Br	79.909	0.860	0.015	67	Ho	164.930	0.778	0.168
36	Kr	83.800	0.858	0.016	68	Er	167.260	0.775	0.177
37	Rb	85.470	0.857	0.018	69	Tm	168.934	0.772	0.185
38	Sr	87.620	0.855	0.020	70	Yb	173.040	0.769	0.194
39	Y	88.905	0.853	0.022	71	Lu	174.970	0.766	0.203
40	Zr	91.220	0.852	0.025	72	Hf	178.490	0.763	0.212
41	Nb	92.906	0.849	0.027	73	Ta	180.948	0.761	0.221
42	Mo	95.940	0.847	0.030	74	W	183.850	0.758	0.230
43	Tc	98.000	0.844	0.033	75	Re	186.200	0.755	0.239
44	Ru	101.070	0.841	0.036	76	Os	190.200	0.752	0.249
45	Rh	102.905	0.839	0.039	77	Ir	192.200	0.749	0.259
46	Pd	106.400	0.836	0.043	78	Pt	195.090	0.746	0.269
47	Ag	107.870	0.834	0.046	79	Au	196.967	0.743	0.279
48	Cd	112.400	0.831	0.050	80	Hg	200.590	0.740	0.289
49	In	114.820	0.828	0.054	81	Tl	204.370	0.737	0.299
50	Sn	118.690	0.826	0.059	82	Pb	207.190	0.734	0.310
51	Sb	121.750	0.823	0.064	83	Bi	208.980	0.731	0.320
52	Te	127.600	0.820	0.068	84	Po	210.000	0.728	0.331
53	I	126.904	0.817	0.073	85	At	210.000	0.725	0.342
54	Xe	131.300	0.815	0.079	86	Rn	222.000	0.722	0.353
55	Cs	132.905	0.812	0.084	87	Fr	223.000	0.719	0.364
56	Ba	137.340	0.809	0.090	88	Ra	226.000	0.716	0.375
57	La	138.910	0.806	0.096	89	Ac	227.000	0.713	0.387
58	Ce	140.120	0.804	0.102	90	Th	232.038	0.710	0.398
59	Pr	140.907	0.801	0.109	91	Pa	231.000	0.706	0.410
60	Nd	144.240	0.798	0.116	92	U	238.030	0.703	0.422

Table A.13 Norrish and Hutton coefficients for silicate analysis, normalized to G1 : W1 (from Norrish & Chappell, 1977).

Line (Kα)	X	Y										
		Fe_2O_3	MnO	TiO_2	CaO	K_2O	SO_3	P_2O_5	SiO_2	Al_2O_3	MgO	Na_2O	Loss
Fe	1.046	−0.027	−0.031	0.146	0.134	0.126	−0.060	−0.060	−0.065	−0.074	−0.090	−0.110	−0.163
Mn	1.045	−0.044	−0.044	0.146	0.135	0.130	−0.037	−0.063	−0.063	−0.074	−0.078	−0.100	−0.163
Ti	0.851	0.081	0.077	0.179	0.647	0.644	0.194	0.181	0.110	0.078	0.069	0.051	−0.132
Ca	0.865	0.090	0.092	0.065	0.110	0.723	0.201	0.182	0.128	0.105	0.068	0.051	−0.134
K	0.897	0.098	0.086	0.017	0.000	0.069	0.182	0.179	0.119	0.101	0.080	0.057	−0.139
S	0.894	0.086	0.074	0.002	−0.023	−0.037	−0.053	0.167	0.131	0.112	0.087	0.063	−0.139
P	0.896	0.108	0.094	−0.020	−0.037	−0.047	−0.059	−0.063	0.127	0.110	0.094	0.046	−0.139
Si	1.014	0.082	0.086	−0.034	−0.042	−0.055	−0.057	−0.061	−0.061	0.122	0.093	0.063	−0.158
Al	1.056	0.112	0.116	−0.032	−0.037	−0.048	−0.056	−0.060	−0.088	−0.072	0.116	0.058	−0.164
Mg	1.050	0.136	0.126	0.010	−0.021	−0.043	−0.046	−0.016	−0.070	−0.078	−0.084	0.080	−0.163
Na	1.032	0.158	0.142	0.041	0.005	−0.016	−0.034	−0.044	−0.053	−0.066	−0.081	−0.103	−0.161

These coefficients are for the weight per cent oxide, not the element. They apply only when samples are prepared according to the Norrish and Hutton 'recipe' described in the text. They are normalized in each case relative to a value of 1,000 for a hypothetical sample consisting of equal (weight) parts of the standard rocks G1 and W1. Similar coefficients can be calculated for other sample preparation procedures, and normalized to other standards (see Table A.14).

Table A.14 Norrish and Hutton coefficients for silicate analysis, normalized to fusion mix (courtesy of Dr Michael Hough).

Line (Kα)	X	Y										
		Fe_2O_3	MnO	TiO_2	CaO	K_2O	SO_3	P_2O_5	SiO_2	Al_2O_3	MgO	Na_2O	Loss
Fe	1.000	−0.026	−0.030	0.140	0.128	0.120	−0.057	−0.057	−0.062	−0.071	−0.086	−0.105	−0.156
Mn	1.000	−0.042	−0.042	0.140	0.129	0.124	−0.035	−0.060	−0.060	−0.071	−0.075	−0.096	−0.156
Ti	1.000	0.095	0.090	0.210	0.760	0.751	0.228	0.213	0.129	0.092	0.081	0.060	−0.155
Ca	1.000	0.104	0.106	0.075	0.127	0.836	0.232	0.210	0.148	0.121	0.079	0.059	−0.155
K	1.000	0.109	0.096	0.019	0.000	0.077	0.203	0.200	0.133	0.113	0.089	0.064	−0.155
S	1.000	0.096	0.083	0.002	−0.026	−0.041	−0.059	0.187	0.147	0.125	0.097	0.070	−0.155
P	1.000	0.121	0.105	−0.022	−0.041	−0.052	−0.066	−0.070	0.142	0.123	0.105	0.051	−0.155
Si	1.000	0.081	0.085	−0.034	−0.041	−0.054	−0.056	−0.060	−0.060	0.120	0.092	0.062	−0.156
Al	1.000	0.106	0.110	−0.030	−0.035	−0.045	−0.053	−0.057	−0.083	−0.068	0.110	0.055	−0.155
Mg	1.000	0.130	0.120	0.010	−0.020	−0.041	−0.067	−0.074	−0.015	−0.044	−0.080	0.076	−0.155
Na	1.000	0.153	0.138	0.040	0.005	−0.016	−0.033	−0.043	−0.051	−0.064	−0.078	−0.100	−0.156

These coefficients are for the weight per cent oxide, not the element. They apply only when samples are prepared according to the Norrish and Hutton 'recipe' described in the text. They are normalized in each case relative to a value of 1,000 for a fused disc containing only the 'recipe' fusion mix and no silicate sample.

Table A.15 Sample Pascal program for calculation of mass absorption coefficients.

```
program MacMAC (input, output);

{program to compute mass absorption coefficient tables for KA, LA and MA emissions, K
    and LI edges}

uses
    SANE;   {MacPascal library, contains xpwri and xpwry functions}

type
    tfile = file of text;
    elsym = array[1..92] of string[2];
    coeff = array[1..10] of double;
    points = array[1..10] of integer;
    heads = array[1..5] of string[70];
    counts = array[1..5] of integer;
    coffs = array[1..16, 1..6] of double;
    drec = record
        p : coeff;
    end;
    symrec = record
        sym : elsym;
    end;
    xafile = file of symrec;
    xfile = file of drec;
```

var

```
Za, Ze, i, j, b, max, dest, mac2, records : integer;    {Za and Ze are atomic numbers
                                                         of absorber and emitter}
min, maxz, emit, lowa, higha : counts;
ptr : points;
lz, K : double;
edge : coeff;
cf, cfmac : coffs;
mac, d : extended;
fname, head, cofile : string;
an : elsym;
txfile : tfile;
xname : xfile;
xaname : xafile;


procedure textwindow;             {sets up Macintosh window for text display}

var
    twindow : rect;

begin
    twindow.top := 30;
    twindow.bottom := 330;
    twindow.left := 10;
    twindow.right := 500;
    settextrect(twindow);
    showtext;
end;   {procedure textwindow}
```

Table A.15 – *continued.*

```
procedure getsymbols (var pan : elsym);

var
   symbols : symrec;
   ppi : integer;

begin
   open(xaname, 'PascfilesB:Symfile');
   with symbols do
      begin
         seek(xaname, 0);
         read(xaname, symbols);
         for ppi := 1 to 92 do
            pan[ppi] := sym[ppi];
      end;
   close(xaname);
end;   {procedure getsymbols}


procedure setpointers (var pptr : points;         {establish pointers for table access
                      and compute ranges}
                      var pemit, pmin, pmaxz, plowa, phigha : counts);
                      {pointers to the absorption edge coefficients in the regression
                      coefficient table}
begin
   pptr[1] := 3;
   pptr[2] := 6;
   pptr[3] := 7;
```

```
pptr[4]  := 8;
pptr[5]  := 11;
pptr[6]  := 12;
pptr[7]  := 13;
pptr[8]  := 14;
pptr[9]  := 15;
pptr[10] := 16;
pemit[1] := 1;   {pointers to KA, LA, MA emission and K, LI edge coefficients in
                  the same table}
pemit[2] := 4;
pemit[3] := 9;
pemit[4] := 3;
pemit[5] := 6;
pmin[1]  := 11;  {minimum Ze to compute for each of the five output tables}
pmin[2]  := 31;
pmin[3]  := 56;  {lower values are below the regression range and subject to
                  extrapolation error}
pmin[4]  := 11;
pmin[5]  := 31;
pmaxz[1] := 50;  {maximum Ze to compute for each of the five output tables}
pmaxz[2] := 92;
pmaxz[3] := 92;  {higher values are beyond the regression range and subject to
                  extrapolation error}
pmaxz[4] := 50;
pmaxz[5] := 92;
plowa[1] := 8;   {minimum Za to compute for each of the five tables}
plowa[2] := 8;
plowa[3] := 8;
plowa[4] := 8;
```

Table A.15 – *continued.*

```
prowa[5] := 50;
phigha[1] := 92;             {maximum Za to compute for each of the five tables}
phigha[2] := 92;
phigha[3] := 92;
phigha[4] := 42;
phigha[5] := 92;
end;   {procedure setpointers}


procedure setup (var pdest, pb : integer;
                 var phead, pfname : string);
                                     {select output table, output device, table
    heading}
var
    i : integer;
    pheading : heads;
begin
    writeln;
    writeln('MacMAC - program to compute tables of mass absorption coefficients');
    writeln("                               K.L.Williams, October, 1985');
    writeln;
    writeln;
    write('Specify emitter series - K (1), L (2), M (3), K edge (4), or LI edge (5) :
         ');
    readln(pb);
    writeln;
```

```
write('Send output to screen (1), printer (2), or disk text file (3) : ');
readln(pdest);
if pdest = 2 then
  begin
    close(output);
    rewrite(output, 'Printer:');
    write(chr(27), 'c', chr(27), 'p');              {initialise Imagewriter printer}
    write(chr(29), 'A@');                            {instruct Imagewriter to skip
                                                     continuous-feed perforations}
    for i := 1 to 58 do
      write('@@');
    write('C@@@@@@@@@@@@@@@@', chr(30));
    write(chr(27), chr(81), chr(27), chr(33));       {select condensed print,
                                                     bold face}
  end;     {pdest = 2}
if (pdest = 3) then      {output can be sent to a disk text file for editing before
                          printing}
  begin
    write('Please specify a name for the text file (Volume:Filename) : ');
    readln(pfname);
    close(output);
    open(output, pfname);
  end;    {pdest = 3}

{specify heading to be printed for the selected table}

pheading[1] := 'Table A.6 :      Mass absorption coefficients for KA emissions';
pheading[2] := 'Table A.7 :      Mass absorption coefficients for LA emissions';
pheading[3] := 'Table A.8 :      Mass absorption coefficients for MA emissions';
```

Table A.15 – *continued*.

```
    pheading[4] := 'Table A.9 :       Mass absorption coefficients for high side of K
      edge';
    pheading[5] := 'Table A.10 :      Mass absorption coefficients for high side of LI
      edge';
    phead := pheading[b];
  end;          {procedure setup}

  procedure get_coefficients (cfname : string;
                  var pn : integer;
                  var cof : coffs);

  var
    xrfile : xfile;
    onerec : drec;
    i, j : integer;

  begin
    open(xrfile, cfname);
    for i := 1 to pn do
      with onerec do
        begin
          seek(xrfile, i - 1);
          read(xrfile, onerec);
          for j := 1 to 6 do
            cof[i, j] := p[j];
        end;    {file read}
    close(xrfile);
  end;          {procedure get_coefficients}
```

```
procedure print_page_head (phead : string;
                          pmin, pmax : integer;
                          pan : elsym);
var
   i : integer;

begin
   writeln(phead);
   writeln;
   writeln;
   writeln('                                            Emitter');
   writeln;
   write('     ');
   for i := pmin to pmax do
      write(i : 6);
   writeln;
   write('     ');
   for i := pmin to pmax do
      write(' ', pan[i], ' ');
   writeln;
   writeln('Absorber');
   writeln;
end;    {procedure print_page_head}

{main program begins here}
```

Table A.15 – *continued*.
```
begin

  textwindow;                              {set up text display window}
  getsymbols(an);                          {get chemical symbols for 92 elements}
  setpointers(ptr, emit, min, maxz, lowa, higha);   {establish element ranges to
      compute}
  setup(dest, b, head, fname);             {specify table to compute, output device, table
      heading}
  cofile := 'PascfilesB:xrcoeffs';
  records := 16;
  get_coefficients(cofile, records, cf);   {get the polynomial coefficients for
      emissions and edges}
  cofile := 'PascfilesB:norrish';
  records := 10;
  get_coefficients(cofile, records, cfmac);  {get the polynomial coefficients for mass
      absorption coeffs}
  max := min[b] - 1;
  while (max < maxz[b]) do       {determine the range of emitters for current page of output,
      21 cols./page}
  begin
    min[b] := max + 1;
    max := min[b] + 20;
    if max > maxz[b] then
      max := maxz[b];
    print_page_head(head, min[b], max, an);

                {preliminaries completed - now start computation for absorber rows and element
      columns}
```

```
for Za := lowa[b] to higha[b] do
begin

  write(Za : 2, ' ', an[Za]);

  {first calculate the wavelengths of the 10 absorption edges K through NI}

  for i := 1 to 10 do
  begin
    lz := ln(Za);
    edge[i] := 0.0;
    for j := 1 to 6 do
      edge[i] := edge[i] + cf[ptr[i], j] * xpwri(lz, j - 1);
    edge[i] := exp(edge[i]);
  end;

  {now calculate the wavelength of the current emitter (i.e. atomic number = Ze)
  and see}
  {where it fits among the 10 absorption edges - when the fit is located,
  calculate the m.a.c.}

  for Ze := min[b] to max do
  begin
    lz := ln(Ze);
    K := 0.0;
    for j := 1 to 6 do
      K := K + cf[emit[b], j] * xpwri(lz, j - 1);
    K := exp(K);
```

Table A.15 – *continued*.

{Search through the calculated absorber edges to find the appropriate bounding pair}
{first - is the emitter wavelength short of the K edge or beyond it? If the latter, }
{assume for the moment that it is between the K and LI edges }

if (K <= edge[1]) **then**
  j := 1
else
  j := 2;

{Now check the L edges - but only if appropriate. For L edges greater than about 25 A, }
{extrapolation of the regressions is too inaccurate. 15 A would be safer, but 25 A }
{includes oxygen KA. }

if Za > 22 **then**
begin
  for i := 2 **to** 3 **do**
  begin
    if (K > edge[i]) **and** (K <= edge[i + 1]) **then**
      j := i + 1;
  end;
  if (K > edge[4]) **then**
    j := 5;
end;

{Now check the M edges, if appropriate - less than about 15 A. M extrapolations are }
{particularly risky. }

if Za > 50 **then**
begin
　for i := 5 **to** 9 **do**
　begin
　　if (K > edge[i]) **and** (K <= edge[i + 1]) **then**
　　　j := i + 1;
　end;
　if (K > edge[10]) **then**
　　j := 11;
end;　　{Za > 50}

{Now calculate the mass absorption coefficient - providing the emission wavelength is }
{short of the NI edge. Extrapolation to regions beyond the NI edge is too inaccurate - }
{set any m.a.c.'s in this region to zero as a warning. }

mac := 0.0;
if (j < 11) **then**
begin
　if (j > 1) **and** (j < 5) **and** (Za > 40) **then**
　　d := cfmac[j, 6]
　else

Table A.15 – *continued*.

```
        d := 0.0;
        for i := 1 to 3 do
            mac := mac + cfmac[j, i] * xpwri(Za, (i - 1));
        mac := xpwri(mac, 3);
        mac := mac * xpwry(K, (cfmac[j, 4] - cfmac[j, 5] * sqrt(Za) - d * (Za -
            40)));
        end;

        {round to nearest integer - optimistic precision!}

        mac2 := round(mac);

        {output the result}

        write(mac2 : 6);
      end;        {Ze := min[b] to max }
        writeln;  {advance output to next absorber row }
      end;        {Za := lowa[b] to higha[b] }
      if (dest = 2) then
        writeln(chr(12));    {for printer output, advance next page }
      end;        {while max < maxz[b] }

    close(output);
    rewrite(output);    {redirect output to screen (default) }
    end.        {main program segment }
```

{This program obtains data from three files stored on a disk with the volume name
'PascfilesB'. }
{These files are:

1. 'xrcoeffs': contains the polynomial coefficients, calculated by regression,
   of the   K, L and M alpha and beta emissions and the K, L, M and
   NI   absorption edges- see Table A.1. These are stored as 16
   records,   each with up to 6 coefficients per parameter; the array
   is large   enough to include 10, for higher-order polynomials.}

2. 'norrish' : contains the coefficients for K. Norrish's polynomials for
   calculation   of mass absorption coefficients as functions of Z and
   wavelength -   see Table A.2 These are stored as 10 records, with room
   for up to   10 coefficients. }

3. 'Symfile' : contains the chemical symbols for the 92 elements, as
   two-character   strings in a 1-dimensional array, stored as a single
   record }

APPENDIX B

Some fundamental statistical concepts

True result: The correct result. In X-ray analysis, the true concentration is the actual concentration in the region analysed. It may never be known absolutely, but in certain cases it is possible to synthesize materials whose composition may be presumed to be known within specified levels of homogeneity, and there is also a range of much-analysed samples about whose compositions there is a reasonable measure of agreement, and which therefore may be taken as having accepted reference levels of concentration that can be used in lieu of true concentrations.

Accuracy: The nearness of a result, or the mean of a group of results, to the true result. As has already been noted, a systematic difference between the experimental and the true results is known as the bias, or the systematic error.

Precision: The closeness of agreement between multiple analyses obtained under the same experimental conditions. Precision can always be evaluated simply by comparing the measured results, but accuracy cannot unless the true result is known (which is seldom the case). It is important to realize that good precision does not necessarily imply good accuracy, as is often assumed – closeness of agreement between duplicate determinations does not eliminate the possibility of systematic error (Fig. B.1).

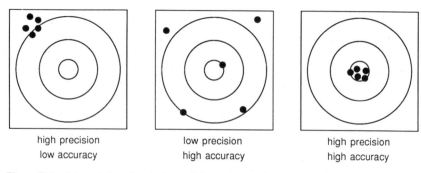

Figure B.1 A target-shooting analogy of the concepts of *accuracy* and *precision*. Clearly, a good analysis must be both accurate and precise.

SOME FUNDAMENTAL STATISTICAL CONCEPTS

Significant figures: The number of significant figures, or significant digits, used in reporting the numerical value of a quantity is an implicit statement of the precision of that value. Thus to quote a quantity as '32.37 per cent', without any other specific indication of error limits, implies that the true result is closer to 32.37 per cent than it is to either 32.36 or 32.38. In other words, the *absolute uncertainty* of the quoted result may be assumed to be 0.01, and the *relative uncertainty* to be $100 \times (0.01 / 32.37)$ or 0.03 per cent.

It follows that there is a significant difference in implicit uncertainty in a quantity variously reported as 1, 1.0 or 1.00 (i.e. with one, two or three significant digits), and analysts should always be careful to report their results with these implications in mind. To eliminate uncertainty, some routinely append the first non-significant digit as a subscript (e.g. '4.1_3'), but this should not be necessary – and can conceivably lead to some confusion, since number bases are reported the same way.

In determining the level of precision to be assigned to a reported value, the magnitude of all possible sources of error must be carefully evaluated. Some of the more serious sources of potential error in XRS analysis are discussed in Chapter 7 and in this Appendix; brief mention should also be made of another aspect which is often overlooked. Whenever a quantity is calculated from two or more other quantities, two fundamental principles should be kept in mind:

(a) In *addition* and *subtraction*, the *absolute uncertainty* in the result must be equal to the largest absolute uncertainty in the components of the calculation. Thus,

$$\begin{array}{ccc} 51.36 & & 51.36 \\ +0.40 & \text{but} & +0.4 \\ \hline 51.76 & & 51.8 \end{array}$$

(b) In *multiplication* and *division*, the *relative uncertainty* in the result must be equal to the largest relative uncertainty in the components.

Thus a common hand calculator will usually show the result of 0.14×9.86 as 1.3804. Few analysts would report such a result with the 'synthetic precision' implied by five significant digits. However, many (if not most) would write it as 1.38 (or possibly even as 1.380 or 1.38_0). The correct product is 1.4, since the relative uncertainty implicit in both the component 0.14 and the result 1.4 is 1/14, or approximately 7 per cent; '1.38' implies a relative uncertainty of 1/138 (or 0.7 per cent), which is false.

Note, however, that in making a sequential calculation it is perfectly proper – indeed mandatory – to carry as many significant digits as are

APPENDIX B

necessary to minimize rounding or truncation errors. 'Synthetic precision' only results if the calculated quantity is *reported* with an unwarranted implicit level of significance.

Average (mean, arithmetic mean, $\bar{x}$): The arithmetic sum of a group of results, divided by the number of results (n):

$$\bar{x} = (\Sigma x_i) / n \qquad (B.1)$$

Median: The middle value of a series of results placed in ascending order, such that there are as many values below the median as above it. If there is an even number of results, the median is taken as the arithmetic mean of the middle pair of results.

Mode: The most frequently occurring value in a series of results.

Frequency distribution: A mathematical description of the way in which the individual members of a set of results are distributed among an arbitrarily defined set of classes or class intervals. Frequency distributions may be described graphically (histograms, frequency curves) or by the mathematical equations which define their frequency curves. There are many common types of frequency distribution, of which two are of particular interest to X-ray spectrometry, viz. the *Poisson* and the *Gaussian* (or *normal*) distributions.

Poisson distribution: The distribution of a set of random values such that:

$$y = \frac{\mu^x}{x!} \times e^{-\mu} \qquad (B.2)$$

where y is the number of events measured in each class interval x, and μ is both the *variance* (see below) and the *population mean*. This distribution has frequently been found to describe the behaviour of relatively rare events; in fact it has been called the 'law of improbable events', applicable to such diverse phenomena as the number of Prussian cavalrymen kicked to death each year by their horses, the number of V-bombs falling on specified areas of London during World War II, the number of defects encountered in certain manufactured products, the number of annual vacancies on the US Supreme Court – and the generation of X-rays. In the specific case of X-ray generation, Equation B.2 may be rewritten as:

$$W(N) = \frac{(N_0)^N}{N!} \times e^{-N_0} \qquad (B.3)$$

SOME FUNDAMENTAL STATISTICAL CONCEPTS

which describes the probability of obtaining a particular random count N in a series of repeated measurements which are distributed around a true value N_0 according to the Poisson distribution.

Relevance of the Poisson distribution depends on several basic assumptions:

(a) the events measured must occur independently (i.e. generation of a particular photon does not depend on the generation of previous or subsequent photons);
(b) the probability of an event occurring is proportional to the length of time since the last event;
(c) the intervals between successive events are, on the average, much longer than the events themselves; and
(d) two events cannot be observed simultaneously.

Gaussian (normal) distribution: For high values of N (in practice, for $N = 50$ counts or more) the Poisson distribution closely approximates a Gaussian or 'normal' distribution (which does not imply that other distributions are in any way abnormal), for which:

$$W(N) = \frac{1}{\sqrt{(2\pi N)}} \times \exp\{-[(N - N_0)^2 / 2N]\} \qquad (B.4)$$

The mathematical definitions of these distributions are not important in the present context. What are important are certain basic properties of the Gaussian and Poisson distributions, on which statistical evaluations of X-ray intensity data are based. In particular, it is important to realize that the form of a Gaussian distribution is completely defined by only two parameters: the mean (μ), defined above, and the standard deviation (σ), defined below. A Poisson distribution is defined by *either* its mean *or* its standard deviation; it has the special property that its standard deviation is equal to the square root of its mean.

Standard deviation: Any distribution, normal or otherwise, has a *standard deviation*, which is defined as the root mean square deviation of a set of observations from their mean:

$$\sigma = \sqrt{\left\{\frac{\Sigma[(x - \mu)^2]}{n}\right\}} \qquad (B.5)$$

The standard deviation (σ) thus defined refers to deviations about the mean (μ) of an infinite number of observations. The symbol S is used for the more practical case of a finite number of observations with a mean of $\bar{x}$:

APPENDIX B

$$S = \sqrt{\left\{\frac{\Sigma\left[(x - \bar{x})^2\right]}{(n - 1)}\right\}} \quad \text{(B.6)}$$

The divisor $(n - 1)$ in Equation B.6 reflects the loss of one degree of freedom in a finite number of observations; specification of the mean and of the first $(n - 1)$ observations of x uniquely determines the nth value of x.

The difference between σ and S may or may not be of practical significance, depending primarily on the value of n.

Variance: The square of the standard deviation; i.e.

$$V = \sigma^2 \quad \text{or} \quad S^2 \quad \text{(B.7)}$$

Coefficient of variation: An expression of the 'spread' of observations about the mean, defined as the ratio of the standard deviation to the mean (usually expressed as a percentage):

$$\varepsilon = (S / \bar{x}) \times 100 \text{ (per cent)} \quad \text{(B.8)}$$

The coefficient of variation is sometimes called the *relative standard deviation*, and it may be expressed as a fraction or a percentage.

Standard error of the mean: If random samples of size n are taken from a population with a mean of μ and a standard deviation of σ, then a fundamental theorem of statistics states that the distribution of the sample means will have the same arithmetic mean as the population (i.e. μ), and its standard deviation will be $\sigma / \sqrt{n}$. The latter is called the *standard error of the mean* – a fundamental measure in statistical inference. It is common practice when dealing with randomly distributed data to determine the mean $\bar{x}$ and to write it in the form $\bar{x} \pm \alpha$, where α is the standard error of the mean. σ (or S) is the standard deviation of the various estimates of μ (or $\bar{x}$), and n is the number of samples used to determine each μ (or $\bar{x}$).

Properties of Poisson and Gaussian distributions: Essential properties of these distributions, of significance to X-ray counting statistics, include the following:

(a) The variance of all values of N is equal to the true value N_0, which approximates to N for all except very low values of N_0 (in general N is known and must be used as an estimator in place of N_0, which is not; the approximation suffices in most practical situations). It follows that the standard deviation $S = \sqrt{N_0}$. This is a property of

SOME FUNDAMENTAL STATISTICAL CONCEPTS

the Poisson distribution, whose equation (B.2) can be written in terms of only one variable, namely the mean µ. The Poisson distribution is a special case of the binomial distribution, from which the relationship $\mu = \sigma^2$ can be derived.

(b) There is a 68.3 per cent probability that a random value will lie within the range of ± one standard deviation (±1σ) about the mean. In the case of standard deviations (S) about the mean of a finite number of observations, the probability is slightly different, but in X-ray counting statistics the difference is negligibly small except at very low count rates.

(c) There is a 95.4 per cent probability that a random value will lie within the range of ± two standard deviations (± 2σ) about the mean.

(d) There is a 99.7 per cent probability that a random value will lie within the range of ± three standard deviations (± 3σ) about the mean.

(e) There is a 99.9 per cent probability that a random value will lie within the range of ± four standard deviations (± 4σ) about the mean.

(b)–(e) are properties of the Gaussian distribution.

The upper and lower limits of the range specified by a certain probability are known as the *confidence limits* for that probability, e.g. 99.7 per cent confidence limits are equivalent to the range ($\bar{x} - 3\sigma$) to ($\bar{x} + 3\sigma$).

Since the relative standard deviation (coefficient of variation) is equal to the standard deviation divided by the mean, it follows that for a finite number of observations

$$\varepsilon = \frac{S}{N_0} \times 100 \text{ per cent}$$

$$= \frac{\sqrt{N_0}}{N_0} \times 100 \text{ per cent}$$

$$= \frac{100}{\sqrt{N_0}} \text{ per cent}$$

$$\approx \frac{100}{\sqrt{N}} \text{ per cent} \qquad (B.9)$$

(since $N \approx N_0$).

APPENDIX B

Estimation of counting errors

Error of a single measurement

If N counts are measured in a single measurement, the standard deviation will be $\sqrt{N_0}$ (a property of the Poisson distribution). If N_0 is sufficiently large, N will be approximately equal to N_0, so that

$$S = \sqrt{N} \tag{B.10}$$

There is a 68 per cent probability that N lies within $\pm S$ of N_0, or conversely that N_0 lies within $\pm S$ of N. Hence if 10 000 counts are measured there is a 68 per cent probability that the true count lies in the range of 10 000 ± 100 counts, a 95 per cent probability that it lies within the range 10 000 ± 200 counts, and so on. Note that it is the *total number of counts* that is significant, not the count rate; in the example quoted it makes no (statistical) difference whether the 10 000 counts were collected in one second, or 1000 cps for 10 seconds – or even 1 cps for 10 000 seconds. Of course, if count times are very long then other possible sources of random error must also be considered.

Propagation of statistical errors .

Suppose that an experimental result w is determined as a function of several independent variables $x, y, z, \ldots$, with each of the latter having an associated statistical error expressed as its variance $V_x, V_y, V_z, \ldots$ (Eqn B.7). That is,

$$w = f(x, y, z \ldots) \tag{B.11}$$

Then,

$$V_x = S_x^2;\ V_y = S_y^2;\ V_z = S_z^2 \ldots \text{etc.}$$

For any given set of observations of $x, y, z \ldots$, each will contribute a component of error to the estimation of w. However, the variance of w (V_w) will in the general case differ from the simple sum of the variances of $x, y, z \ldots$, because of the probability that a random high estimate of x will be wholly or partly compensated by a low estimate of y, and so on.

A useful approximate estimate of V_w is given by

$$V_w = \left(\frac{\partial w}{\partial x}\right)^2 \cdot V_x + \left(\frac{\partial w}{\partial y}\right)^2 \cdot V_y + \left(\frac{\partial w}{\partial z}\right)^2 \cdot V_z + \ldots \tag{B.12}$$

SOME FUNDAMENTAL STATISTICAL CONCEPTS

This equation applies equally well to all arithmetic manipulations of x, y, z ... used to determine w (addition, subtraction, multiplication, division etc.), although the solutions of the partial differentials will obviously vary according to the actual manipulations involved in a particular case. Three of the simplest but most important possibilities are as follows:

ERROR OF A DIFFERENCE

Consider a simple example, in which

$$w = ax + by$$

in which $a = 1$ and $b = -1$ (i.e. $w = x - y$).
In this case, Equation B.12 takes the form

$$\begin{aligned} V_w &= a^2 \cdot V_x + b^2 \cdot V_y \\ &= 1^2 \cdot V_x + (-1)^2 \cdot V_y \\ &= V_x + V_y \end{aligned}$$

Since $V = \sigma^2$,

$$\sigma_w^2 = \sigma_x^2 + \sigma_y^2$$
$$\sigma_w = \sqrt{(\sigma_x^2 + \sigma_y^2)} \tag{B.13}$$

which is a statement of the simplest form of the so-called *rule of variance*. In this case, the relative standard deviation (ε_w) is given by

$$\begin{aligned} \varepsilon_w &= \frac{\sigma_w}{w} \\ &= \frac{\sqrt{(\sigma_x^2 + \sigma_y^2)}}{(ax - by)} \\ &= \frac{\sqrt{[(x\,\varepsilon_x)^2 - (y\,\varepsilon_y)^2]}}{(ax - by)} \end{aligned} \tag{B.14}$$

(since $\varepsilon_x = \sigma_x / x$; Eqn B.8).

For example, a net count rate C_{net} may be estimated as the difference between a peak count rate C_p and a background count rate C_b:

$$C_{net} = C_p - C_b$$

In this case, $a = 1$ and $b = -1$. From Equation B.13,

APPENDIX B

$$\sigma_{net} = \sqrt{(\sigma_{C_p}^2 + \sigma_{C_b}^2)}$$

If N_p and N_b are the numbers of quanta detected in t_p and t_b seconds, on peak and background respectively, then $C_p = N_p / t_p$ and $C_b = N_b / t_b$. Provided that N_p and N_b are sufficiently large, there will be no practical difference between σ_p and S_p, or between σ_b and S_b. Hence it follows that

$$\sigma_p = \sqrt{N_p} = \sqrt{(C_p\, t_p)}$$

and

$$\varepsilon_p = \sigma_p / N_p = \sqrt{N_p} / N_p = 1 / \sqrt{N_p} = 1 / \sqrt{(C_p \times t_p)}$$

Similarly,

$$\varepsilon_b = 1 / \sqrt{(C_b \times t_b)}$$

Substituting in Equation B.14,

$$\varepsilon_{net} = \frac{\sqrt{\left[\left(\frac{C_p}{\sqrt{(C_p\, t_p)}}\right)^2 + \left(\frac{C_b}{\sqrt{(C_b\, t_b)}}\right)^2\right]}}{C_p - C_b}$$

$$= \frac{\sqrt{\left[\left(\frac{C_p}{t_p}\right) + \left(\frac{C_b}{t_b}\right)\right]}}{C_p - C_b} \tag{B.15}$$

ERROR OF A RATIO

Suppose that a parameter w is given by

$$w = x / y$$

From Equation B.12,

$$V_w = \left(\frac{\partial w}{\partial x}\right)^2 V_x + \left(\frac{\partial w}{\partial y}\right)^2 V_y$$

$$= \left(\frac{1}{y}\right)^2 V_x + \left(\frac{-x}{y^2}\right)^2 V_y$$

$$= \frac{x^2}{y^2}\left(\frac{V_x}{x^2} + \frac{V_y}{y^2}\right)$$

so that

SOME FUNDAMENTAL STATISTICAL CONCEPTS

$$\sigma_w = \frac{x}{y} \sqrt{\left[\left(\frac{\sigma_x}{x}\right)^2 + \left(\frac{\sigma_y}{y}\right)^2\right]}$$

and

$$\varepsilon_w = \frac{\sigma_w}{(x/y)}$$

$$= \sqrt{\left[\left(\frac{\sigma_x}{x}\right)^2 + \left(\frac{\sigma_y}{y}\right)^2\right]}$$

$$= \sqrt{(\varepsilon_x^2 + \varepsilon_y^2)}$$

If $x = C_{sam}$ and $y = C_{std}$,

$$\sigma_{sam} = \sqrt{N_{sam}} = \sqrt{(C_{sam} t_{sam})}$$
$$\varepsilon_{sam} = \sqrt{N_{sam}}/N = 1/\sqrt{(C_{sam} t_{sam})}$$
$$\varepsilon_{std} = 1/\sqrt{(C_{std} t_{std})}$$
$$\varepsilon_{ratio} = \sqrt{(\varepsilon_{sam}^2 + \varepsilon_{std}^2)}$$

$$= \sqrt{\left(\frac{1}{C_{sam} t_{sam}} + \frac{1}{C_{std} t_{std}}\right)} \qquad (B.16)$$

ERROR OF A RATIO OF DIFFERENCES

The above reasoning can readily be extended to other cases encountered in X-ray spectrometric analysis. For example, intensity ratios are often calculated from net count rates (peak − background) measured on a sample and a standard, i.e.

$$\text{intensity ratio} = (C_{p(sam)} - C_{b(sam)}) / (C_{p(std)} - C_{b(std)})$$

The relative standard deviation of the intensity ratio may be calculated by first determining the relative standard deviations of the two net count rates (sample and standard), using Equation B.15. Then, from the derivation of Equation B.16,

$$\varepsilon_{ratio} = \sqrt{(\varepsilon_{sam}^2 + \varepsilon_{std}^2)}$$

For more complex cases, appropriate solutions may be sought for Equation B.12. However, this is seldom necessary, since it is usually possible to estimate the variances of independent components of the final estimate and combine them according to the principles outlined above.

Bibliography

Beaman, D. R. and J. A. Isasi 1972. Electron beam microanalysis. *American Society for Testing and Materials, Special Technical Publication* **506**.

Bence, A. E. and A. L. Albee 1968. Empirical correction factors for the electron microanalysis of silicates and oxides. *Journal of Geology* **76**, 382.

Birks, L. S. 1963. *Electron probe microanalysis*. New York: Interscience.

Birks, L. S. 1969. *X-ray spectrochemical analysis*, 2nd edn. New York: Wiley.

Bishop, H. E. 1974. Prospects for an improved absorption correction in electron probe microanalysis. *Journal of Physics* **D7**, 2009.

Burhop, E. H. S. and W. N. Asaad 1972. The Auger effect. *Advances in Atomic and Molecular Physics* **8**, 163.

Castaing, R. 1951. *Application des sondes électroniques a une méthode d'analyse ponctuelle chimique et cristallographique*. Thesis, University of Paris.

Castaing, R. and R. Guinier 1949. Application des sondes électroniques a l'analyse metallographique. *Proceedings of the Conference on Electron Microscopy, Delft*, p. 60. The Hague: Martinus Nijhoff.

Chappell, B. W. and A. J. R. White 1968. The X-ray spectrographic determination of sulphur coordination in scapolite. *American Mineralogist* **53**, 1735.

Colby, J. W. 1968. *MAGIC IV – a computer program for quantitative electron microprobe analysis*. Bell Telephone Labs., Allentown, Pa. (See also *Advances in X-ray Analysis* **11**, 287).

Criss, J. W. and L. S. Birks 1968. Calculation methods for fluorescent X-ray spectrometry: empirical coefficients vs. fundamental parameters. *Analytical Chemistry* **40**, 1080.

Duncumb, P. and S. J. B. Reed 1968. The calculation of stopping power and backscatter effects in electron probe microanalysis. In *Quantitative electron probe microanalysis*, K. F. J. Heinrich (ed.), 133–54, National Bureau of Standards Special Publication 298.

Frazer, J. Z. 1967. *A computer fit to mass absorption coefficient data*. Institute for the Study of Matter, University of California, La Jolla, Publication 67–29.

Goldstein, J. I., D. E. Newbury, P. Echlin, D. C. Joy, C. Fiori and E. Lifshin 1981. *Scanning electron microscopy and X-ray microanalysis*. New York: Plenum.

Heinrich, K. F. J. 1966. X-ray absorption uncertainty. In *The electron microprobe*, T. D. McKinley, K. F. J. Heinrich and D. B. Wittry (eds), 296. New York: Wiley.

Heinrich, K. F. J. (ed.) 1968. *Quantitative electron probe microanalysis*. National Bureau of Standards Special Publication 298.

Heinrich, K. F. J. 1981. *Electron beam X-ray microanalysis*. New York: Van Nostrand Reinhold.

Hillier, J. 1947. *Electron probe analysis employing X-ray spectrography*. US Patent No. 2 418 029.

Jenkins, R. 1974. *An introduction to X-ray Spectrometry*. London: Heyden.

Jenkins, R. and J. L. de Vries 1969. *Practical X-ray spectrometry*, 2nd edn. London: Macmillan.

Kelly, T. K. 1966. Mass absorption coefficients and their relevance in electron probe

BIBLIOGRAPHY

microanalysis. *Transactions of the Institute of Mining and Metallurgy* **75**, B59–B73.

Kramers, H. A. 1923. On the theory of X-ray absorption and the continuous X-ray spectrum. *Philosophical Magazine* **46**, 836.

Long, J. V. P. 1977. Electron probe microanalysis. In *Physical methods in determinative mineralogy*, J. Zussman (ed.), 273–342. London: Academic Press.

McKinley, T. D., K. F. J. Heinrich and D. B. Wittry (eds) 1966. *The electron microprobe*, Proceedings of the Symposium of the Electrochemical Society, Washington, 1964. New York: Wiley.

Norrish, K. and B. W. Chappell 1977. X-ray fluorescence spectrometry. In *Physical methods in determinative mineralogy*, J. Zussman (ed.), 201–72. London: Academic Press.

Norrish, K. and J. T. Hutton 1969. An accurate X-ray spectrographic method for the analysis of a wide range of geological samples. *Geochemica et Cosmochimica Acta* **33**, 431.

Pattee, H. H., V. E. Cosslett and A. Engstrom (eds) 1963. *X-ray optics and X-ray microanalysis*. (Proceedings of the Stanford Conference). New York: Academic Press.

Philibert, J. 1963. A method for calculating the absorption correction in electron-probe microanalysis. In *X-ray optics and X-ray microanalysis*, H. H. Pattee, V. E. Cosslett and A. Engstrom (eds), p. 379. New York: Academic Press.

Pik, A. J., J. M. Eckert and K. L. Williams 1981. The determination of dissolved chromium (III) and chromium (VI) and particulate chromium in waters at $\mu g\ l^{-1}$ levels by thin-film X-ray fluorescence spectrometry. *Analytica Chimica Acta*, **124**, 351–6.

Reed, S. J. B. 1965. Characteristic fluorescence corrections in electron-probe microanalysis. *British Journal of Applied Physics* **16**, 913.

Reed, S. J. B. 1975. *Electron microprobe analysis*. Cambridge: Cambridge University Press.

Smith, D. G. W. (ed.) 1976a. *Microbeam techniques*. Short Course Handbook, Vol. 1. Edmonton: Mineralogical Association of Canada.

Smith, D. G. W. 1976b. Quantitative energy dispersive microanalysis. In *Microbeam techniques*, Short Course Handbook Vol. 1, D. G. W. Smith (ed.), 63–106. Edmonton: Mineralogical Association of Canada.

Springer, G. 1967. The correction for 'continuous fluorescence' in electron-probe microanalysis. *Neues Jahrbuch für Mineralogie Abhandlung* **106**, 241.

Springer, G. 1974. Quantitative electron-probe microanalysis: review of iterative procedures and evaluation of electron backscattering. *Proceedings of the Ninth Annual Conference Microbeam Analytical Society, Ottawa*, 38A.

Springer, G. 1976. Correction procedures in electron-probe analysis. In *Microbeam techniques*, Short Course Handbook, Vol. 1, D. G. W. Smith (ed.), 45–62. Mineralogical Association of Canada.

Stephenson, D. A. 1971. Multivariable analysis of quantitative X-ray emission data. The system zirconium oxide – aluminium oxide – silicon oxide – calcium oxide – cerium oxide. *Analytical Chemistry* **43**, 310.

Tertian R. and F. Claisse 1982. *Principles of quantitative X-ray fluorescence analysis*. London: Heyden.

White, E. W. and G. G. Johnson 1972. *X-ray and absorption wavelengths and two-theta tables*, 2nd edn. American Society for Testing and Materials.

BIBLIOGRAPHY

Yakowitz, H., R. L. Myklebust and K. F. J. Heinrich 1973. *FRAME: an on-line correction procedure for quantitative electron microprobe analysis*. National Bureau of Standards, Technical Note 796.

Ziebold, T. O. and R. E. Ogilvie 1966. An empirical method for electron microanalysis. *Analytical Chemistry* **36**, 322–7.

Zussman, J. (ed.) 1977. *Physical methods in determinative mineralogy*, 2nd edn. London: Academic Press.

Index

aberrations, electron beam 132
absorbed electrons 237
absorptiometry 41
absorption 34, 107, 115, 136, 185, 189, 199–210, 227, 253, 270
   edge 38–42, 147, 171, 186, 255, 256, 262
   factor 243–51
   jump ratio 41, 253, 255, 257, 261
   of fluorescent radiation 261
accuracy 142, Fig. B.1
adsorption traps 135
airlock, XRF 125
   electron probe 133
alpha-factors 194, 196–9, 210–14, 268–70
amplifier 11, 127, 137, 141
analog-to-digital converter 111
analysis totals 214
analytical strategy 144, 159, 174
Ångstrom unit 1, 11
aperture, electron beam 132
asymmetric line profiles 104
atomic number 223, 228, 253, Fig. A.2
   factor 228, 229–43
atomic weight 232, 252
attenuation, X-ray *see* absorption
attenuator 141
Auger electron 21, 91, 229
average Fig. B.3
Avogadro's number 233

background 118, 140, 169–79, 217–18, 222–4
   non-linear 173, 177
backing pump 135
backlash 146, 154
backscatter ionization loss 237–8
backscattering 230, 234, 245
band gap 29, 99
baseline restoration 118
   discriminator 140
beam current 137, 237
blanking, electron beam 132
borate fusion 193
Bragg angle 47, 101, 105, 145
Bragg equation 4

calibration curve *see* working curve
calibration standard 144
Cassegrainian objective 132
channel, MCA 113
characteristic fluorescence 252–9
characteristic spectrum 3–4, 11, 15–16, 203, 229

charge carrier 28, 92
chart recorder 113, 128, 140
coefficient of variation *see* relative standard deviation
coherent scattering 34, 43, 203
collimator 27, 87, 98, 102, 124, 126, 145, 146
Compton scattering *see* incoherent scattering
computer 7, 113, 138, 139, 140, 213, 237, 242, 253, 270, Fig. A.1
condensing lens 131
conduction band 29, 91
conductor 29
confidence limits Fig. B.9
contamination spot 114, 132, 134
continuous spectrum *see* continuum
continuum 3, 11, 12–15, 121, 147, 179, 229, 252, 253, 260
   fluorescence 253, 254, 259–68
control system 137; *see also* computer
counter 127; *see also* detector
counting statistics 157
critical excitation potential 15, 40, 229
cryostat 96
crystal 4, 47, 49, 146
   fluorescence 56
   interchange 108, 136
   dispersing 126, 136
   flexible 106
   'ideally imperfect' 55
   thermal drift 52, 57

defocussing 114
dead layer 96, 115
dead time 81–5, 113, 115, 118, 153
   measurement 85–7
deconvolution *see* peak stripping
degree of freedom Fig. B.6
demountable tubes 4
depletion layer 93
depth function 242
   measurement 246
depth generation factor 186
detection limits 2, 139, 165–6
detector 28, 47, 58–99, 126, 136, 146
   filling gas 60
   proportionality 65
   resolution 73–6, 96–8
   voltage 148
   window 61–2, 99
   electron 137
   gas flow 62

INDEX

gas-filled 59–87, 109, 115
  proportional response 115
  scintillation 59, 88–90
  semiconductor *see* detector, solid-state
  Si(Li) 92–9, 115
  solid-state 28, 59, 91–9, 111
differential matrix factor 209
differential mode 77
diffracting crystal *see* crystal
diffraction 44–6
diffusion pump 135
dilution 192–3, 212
dispersing crystals *see* crystal
dispersion 3, 46, 53, 100, 145
divergence 101
doped semiconductor 31-3
double dilution 207–9
drift 155, 158

elastic scattering 229
electron energy loss 227, 229–31
  hole 30
  microscope 131
  microscope, scanning 113, 137
  optics 131
  backscattered 137
  secondary 137, 229
electron-hole pairs 92
emission current 131
energy, X-ray 10
energy, critical excitation 2, 11
  effective excitation 186–8
energy-wavelength relation 10
enhancement, XRF 191–2
enhancement, electron probe 225, 251–68
error of a difference Fig. B.11
  of a ratio Fig. B.14
  of a ratio of difference Fig. B.16
errors, operator 143
  statistical 155
escape peaks 80–1
extrinsic semiconductor 31

fano factor 71, 76, 97
Faraday cage 155
Fermi level 16, 29
FET *see* field-effect transistor
frequency distribution Fig. B.4
field-effect transistor 96
filament 131, 134
filters, diffraction 41
finite depth factor 188–90
fixed count mode 159, 162
fixed time mode 160, 161
fluorescence 6
  efficiency *see* fluorescent yield
  factor 228, 251–68
  yield *see* fluorescent yield

characteristic 252–60
  secondary 183, 187, 191, 195, 210, 225
  third-element 183, 253
fluorescent yield 21, 233, 252, 255
fundamental parameters method 187
fusion 151, 192-4, 212, 214–20
FWHM 75, 97
FWTM 76

gamma-rays 1
gas amplification 64
Gaussian distribution 69, 156, Fig. B.5
Geiger counter 7, 65
  region 65
generation factor 228
generator, X-ray 121
germanium detector 98
goniometer 57, 101, 141
gun brightness 137
gun, electron 129, 137
  field emission 131
  lanthanum hexaboride 131

heavy absorber 193, 215
helium path 126

images, X-ray, scanning 114
incoherent scattering 34, 43, 141, 204
inelastic scattering 229
infinite depth *see* infinite thickness
infinite thickness 182, 202, 225, 243
influence coefficients 194, 195–9, 210–14, 227, 269–70
insulator 29
integral mode 77, 141
interelement effects, XRF 179–220
  electron probe 222, 224–7
interference 44, 115, 123, 145, 169–79, 222–4
  harmonic 119
  high-order *see* interference, harmonic
internal standard 206–7
intrinsic semiconductor 30, 91
ion pair 63
ion pump 135
ionization cross-section 233
ionization energy 68
iteration 184, 187, 216, 241, 251, 258, 269

Johann optics 104
Johannson optics 103
jump ratio *see* absorption jump ratio

lanthanum oxide addition 215
Laplace transform 245
lateral intensity function 245
Lenard coefficient 246, 256
light optics, electron probe 132

INDEX

linear absorption coefficient 35, 150, 188
live time 113, 117, 118
LN2 94, 99, 115
lower limit of detection 165
lower limit of determination 166

macroprobe, X-ray 6, 106
magnetic data storage 113
mass absorption coefficient 37, 184, 190, 195–209, 216–18, 226, 228, 244, 249, 252, 255, 261, 268, Fig. A.1
mass attenuation coefficient *see* mass absorption coefficient
mass scattering coefficient 203
mass thickness 200, 235
MCA *see* multi-channel analyser
mean Fig. B.3
mean excitation energy (J) 236
mean mass depth 247
median Fig. B.3
microabsorption effects 150–2, 192, 206, 209, 212, 214
mode Fig. B.3
monitor standard 155
Monte Carlo technique 237
mosaic imperfections 56
multi-channel analyser 95, 112
multiple regression 194, 212
Mylar 61

noise 96, 97, 115
normal distribution 156, Fig. B.5
Norrish and Hutton method 194, 214–20, 270

objective lens 131
operator errors 143
optimised time mode 162
order of reflections 49, 145
oscilloscope 113
overvoltage 234, 238, 253

particle-size effects *see* microabsorption effects
Pauli exclusion principle 17
peak asymmetry 145
  broadening 118
  location 145
  shape 119, 145
  shifts 118
  stripping 119
peak-seeking 146
peak-to-background ratio 114, 118, 170
PHA *see* pulse height discrimination
PHD *see* pulse height discrimination
Philibert expression 246, 270
photodetectors 30
photoelectric absorption 34

photoelectron 20, 91, 137
photomultiplier 88
photon 10
Pirani gauge 127
Poisson distribution Fig. B.4
polynomial coefficients Fig. A.1–A.4
power supplies 127, 128, 137
preamplifier 97, 111, 127
precision 159, 160, Fig. B.1
preset count mode 128
preset time mode 128
pressed powder samples 151
printer 113
propagation of errors 160, Fig. B.10
pseudo-crystals 52
pulse amplifier 68
  amplitude 54, 68
  amplitude depression 77–80, 118
  amplitude discrimination 72–6
  amplitude shift 147, 153
  discrimination 117, 140, 145, 148, 170
  height analysis *see* pulse amplitude discrimination
  height discrimination *see* pulse amplitude discrimination
  pile-up 115–17
  shaping 111
pump baffles 135

quantum numbers 16
quench gas 66

radio-isotope source 110
radiography 2
random errors 156
raster scan 132
ratemeter 113, 127, 140
Rayleigh scattering *see* coherent scattering
reflection efficiency 54
  order *see* order of reflections
reflections, abnormal 58
regression analysis Fig. A.3
relative standard deviation 71, 74, Fig. B.7
replicate analyses 214
resolution, detector 115
  electron beam 137
  spatial 2, 6
  spectrometer 49, 102, 105, 109, 111, 140
resolving power *see* resolution, dispersion
  rise-time 116
rocking curve 56
rotary pump 135
Rowland circle 103, 108
roughing pump 135

safety 124
sample chamber, XRF 124
  electron probe 133

369

INDEX

current 237
dilution 151
homogeneity 125, 149
preparation errors 149
preparation, XRF 149
preparation, electron probe 152
turret, XRF 125
sampling errors 149
satellite line 21–2, 139, 145, 174
SCA *see* single-channel analyzer
scaler 113, 128
scanning coils 132
scanning images 132
scapolite, sulphur in 25
scattering 34, 40, 42, 203
scattering, electron 229
self-absorption, tube 121
semiconductor 29
sensitivity 2, 127, 166
significant figures Fig. B.2
Soller collimator 126
standard error of mean Fig. B.7
simplified theory, XRF 186–92
simplified theory, electron probe 227–68
sin-θ potentiometer 54, 141, 146, 148
single-channel analyzer 111
slit 98, 103, 107, 109, 146
specimen current *see* sample current
spectral series 3, 18–24, 252
    intensity ratio 257
    stripping 140
spectrography 6
spectrometer 1, 8
    Bragg 47
    curved crystal *see* spectrometer, focussing
    electron probe 135
    electron traps 223
    energy dispersive 7, 100, 110–20, 135, 139, 140, 224
    fixed 100
    flat crystal 102
    focussing 102, 135
    linear focussing 107, 110
    scanning 100, 110
    sequential 101
    wavelength dispersive 7, 100–10, 135
    portable 110, 120
spiking 207
spinner 125
spreadsheet 242
square model (Bishop) 247
standard deviation Fig. B.6
statistical errors 155
stigmator 132

stopping number 236
stopping power 234, 236
strategy, analytical 144
systematic errors 156, 157

tail effects 177
take-off angle 106, 136, 249, 260, 270
tertiary fluorescence 183, 185, 187, 191
thermal effects (microprobe) 114
thermistor 30
thin-film method 185, 194–5
third-element fluorescence 183, 253
time constant 117, 128
timer 113, 128
tracer element 246
transition levels 17
    nomenclature 19
    selection rules 17
trapped carriers 96
triode electron gun 129
true result Fig. B.1
tube (XRF) 121
    anode 121, 147
    continuum 121
    life 124
    lines 121, 141
    operating conditions 124
    power 121
turbo-molecular pump 135

uncertainty Fig. B2.

vacuum path 126
    pump, rotary 127
    system, electron probe 134
    system, XRF 126
valence band 28
variance Fig. B.7
visual display unit 113

wavelength, effective 188, 195, 214
    shifts 24–7
Wehnelt cylinder 131
window, X-ray tube 121
    discriminator 140
working curve 180, 195

X unit 11
X-ray 10–46
    diffraction 2–3, 42–7
    generator 121

ZAF corrections 227–68, 270